Governors State University
Library
Hours:
Monday thru Thursday 8:30 to 10:30
Friday and Saturday 8:30 to 5:00
Sunday 1:00 to 5:00 (Fall and Winter Trimester Only)

DEMCO

Alcohol, Tobacco and Cancer

Alcohol, Tobacco and Cancer

Editors

Chi Hin Cho   *Hong Kong*
Vishnudutt Purohit   *Bethesda, Md.*

43 figures, 2 in color and 11 tables, 2006

Basel · Freiburg · Paris · London · New York ·
Bangalore · Bangkok · Singapore · Tokyo · Sydney

· ·

Prof. Chi Hin Cho

Department of Pharmacology
Li Ka Shing Faculty of Medicine
The University of Hong Kong
21 Sassoon Road
Hong Kong, China

Dr. Vishnudutt Purohit

Division of Metabolism and Health Effects
National Institute of Alcohol
Abuse and Alcoholism
National Institute of Health
5635 Fishers Lane
Bethesda, MD, USA

Library of Congress Cataloging-in-Publication Data

Alcohol, tobacco, and cancer / editors, Chi Hin Cho, Vishnudutt Purohit.
      p. ; cm.
   Includes bibliographical references and index.
   ISBN 3-8055-8107-6 (hard cover : alk. paper)
   1. Cancer–Etiology.   2. Alcohol–Carcinogenicity.   3. Smoking–Health
aspects.   I. Cho, C. -H. (Chi-Hin)   II. Purohit, Vishnudutt.
   [DNLM: 1. Neoplasms–etiology.   2. Alcohol Drinking–adverse effects.
3. Ethanol–adverse effects.   4. Risk Factors.   5. Smoking–adverse
effects.   6. Tobacco–adverse effects.   QZ 202 A3547 2006]
RC268.48.A43 2006
616.99′4071–dc22

      2006013729

Bibliographic Indices. This publication is listed in bibliographic services, including Current Contents® and Index Medicus.

© Copyright 2006 by S. Karger AG, P.O. Box, CH–4009 Basel (Switzerland)
www.karger.com
ISBN-10: 3–8055–8107–6
ISBN-13: 978–3–8055–8107–3

Contents

Preface

Seven million deaths have been attributed to cancer in 2001. Of these 35% were attributable to nine potentially modifiable risk factors, including smoking and excessive alcohol consumption. Alcohol consumption is a leading risk factor for cancer of the upper aerodigestive tract (UADT, esophagus, pharynx, larynx and oral cavity), liver and colon, while smoking increases the risk of mouth, oropharyngeal, and esophageal cancers as well as cancer of the trachea, bronchus, lung, liver, stomach and urinary bladder. This book highlights general mechanisms leading to cancer, and provides the latest information on the underlying mechanisms, whereby alcohol consumption and tobacco use initiate and/or promote carcinogenesis.

Concurrent smoking and drinking, which is common, further increases the risk for cancers of the UADT. Several metabolic and genetic factors interact to increase the risk of carcinogenesis due to smoking and drinking. Acetaldehyde, which is the first product of alcohol metabolism (by cytosolic alcohol dehydrogenase [ADH], and, at high concentrations of alcohol, by CYP2E1) is also naturally present in tobacco. Acetaldehyde is carcinogenic by forming adducts with DNA and by inhibiting the DNA repair processes. Furthermore, procarcinogens present in tobacco smoke, such as aromatic and heterocyclic amines, are metabolized by CYP2E1, which is induced by alcohol, and by CYP1A2. Genetic variations in alcohol metabolizing enzymes that result in acetaldehyde accumulation (high-activity ADH variants, or the low-activity aldehyde dehydrogenase [ALDH] variant which is unable to metabolize acetaldehyde) contribute to the risk for various cancers. In addition, both smoking and drinking result in the formation of reactive oxygen species (ROS) which could result in DNA damage. Genetic factors,

VII

including variations in ADH, ALDH, and CYP2E1, which increase the risk for cancers are addressed in this book. In addition, smoking and drinking interfere with retinoid metabolism and signaling, which can result in uncontrolled cell proliferation and cancer.

Several chapters address in detail the correlations between alcohol and carcinogenicity, and focus on specific organs such as UADT, liver, colon, pancreas and breast. Of particular interest is the notion that the risk for breast cancer appears to be significant even with moderate alcohol consumption. The impairment of methionine-folate metabolism and DNA methylation by alcohol, as well as the use of s-adenosylmethionine to prevent cancer, are addressed in a comprehensive manner. Also, the synergistic effects of ROS and iron in producing liver cancer are addressed.

Over 1.3 billion people use tobacco worldwide and cigarette smoking alone causes approximately 20% of cancer deaths globally. Several chapters are dedicated to the role of tobacco in lung and digestive tract cancers, and equally important to the prevention of tobacco-induced cancers, especially phytochemicals. Nicotine, a major alkaloid of tobacco, is responsible for smoking-related malignancies. The mutagenic, mitogenic, pre-carcinogenic, anti-apoptotic, and immunosuppressive properties of nicotine which are involved in the development of cancer, and the possible use of nicotine vaccine in the prevention of mortality associated with tobacco use are discussed.

In summary, this book contains a wide array of the state-of-science knowledge and addresses an important topic that highlights the association between cancer and smoking and/or drinking, two modifiable risk factors.

Ting Kai Li, M.D.
Director
National Institute on Alcohol
Abuse and Alcoholism
National Institutes of Health
Bethesda, Maryland, USA

Cho CH, Purohit V (eds): Alcohol, Tobacco and Cancer.
Basel, Karger, 2006, pp 1–12

....

General Mechanisms of Cancer

Annie On On Chan, Benjamin Chun Yu Wong

Department of Medicine, University of Hong Kong, Hong Kong, China

Abstract

In the past decade, we are gaining more insight of cancer mechanism. The association between environmental factors, such as diet, irradiation, infection, and cancer have long been recognized through epidemiology studies. Host susceptibility to cancer has been increasingly recognized through the studies of familial cancer syndromes, single nucleotide polymorphism etc. Carcinogenesis probably results from the interaction between environmental factors and host susceptibility, and is a multi-steps process, involving a number of somatic genetic alterations through a series of morphological changes. These include activation of oncogenes, silencing of tumor suppressor genes through mutation, loss of heterozygosity or CpG island methylation. The epigenetic change, CpG island methylation, is increasingly being recognized to be an early and important mechanism for tumor suppressor gene silencing. Epigenetic silencing does not involve the changes of nucleotide sequence, hence it is potentially reversible and may potentially offer early chemoprevention. These mechanisms are discussed in detail in this review.

Ideas about the causes of cancer began in Greece, with some of the earliest reports ascribed to Hippocrates. Tumors were thought to result from an accumulation of black bile, or 'melancholic humor.' It was not until the invention of the microscope in 1830 that cells were identified as the fundamental unit in tumor tissue, and cell physiology became an active field of study. Soon thereafter, Rudolf Virchow first expressed the theory that all cells derive from cells, a maxim that became the foundation for the modern cellular approach to cancer.

The notion that cancer might be caused by genetic abnormalities originated in the early nineteenth century when it was noted that predisposition to cancer seemed to run in families. However, only when the necessary technologies became available in the early 1970s, tumor formation was related to the action of specific genes. The advances of molecular technology in the last two decades enable the understanding of cancer as a genetic disease. The identification of germline mutation in the adenomatous polyposis coli (*APC*) gene in patients with familial adenomatous polyposis of colorectal cancer by Vogelstein et al. [1] further corroborated the genetic point of view in carcinogenesis and opened up a new page in the study of genetic alterations in cancers. Much of the focus of molecular biologic research has concentrated on investigating the role of genetic changes, i.e. direct alterations of DNA base sequence through mutation, deletion, or insertion, and their effect on subsequent gene expression and cell behavior. Thereafter, much advances and insight were gained in the underlying mechanisms of carcinogenesis. In this chapter, we shall discuss and summarize the general mechanisms in carcinogenesis.

Environmental Causes

For more than two centuries, the focus of the causes of cancer has been on external causes. Chemical and physical agents were probably the earliest recognized carcinogenic substances. Percival Pott (1775) reported that the chimney sweeps of Britain were prone to scrotal cancer. Beginning in the late 19th century, industrial exposures to large quantities of noxious agents including paraffin oils, mining dust, arsenic, and aniline dyes, as the cause for cancer became apparent. The uncontrolled use of asbestos in the European construction industry in mid 20th century led to a rapid increase in mesothelioma and lung cancer [2].

Cigarette smoking, as one of the most important environmental causes for cancer, is the direct and avoidable cause of an enormous cancer burden. No other known single environmental factor has anything like the same degree of importance for cancer in the developed world. The large increase in male cigarette smoking in Britain during and after the First World War caused an unprecedented epidemic of lung cancer among men born around 1900, and by 1955 the rate in British men aged under 55 was the highest in the world [3]. It not only causes lung cancer, it also increases risk for heart attack and for cancers of the mouth, throat, bladder, colon, rectum, pancreas and cervix.

Radiation carcinogenesis is a twentieth century problem. One of the most important pieces of information comes from the long-term follow-up of survivors of the atomic bombs dropped in Nagasaki and Hiroshima. Studies of occupational exposure of diagnostic radiologists, uranium miners, and workers in the nuclear industries have also provided valuable information. More has come from the analysis of cancer incidence in patients exposed to radiation for medical purposes, either for diagnosis or for treatment of non-malignant conditions. Cancers induced by ionizing radiation are indistinguishable from most cancers arising from other causes. Common cancers resulting from radiation are leukemia, lymphoma and brain tumor, and skin cancer from UV light exposure.

The first definite connection between viruses and cancer came from the pioneering work of Peyton Rous, who in 1900s, described a virus that could induce sarcomas in chickens and could transmit the disease in the manner of an infectious agent. Today, several kinds of viruses have been recognized to play important roles in cancer, including Epstein–Barr virus in Burkitt's lymphoma and nasopharyngeal cancer [4], human papillomavirus in cervical and other genitourinary cancer [5, 6], human T-cell lymphotropic virus type 1 in some T-cell leukemias and lymphomas [7], and hepatitis B virus in hepatocellular carcinoma [8]. *Helicobacter pylori*, a chronic gastric bacterial infection which was discovered 20 years ago, is now thought to be a major factor in the development of gastric cancer [9]. Schistosomiasis has been implicated in bladder and colon cancer, and liver flukes in cholangiosarcoma [9].

In addition, the above environmental factors that cause cancers directly, diet has been consistently reported to be predisposing factor in many types of cancers, especially gastrointestinal tumors. For example, a number of studies have found an association between red meat consumption and colorectal cancer. A prospective study found that women with the highest ratio of red meat to chicken and fish intake had two and one-half times the colon cancer risk of those with the lowest ratio [10]. A prospective cohort study of nearly 50,000 U.S. male health professionals showed that men who ate beef, pork, or lamb as a main dish, compared to those consuming such foods less than once a month, had over three times the risk of colon cancer [11]. Cooked meats have been found to contain compounds, including a class of heterocyclic amines, which are mutagenic and carcinogenic in animal models. They are produced during high-temperature cooking such as broiling or frying [12]. The association between N-nitroso compounds and gastric cancer has been summarized by Bartsch et al. [13]. The risk of gastric cancer induced by N-nitroso compounds has been demonstrated in animal experiments [14–16]. An increase in gastric nitrite was observed in patients with intestinal metaplasia, dysplasia, and gastric cancer [17–19]. The uses of nitrate-based fertilizers [18, 20, 21] and pickled foods that contain nitrosated products [22, 23] have been shown to positively correlate with gastric cancer.

Countries and regions with the highest per capita dietary fiber consumption tend to have the lowest colorectal cancer rates. A meta-analysis of 16 case-control studies found nearly a 35% reduction in the relative risk of colorectal cancer for those in the highest, compared to the lowest, category of dietary fiber intake [24]. In the cohort study of polyps in male health professionals, men in the highest, relative to those in the lowest, category of dietary fiber intake had half the risk of developing an adenomatous polyp of the large bowel [25]. The large majority of case-control studies of colorectal cancer that assess vegetable intake find it to be protective [26]. Several case-control studies of large bowel cancer have shown an inverse association for fruit intake, but in general the analytic epidemiologic findings are not as consistent for fruits as for vegetables [27]. However, later evidence from prospective studies indicates a weaker overall association with high consumption of fruits and vegetables [28].

Folic acid supplements have been associated with a preventive role and reduced risks of colorectal cancer and adenomas in both case-control and cohort studies [28]. A particularly striking finding came from the Nurses' Health Study: consumption of multiple vitamins containing folic acid for more than 15 years was associated with a risk of colorectal cancer that was 75% lower than that in non-users [29]. A preventive role of folic acid intake is further supported by animal studies, as well as by mechanistic investigations. Findings from the latter demonstrate that low folic acid concentrations greatly increase the substitution and incorporation of uracil for thymine into DNA due to the depletion of essential methyl donors in the cell.

Inherited Susceptibility to Cancer

Most cancers seem to be induced in ostensibly normal individuals by exposure to exogenous agents. Although cancer-prone families are uncommon, examination of such families has uncovered specific genetic loci that serve as targets of carcinogenesis. The cells of each person contain two copies of each gene, one inherited from the mother and one from the father. Based on the differences between familial and sporadic diseases, Knudson proposed the 'two-hit' theory. He argued that whether familial or sporadic, the cancer requires two distinct genetic alterations [30]. The first alteration in familial cancers would be a mutant gene inherited from a parent (1st hit). The second genetic change would be somatic (2nd hit). Whereas in sporadic cancers, two independent somatic mutation would be required.

One of the well know example is familial adenomatous polyposis, which is the most common polyposis syndrome but accounts for <1% of colorectal cancer cases per year. Familial adenomatous polyposis is inherited in an autosomal dominant fashion and is associated with a germline mutation in the *APC* gene [31]. Patient with familial diffuse type of gastric cancer was found to have germline mutation at *E-cadherin* gene [32, 33]. Thereafter, promoter CpG hypermethylation at *E-cadherin* gene was found to be the second 'genetic hit' in abrogating *E-cadherin* expression in two other kindreds with familial gastric cancer with germline *E-cadherin* mutation as the first 'genetic hit' [34].

In addition to the harboring of a germline mutation, there are increasing studies showing that certain genetic polypmorphisms may predispose an individual in developing cancer. One of the most well known examples is gastric cancer. *H. pylori* infection is associated with increase risk of developing gastric cancer. However, not everyone infected with the bug will develop gastric cancer eventually. El-Omar et al. [35] have studied the human interleukin 1 beta (*IL-1β*) gene and that the gene is the most important candidate gene in the host that could affect the clinical outcome of *H. pylori* infection. *IL-1β* is upregulated by the infection, is profoundly proinflammatory, and is the most powerful acid inhibitor known. Polymorphisms in *IL-1β* gene (carriers of *IL-1β*-511*T) and in the *IL-1* receptor antagonist gene (*IL-1RN*2/*2*) were found to be associated with an increased risk of gastric cancer in patients with *H. pylori* infection [34]. Figueiredo et al. [36] had further suggested *IL-1β*-511*T carriers (*IL-1β*-511*T/*T or *IL-1β*-511*T/*C) homozygous for the short allele of *IL-1RN* (*IL-1RN*2/*2*) associating with vacAs1-, vacAm1-, and cagA-positive strains of *H. pylori* had an increased gastric carcinoma risk.

Unifying Themes in Carcinogenesis

Only 5–10% of cancers are hereditary. Most cancers develop through genetic alterations due to damage by environmental factors at a background of host susceptibility. These genetic changes are called acquired/somatic genetic alterations, which usually take many years before they develop into a cancerous cell. An example of the interactions of environmental factors and host susceptibility on genetic alterations in gastric carcinogenesis is shown in figure 1. Regardless of the causes of the cancer (i.e. environmental exposure to chemicals, physical, viral agents, or familial cancers), some common themes emerge. One is that cancer development is a multi-step, progressive phenomenon that arises because a cell accumulates a series of genetic changes that cause its growth pattern to become progressively more abnormal. Another theme is that a limited number of genetic changes in the cancer cell account for at least some of these alterations, and that more than one genetic modification is required. Loss of genome stability and integrity in the affected cells is a hallmark of the multi-step process of cancer, which progressively leads to uncontrolled cell growth and metastasis. This loss of normal control mechanisms arises from the acquisition of alterations in three broad categories of genes:

(1) Proto-oncogenes, the normal products of which are components of signaling pathways that regulate proliferation and which, in their mutated form, become dominant oncogenes;

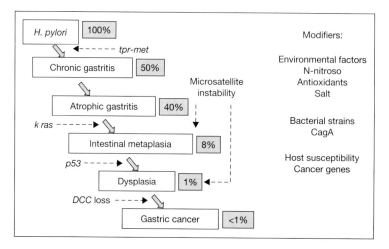

Fig. 1. The interactions of environmental factors including *H. pylori* infection, host susceptibility and the resulted somatic genetic changes in gastric carcinogenesis.

(2) Tumor suppressor genes, which generally exhibit recessive behavior, the loss of function of which in cancers leads to deregulated control of cell cycle progression, cellular adhesion etc., and

(3) DNA repair enzymes, which promote genetic instability when mutated or epigenetically silenced.

Though a great deal of information on changes in DNA structure and their relationships to neoplastic environment is available, the knowledge of the control of gene expression is much more limited. The importance of epigenetic mechanisms, which are defined as a heritable trait not based upon the primary DNA sequence [37], is only recognized in the recent two decades. Methylation is one the most widely studied and important epigenetic mechanism.

Methylation and Carcinogenesis

Altered methylation patterns are known to occur in the DNA of cancer cells. Two patterns have been observed: wide areas of global hypomethylation along the genome, and localized areas of hypermethylation at certain specific sites, the CpG islands, within the gene promoter regions [38, 39].

Based on these patterns, several theories have emerged to implicate DNA methylation mechanisms in carcinogenesis [40]. In fundamental genetic models of cancer, the amplification of protooncogenes, or the silencing of tumor suppressor genes, disrupts the balance that normally controls cell proliferation and drives it through the succession of events leading to full malignant status. Thus, in theory, decreased methylation, and hence relief of transcriptional silencing, may allow the expression of previously quiescent protooncogenes to become active and induce the cell proliferation events. Alternatively, increased methylation at previously unmethylated sites, such as the promoter regions of a tumor suppressor gene, may result in their silencing through inhibition of transcription and their inability to suppress cell proliferation.

De Novo Methylation of CPG Islands

Jones et al. [41] demonstrated that the CpG island of the mouse MyoD1 muscle determination gene acquired de novo methylation in all immortalized cell lines examined, although unmethylated in adult tissues and low passage fibroblasts. Interestingly, the island became progressively more methylated when cells were oncogenically transformed with chemicals such as 3-methylcholanthrene or by transfection with an activated *ras* gene [42]. This comes from the evidence of the study of *RB* [43–45] and *VHL* gene [46]. Approximately half of the tumor suppressor genes that are responsible for familial cancer syndromes such as *p16* [47], *E-cadherin* [48], *BRCA1* [49] and *hMLH1* [50] have also been reported to be methylated in sporadic cancers. It is now proposed that such hypermethylation may be considered to be a third pathway causing loss of function which is equivalent to mutation and deletions in Knudson's [30, 51] two-step model of inactivation of tumor suppressor gene, e.g. the silencing of *hMLH1* and *APC* [52].

The evident alterations in levels and patterns of DNA methylation in tumors may be the result of a deregulation of the enzymes involved in the methylation process. Kautiainen and Jones [53] observed a 2–3000-fold higher level of extractable DNA methyltransferase activity in cultured tumorigenic cells compared to non-tumorigenic cells. El-Deiry et al. [54] studying DNA methyltransferase in colon cancer demonstrated a 60-fold increase in DNA methyltransferase expression in benign colon polyps, and a greater than 200-fold increase in colon carcinomas compared to normal human colon mucosa. In addition, it is shown that the normal mucosa from patients with benign polyps or colon cancers possesses a 15-fold increase in DNA methyltransferase expression.

Hypomethylation

It has been shown, more than a decade ago, that cancers cells have lower DNA methylation than in normal cells [55–57]. The overall level of genomic

DNA methylation is reduced by approximately 10% in colonic neoplasia [58]. Hypomethylation may also become more pronounced in the later stages of tumor progression. Bedford and van Helden [59] observed decreases in methylated-cytosine content in metastatic prostate carcinomas but not in non-metastatic variants. In addition, Gama-Sosa et al. [57] have observed hypomethylation associated with metastatic tumor states compared to benign neoplasms. However, demethylation does not appear to be a universal phenomenon of all tumors and tumor-derived cell lines since Pfeifer et al. [60] failed to detect changes in methylation levels in acute human leukemia. Hypomethylation in mutant mouse embryonic stem cells, due to the lack of a functional copy of the DNA methyltransferase DNMT-1, have been shown to have a 10-fold increase in mutation rate involving gene rearrangements [61]. Interaction seems to exist between hypomethylation and hypermethylation. Experiments in plants have suggested that global hypomethylation might induce local de novo methylation. Hypomethylation may trigger the transposon promoter, which results in aberrant mRNAs containing the transposon and exon sequences [62]. The aberrant mRNAs produced were shown to cause methylation and silencing of the cellular gene in plant [63].

DNA Methylation and Mutation

DNA methylation can predispose to mutations through cytosine deamination, which converts cytosine directly to thymine while the unmethylated cytosine is converted to uracil, which is recognized and repaired more efficiently [64]. The increase mutability of 5-methylcytosine versus cytosine is influenced by three factors: differential repair efficiency, rate of spontaneous deamination, and rate of cell division. The mutagenicity of methylated cytosine in biological systems was first demonstrated by the observation that sites of cytosine methylation corresponded to mutational hotspots in the LacI repressor gene of *E. Coli* [65]. The increased frequency of mutation driven by methylation-mediated deaminations of cytosine at these sites may be responsible for the under-representation of the CpG sequence in the vertebrate genome. CpG dinucleotides are present at a frequency of 20% of that expected in the total genome, and 37% of that expected in coding regions [66, 67]. Sved and Bird [68] estimate that an enhanced mutation frequency at CpG dinucleotides, of 12 times the normal frequency of transitions, resulting in CpG to TpG or CpA base changes, can explain the depletion of CpG during the course of vertebrate evolution.

The role of methylated cytosine in human carcinogenesis is particularly evident in the analysis of inactivating point mutations of the *p53* tumor suppressor gene in somatic cells. Abnormalities associated with the *p53* gene represent the most common known genetic alteration in human neoplastic processes [69]. About 28% of point mutations in the *p53* gene are due to C to T transitions at CpG dinucleotides [70]. Hence, epigenetic silencing of tumor suppressor gene by methylation at the CpG island would simultaneously increase the likelihood of the gene to sustain point mutations at the CpG dinucleotides. This would be particularly important for genes, like *p16*, that have CpG islands extending into the protein-encoding region of the gene.

Several lines of evidence points to a causal role of methylation in tumorigenesis. The first is that reduced methylation suppresses the formation of intestinal polyps in APC$^{min/+}$ mice [71]. Secondly, promoter-region methylation of the retinoblastoma (*pRB*) gene and the von Hippel Lindau (*VHL*) gene has been demonstrated in familial cases of unilateral retinoblastoma [72] and renal cancer [46], respectively. Thirdly, studies of sporadic cases of colorectal cancers exhibiting microsatellite instability demonstrate a high frequency of promoter-region hypermethylation of the mismatch repair gene (MMR) hMLH1. Treatment of colon cell lines containing a hypermethylated hMLH gene with 5'-aza-2'-deoxycytidine results in re-expression of hMLH1 and restoration of MMR ability indicating that hypermethylation of the hMLH1 CpG island is the inactivating event [73].

Conclusion

We are gaining more insight and understanding to the underlying mechanism for carcinogeneis. Early detection or screening is now possible for a number of cancers. We shall aim at primary cancer prevention by leading a healthy life style.

References

1   Vogelstein B, Fearon ER, Hamilton SR, Kern SE, Preisinger AC, Leppert M, Nakamura Y, White R, Smits AM, Bos JL: Genetic alterations during colorectal-tumor development. N Engl J Med 1988;319:525–532.
2   Peto J, Decarli A, La Vecchia C, Levi F, Negri E: The European mesothelioma epidemic. Br J Cancer 1999;79:666–672.
3   Peto R, Lopez AD, Boreham J, Heath C, Thun M: Mortality from Tobacco in Developed Countries, 1950–2000. Oxford, Oxford University Press, 1994.
4   IARC: Epstein-Barr Virus and Kaposi's Sarcoma Herpesvirus/Human Herpesvirus 8 (IARC Monographs on the Evaluation of the Carcinogenic Risks to Humans, 70). IARC, Lyon, 1997.
5   Walboomers JM: Human papillomavirus is a necessary cause of invasive cervical cancer worldwide. J Pathol 1999;189:12–19.
6   IARC: Human Papillomaviruses (IARC Monographs on the Evaluation of Carcinogenic Risks to Humans, 64). IARC, Lyon, 1995.

7   IARC: Human Immunodeficiency Viruses and Human T-cell Lymphotropic Viruses (IARC Monographs on the Evaluation of Carcinogenic Risks to Humans, 67). IARC, Lyon, 1996.

8   IARC: Hepatitis Viruses (IARC Monographs on the Evaluation of Carcinogenic Risks to Humans, 59). IARC, Lyon, 1994.

9   IARC: Schistosomes, Liver Flukes and *Heliobacter pylori* (IARC Monographs on the Evaluation of Carcinogenic Risks to Humans, 61). IARC, Lyon, 1994.

10 Willett WC, Stampfer MJ, Colditz GA: Relation of meat, fat, and fiber intake to the risk of colon cancer in a prospective study among women. New Engl J Med 1990;323:1664–1672.

11 Giovannucci E, Rimm EB, Stampfer MJ: Intake of fat, meat, and fiber in relation to risk of colon cancer in men. Cancer Res 1994;54:2390–2397.

12 Sugimura T: Past, present, and future of mutagens in cooked foods. Environ Health Perspect 1986;67:5–10.

13 Bartsch H, O'Neill I, Hermann R: The relevance of N-nitroso compounds to human cancer. Exposures and mechanisms. IARC Scientific Publications No 84. Lyon, France, International Agency for Research on Cancer 1987;84:1–663.

14 Magee PN, Montesano R, Preussmann R: N-Nitroso compunds and related carcinogens; in Searle CE (ed): Chemical Carcinogens. Am Chem Soc Monogr 173. Washington, DC, American Chemical Society, 1976, pp 491–625.

15 Druckrey H: Chemical carcinogenesis on N-nitroso derivatives. Gann Monogr 1975;17:107–132.

16 Bulay O, Mirvish SS, Garcia H, et al: Carcinogenicity test of six nitrosamides and a nitro-cyanamide administered orally to rats. J Natl Cancer Inst 1979;62:1523–1528.

17 Ruddell WS, Bone ES, Hill MJ, et al: Pathogenesis of gastric cancer in pernicious anaemia. Lancet 1978;1:521–523.

18 Jones SM, Davies PW, Savage A: Gastric-juice nitrite and gastric cancer. Lancet 1978;1:1355.

19 Stewart HL: Experimental alimentary tract cancer. NCI Monogr 1967;25:199–217.

20 Schlag P, Bockler R, Ulrich H, et al: Are nitrite and N-nitroso compounds in gastric juice risk factors for carcinoma in the operated stomach? Lancet 1980;1:727–729.

21 Frazer P, Chilvers C, Beral V, et al: Nitrate and human cancer: a review of the evidence. Int J Epidemiol 1980;9:3–11.

22 Haenszel W, Kurihara M, Segi M, et al: Stomach cancer among Japanese in Hawaii. J Natl Cancer Inst 1972;49:969–988.

23 Sato T, Fukuyama T, Suzuki T, et al: Studies of the causation of gastric cancer. 2. The relation between gastric cancer mortality rate and salted food intake in several places in Japan. Bull Inst Public Health 1959;8:187–198.

24 Trock B, Lanza E, Greenwald P: Dietary fiber, vegetables, and colon cancer: critical review and meta-analyses of the epidemiologic evidence. J Natl Cancer Inst 1990;82:650–661.

25 Giovannucci E, Stampfer MJ, Colditz GA: Relation of diet to risk of colorectal adenoma in men. Am J Epidemiol 1990;132:783.

26 Potter JD, Slattery ML, Bostick RM: Colon cancer: a review of the epidemiology. Epidemiol Rev 1993;15:499–545.

27 Slattery ML, Sorenson AW, Mahoney AW: Diet and colon cancer: assessment of risk by fiber type and food source. J Natl Cancer Inst 1988;80:1474–1479.

28 Giovannucci E, Stampfer MJ, Colditz GA, Rimm EB, Trichopolous D, Rosner BA, Speizer FE, Willett WC: Folate, methionine, and alcohol intake and risk of colorectal adenoma. J Natl Cancer Inst 1993;85:875–884.

29 Giovannucci E, Stampfer MJ, Colditz GA, Hunter DJ, Fuchs C, Rosner BA, Speizer FE, Willett WC: Multivitamin use, folate, and colon cancer in women in the Nurses' Health Study. Ann Intern Med 1998;129:517–524.

30 Knudson AG: Mutation and cancer: statistical study of retinoblastoma. Proc Natl Acad Sci USA 1971;68:820.

31 Rustgi AK: Hereditary gastrointestinal polyposis and non-polyposis syndromes. N Engl J Med 1994;331:1694–1702.

32 Guilford P, Hopkins J, Harraway J, McLeod M, McLeod N, Harawira P, Taite H, Scoular R, Miller A, Reeve AE: E-cadherin germline mutations in familial gastric cancer. Nature 1998;392:402–405.

33   Gayther SA, Gorringe KL, Ramus SJ, Huntsman D, Roviello F, Grehan N, Machado JC, Pinto E, Seruca R, Halling K, MacLeod P, Powell SM, Jackson CE, Ponder BA, Caldas C: Identification of germ-line E-cadherin mutations in gastric cancer families of European origin. Cancer Res 1998; 58:4086–4089.

34   Grady WM, Willis J, Guilford PJ, et al: Methylation of the CDH1 promoter as the second genetic hit in hereditary diffuse gastric cancer. Nat Genet 2000;26:16–17.

35   El-Omar EM, Carrington M, Chow WH, McColl KE, Bream JH, Young HA, Herrera J, Lissowska J, Yuan CC, Rothman N, Lanyon G, Martin M, Fraumeni JF Jr, Rabkin CS: Interleukin-1 polymorphisms associated with increased risk of gastric cancer. Nature 2000;404:398–402.

36   Figueiredo C, Machado JC, Pharoah P, Seruca R, Sousa S, Carvalho R, Capelinha AF, Quint W, Caldas C, van Doorn LJ, Carneiro F, Sobrinho-Simoes M: *Helicobacter pylori* and interleukin 1 genotyping: an opportunity to identify high-risk individuals for gastric carcinoma. J Natl Cancer Inst 2002;94:1680–1687.

37   Ridds AD, Martienssen RA, Russo VEA: Introduction; in Epigenetic Mechanisms of Gene Regulation. Cold Spring Harbor, Cold Spring Harbor Laboratory Press, 1996, pp 1–4.

38   Feinberg AP, Vogelstein B: Hypomethylation distinguishes genes of some human cancers from their normal counterparts. Nature 1983;301:89–92.

39   Baylin SB, Hoppener JW, de Bustros A, Steenbergh PH, et al: DNA methylation patterns of the calcitonin gene in human lung cancers and lymphomas. Cancer Res 1986;46:2917–2922.

40   Laird PW: Oncogenic mechanisms mediated by DNA methylation. Mol Med Today 1997;3: 223–229.

41   Jones PA, Wolkowicz MJ, Rideout WM 3rd, Gonzales FA, Marziasz CM, Coetzee GA, Tapscott SJ: De novo methylation of the MyoD1 CpG island during the establishment of immortal cell lines. Proc Natl Acad Sci USA 1990;87:6117–6121.

42   Jones PA, Buckley JD: The role of DNA methylation in cancer. Adv Cancer Res 1990;54:1–23.

43   Greger V, Passarge E, Hopping W, Messmer E, Horsthemke B: Epigenetic changes may contribute to the formation and spontaneous regression of retinoblastoma. Hum Genet 1989;83:155–158.

44   Ohtani-Fujita N, Fujita T, Aoike A, Osifchin NE, Robbins PD, Sakai T: CpG methylation inactivates the promoter activity of the human retinoblastoma tumor-suppressor gene. Oncogene 1993;8:1063–1067.

45   Sakai T, Toguchida J, Ohtani N, Yandell DW, Rapaport JM, Dryja TP: Allele-specific hypermethylation of the retinoblastoma tumor-suppressor gene. Am J Hum Genet 1991;48:880–888.

46   Herman JG, Latif F, Weng Y, Lerman MI, Zbar B, Liu S, Samid D, Duan DS, Gnarra JR, Linehan WM: Silencing of the VHL tumor-suppressor gene by DNA methylation in renal carcinoma. Proc Natl Acad Sci USA 1994;91:9700–9704.

47   Merlo A, Herman JG, Mao L, Lee DJ, Gabrielson E, Burger PC, Baylin SB, Sidransky D: 5′ CpG island methylation is associated with transcriptional silencing of the tumour suppressor p16/CDKN2/MTS1 in human cancers. Nat Med 1995;1:686–692.

48   Kanai Y, Ushijima S, Hui AM, Ochiai A, Tsuda H, Sakamoto M, Hirohashi S: The E-cadherin gene is silenced by CpG methylation in human hepatocellular carcinomas. Int J Cancer 1997;71: 355–359.

49   Dobrovic A, Simpfendorfer D: Methylation of the BRCA1 gene in sporadic breast cancer. Cancer Res 1997;57:3347–3350.

50   Kane MF, Loda M, Gaida GM, Lipman J, Mishra R, Goldman H, Jessup JM, Kolodner R: Methylation of the hMLH1 promoter correlates with lack of expression of hMLH1 in sporadic colon tumors and mismatch repair-defective human tumor cell lines. Cancer Res 1997;57: 808–811.

51   Knudson AG Jr: Contribution and mechanisms of genetic predisposition to cancer: hereditary cancers and anti-oncogenes. Prog Clin Biol Res 1983;132C:351–360.

52   Hiltunen MO, Alhonen L, Koistinaho J, Myohanen S, Paakkonen M, Marin S, Kosma VM, Janne J: Hypermethylation of the APC (adenomatous polyposis coli) gene promoter region in human colorectal carcinoma. Int J Cancer 1997;70:644–648.

53   Kautiainen TL, Jones PA: DNA methyltransferase levels in tumorigenic and nontumorigenic cells in culture. J Biol Chem 1986;261:1594–1598.

54   El-Deiry WS, Tokino T, Velculescu VE, Levy DB, Parsons R, Trent JM, Lin D, Mercer WE, Kinzler KW, Vogelstein B: Waf1, a potential mediator of p53 tumor suppression. Cell 1993;75: 817–825.

55   Lapeyre JN, Becker FF: 5-Methylcytosine content of nuclear DNA during chemical hepatocarcinogenesis and in carcinomas which result. Biochem Biophys Res Commun 1979;87:698–705.

56   Diala ES, Cheah MS, Rowitch D, Hoffman RM: Extent of DNA methylation in human tumor cells. J Natl Cancer Inst 1983;71:755–764.

57   Gama-Sosa MA, Slagel VA, Trewyn RW, Oxenhandler R, Kuo KC, Gehrke CW, Ehrlich M: The 5-methylcytosine content of DNA from human tumors. Nucleic Acids Res 1983;11:6883–6894.

58   Feinberg AP, Gehrke CW, Kuo KC, Ehrlich M: Reduced genomic 5-methylcytosine content in human colonic neoplasia. Cancer Res 1988;48:1159–1161.

59   Bedford MT, van Helden PD: Hypomethylation of DNA in pathological conditions of the human prostate. Cancer Res 1987;47:5274–5276.

60   Pfeifer GP, Steigerwald S, Boehm TL, Drahovsky D: DNA methylation levels in acute human leukemia. Cancer Lett 1988;39:185–192.

61   Chen RZ, Pettersson U, Beard C, Jackson-Grusby L, Jaenisch R: DNA hypomethylation leads to elevated mutation rates. Nature 1998;395:89–93.

62   Walsh CP, Chaillet JR, Bestor TH: Transcription of IAP endogenous retroviruses is constrained by cytosine methylation. Nat Genet 1998;20:116–117.

63   Mette MF, van der Winden J, Matzke MA, Matzke AJ: Production of aberrant promoter transcripts contributes to methylation and silencing of unlinked homologous promoters in trans. EMBO J 1999;18:241–248.

64   Duncan BK, Miller JH: Mutagenic deamination of cytosine residues in DNA. Nature 1980;287:560–561.

65   Coulondre C, Miller JH, Farabaugh PJ, Gilbert W: Molecular basis of base substitution hotspots in *Escherichia coli*. Nature 1978;274:775–780.

66   Nussinov R: The universal dinucleotide asymmetry rules in DNA and the amino acid codon choice. J Mol Evol 1981;17:237–244.

67   Beutler E, Gelbart T, Han JH, Koziol JA, Beutler B: Evolution of the genome and the genetic code: selection at the dinucleotide level by methylation and polyribonucleotide cleavage. Proc Natl Acad Sci USA 1989;86:192–196.

68   Sved J, Bird A: The expected equilibrium of the CpG dinucleotide in vertebrate genomes under a mutation model. Proc Natl Acad Sci USA 1990;87:4692–4696.

69   Vogelstein B. Cancer. A deadly inheritance. Nature 1990;348:681–682.

70   Hainaut P, Hernandez T, Robinson A, Rodriguez-Tome P, Flores T, Hollstein M, Harris CC, Montesano R: IARC Database of p53 gene mutations in human tumors and cell lines: updated compilation, revised formats and new visualisation tools. Nucleic Acids Res 1998;26:205–213.

71   Laird PW, Jackson-Grusby L, Fazeli A, Dickinson SL, Jung WE, Li E, Weinberg RA, Jaenisch R: Suppression of intestinal neoplasia by DNA hypomethylation. Cell 1995;81:197–205.

72   Stirzaker C, Millar DS, Paul CL, Warnecke PM, Harrison J, Vincent PC, Frommer M, Clark SJ: Extensive DNA methylation spanning the Rb promoter in retinoblastoma tumors. Cancer Res 1997;57:2229–2237.

73   Herman JG, Baylin SB: Methylation specific PCR; in Current Protocols in Human Genetics, 1998.

Prof. Benjamin C.Y. Wong
Department of Medicine
University of Hong Kong
Queen Mary Hospital
Pokfulam Road, Hong Kong (China)
E-Mail bcywong@hku.hk

Cho CH, Purohit V (eds): Alcohol, Tobacco and Cancer.
Basel, Karger, 2006, pp 13–28

....

Epidemiology of Alcohol-Associated Cancers

Linda Morris Brown

Division of Cancer Epidemiology and Genetics, National Cancer Institute, National
Institutes of Health, Department of Health and Human Services, Bethesda, Md., USA

Abstract

Background: The International Agency for Research on Cancer concluded in 1988
that 'Alcoholic beverages are carcinogenic to humans'. **Methods:** Descriptive patterns and
time trends in the United States were examined for alcohol use and for the seven alcohol-
associated tumors (oral cavity and pharynx, squamous cell carcinoma of the esophagus, larynx, liver, colorectum, breast, and pancreas). The epidemiologic evidence on the relation
between alcohol consumption and each tumor type is reviewed. **Results:** Alcohol, especially
in combination with smoking, is a well-established risk factor for cancers of the oral cavity
and pharynx, esophagus, and larynx. Rates of these cancers have been decreasing in recent
years possibly due to reductions in cigarette smoking and alcohol use. Chronic/heavy alcohol
consumption has been linked with increased risk of liver cancer, but rising rates most likely
reflect increases in hepatitis B and C infections. Recent epidemiologic evidence has linked
light to moderate intake of alcohol to cancers of the colorectum and female breast; a role of
heavy alcohol consumption in pancreas cancer is possible. **Conclusions:** Although rates of
smoking and drinking are declining, they remain the major risk factors for cancers of the oral
cavity and pharynx, larynx, and squamous cell esophageal cancer. Chronic/heavy alcohol use
contributes to the risk of liver cancer and moderate levels of alcohol use are associated with
modest increases in the rates of cancers of the colorectum and breast. Heavy intake of alcoholic beverages may be related to a modest increase in pancreas cancer risk.

Alcohol has been consumed by humans for thousands of years. Ancient
Sumerian tablets document the production of beer from barley at least 6,000
years ago [1]. Wine may even be older than beer with our Neolithic ancestors
enjoying the juice of fermented wild grapes and berries [2].

The earliest report of an association between alcohol use and cancer was
made in 1836 by a Boston surgeon, Warren [3] who described a case of tongue

cancer in a tobacco chewer with a 'predisposition' due to chronic use of spirits. An excess of esophageal cancer among alcoholics in Paris in 1910 was the earliest reported 'association' [4]. After a thorough review of the epidemiologic data, the International Agency for Research on Cancer concluded in 1988 that 'Alcoholic beverages are carcinogenic to humans' and that consumption of alcoholic beverages was causally related to the occurrence of cancers of the upper aerodigestive tract (UADT) (i.e. oral cavity, pharynx, larynx, esophagus) and liver [5]. Although the IARC working group found insufficient evidence to implicate alcohol in the etiology of cancers of the breast and colorectum, more recent evidence supports a causal role [6]. The question remains unresolved for cancer of the pancreas.

This chapter reviews the epidemiology of seven alcohol-associated cancers: oral cavity and pharynx, squamous cell carcinoma of the esophagus, larynx, liver, colorectum, breast, and pancreas. It examines similarities and differences in the descriptive patterns and time trends for alcohol use in the United States and for the seven alcohol-associated tumors. It also summarizes the epidemiologic evidence on the relation between alcohol consumption and each tumor type and discusses measurement and other issues related to the collection of alcohol intake data.

United States Trends in Cancer Incidence

Incidence data were obtained from nine Surveillance, Epidemiology, and End Results (SEER) registries (Atlanta, Connecticut, Detroit, Hawaii, Iowa, New Mexico, San Francisco-Oakland, Seattle-Puget Sound, and Utah) that cover approximately 10% of the U.S. population [7]. The SEER*Stat statistical software package was used to calculate age-adjusted (to the 2000 U.S. standard population) incidence rates by sex, race, and 5-year time period of diagnosis (1973–1977, 1978–1982, 1983–1987, 1988–1992, 1993–1997, 1998–2002) [8]. Figures were prepared so that a slope of 10 degrees represents an annual change of 1% by using the ratio of 1 log cycle to 40 years = 1 [9].

Oral and Pharyngeal Cancer
Incidence trends for oral and pharyngeal cancer are displayed in figure 1. Incidence rates in black men show a peak of 28.0/100,000 in 1983–1987 and then a downward trend in recent years dropping 30.4% to a rate of 19.5/100,000 during 1998–2002. There was a 24.5% decrease in incidence for white men from 20.8/100,000 in 1973–1977 to 15.7/100,000 in 1998–2002. Rates for females are 2–3 times lower than for males (in the range of 6.0/100,000–8.9/100,000)

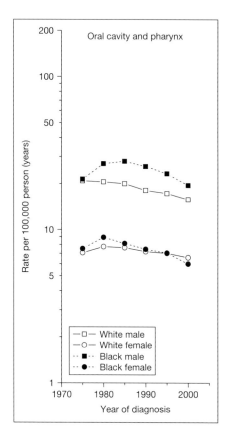

Fig. 1. Age-adjusted (2000 U.S. standard) incidence rates for cancer of the oral cavity and pharynx by race and gender, in 9 SEER registries, 1973–2002.

and also show a downward trend in recent years. It is of interest that rates for white females (6.6/100,000) exceed rates for black females (6.0/100,000) for the first time in 1998–2002.

Squamous Cell Esophageal Cancer

The downward trends in squamous call carcinoma of the esophagus (fig. 2) are evident for all race-sex groups. There was a striking 55.6% decrease in incidence rates for black men over the past two decades from a high of 19.6/100,000 in 1983–1987 to a low of 8.7/100,000 in 1998–2002. White men exhibited a dramatic 47.4% drop in incidence from 3.8/100,000 in 1973–1977 to 2.0/100,000 in 1998–2002. Rates in black women peaked in 1978–1982 at 5.4/100,000 before dropping 42.6% to 3.1/100,000 in the most recent time period. Rates for white females declined 33.3% from a high of 1.5/100,000 in 1983–1987 to 1.0/100,000 in the most recent time period.

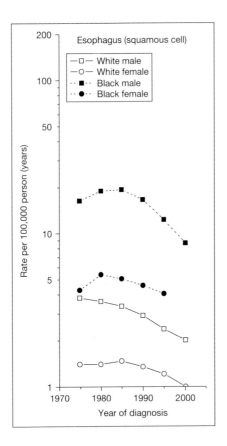

Fig. 2. Age-adjusted (2000 U.S. standard) incidence rates for squamous cell carcinoma of the esophagus by race and gender, in 9 SEER registries, 1973–2002.

Larynx Cancer

Larynx cancer incidence rates among black men peaked at 15.0/100,000 in 1988–1992 and exhibited a downward trend dropping 22.0% in the last two time periods to 11.7/100,000 in 1998–2002 (fig. 3). Rates for white men have shown a 32.6% decline from a high of 9.8/100,000 in 1978–1982 to a low of 6.6/100,000 in 1998–2002. Similar to black men, rates in black women peaked in 1988–1992 at 3.2/100,000 before dropping to 2.6/100,000 in 1998–2002. Trends in white women are less consistent and rates are the lowest of the four race-sex groups ranging from 1.4/100,000 to 1.8/100,000.

Liver Cancer

There have been notable increases in the rates of liver cancer for all race-sex groups since 1978–1982 with blacks having significantly higher rates than whites and males having significantly higher rates than females over the entire

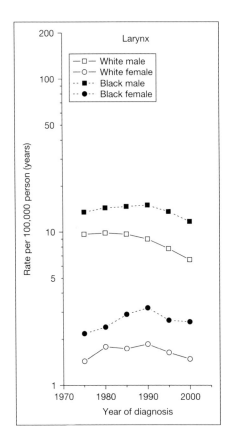

Fig. 3. Age-adjusted (2000 U.S. standard) incidence rates for laryngeal cancer by race and gender, in 9 SEER registries, 1973–2002.

time period (fig. 4). The percent increase from 1978–1982 to 1998–2002 was 83.3% for black men (from 6.0/100,000 to 11.0/100,000), 81.2% for white men (from 3.2/100,000 to 5.8/100,000), 50.0% for black women (from 2.2/100,000 to 3.3/100,000), and 42.9% for white women (from 1.4/100,000 to 2.0/100,000).

Colorectal Cancer

Over the 30 year time period, there was a 12.0% decrease in colorectal cancer incidence for white men (from 71.5/100,000 to 62.9/100,000) and an 18.0% increase in incidence for black men (from 61.8/100,000 to 72.9/100,000) which resulted in a crossing over of the race-specific rates for men around 1990 (fig. 5). The incidence rates for black women exceeded those for white women after 1978–1982. The rate for white women decreased 17.0% from a high of 55.9/100,000 in 1978–1982 to a low of 46.4 in 1998–2002; the rate of decrease

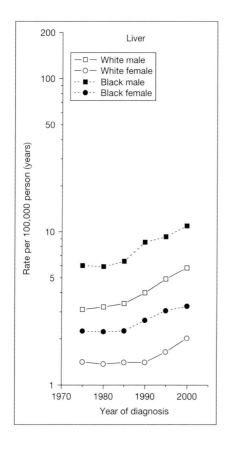

Fig. 4. Age-adjusted (2000 U.S. standard) incidence rates for liver cancer by race and gender, in 9 SEER registries, 1973–2002.

was smaller for black women, 4.9% (from 58.5/100,000 to 56.2/100,000), during the same time period.

Breast Cancer

Breast cancer, the number one cancer among women [10], is the alcohol-associated cancer with the highest incidence rates. It is also the only cancer among this group, where rates for whites exceed those for blacks over the whole time period (fig. 6). Rates for white women increased 34.3% from 106.5/100,000 in 1973–1977 to 143.0/100,000 in 1998–2002; rates for black women increased 38.1% from a low of 88.6/100,000 in 1973–1977 to a high of 122.3/100,000 in 1993–1997 before falling slightly to 118.7/100,000 in the most recent time period.

Pancreas Cancer

Compared to the other alcohol-associated cancers, trends in pancreas cancer incidence have been relatively stable over the past three decades with rates

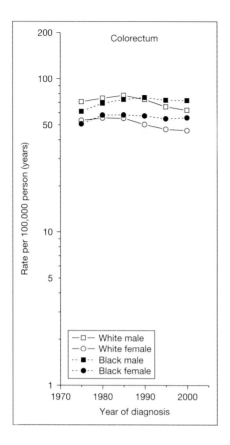

Fig. 5. Age-adjusted (2000 U.S. standard) incidence rates for colorectal cancer by race and gender, in 9 SEER registries, 1973–2002.

for blacks of both sexes exceeding rates for whites since the early 1980s (fig. 7). From 1973 to 2002 incidence rates decreased 18.5% for black men (from 20.3/100,000 to 16.6/100,000) and 14.7% for white men (from 14.9/100,000 to 12.7/100,000). Rates for both black and white women peaked at 15.9/100,000 and 10.1/100,000, respectively in 1983–1987 before dropping to 14.0/100,000 and 9.5/100,000, respectively, in 1998–2002.

United States Trends in Alcohol and Tobacco Use

Data on per capita alcohol consumption obtained from the National Institute on Alcohol Abuse and Alcoholism is presented in figure 8 [11]. Per capita alcohol consumption peaked in the United States in 1980 and 1981 at 2.76 gallons and declined 22.5% to a low of 2.14 gallons in 1997 and 1998 because of lower rates of heavy drinking among whites and higher rates of

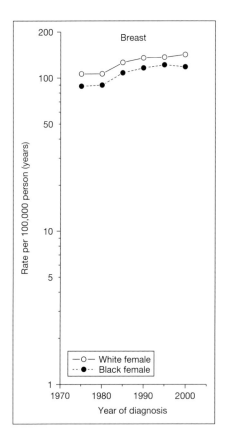

Fig. 6. Age-adjusted (2000 U.S. standard) incidence rates for female breast cancer by race, in 9 SEER registries, 1973–2002.

abstinence among blacks and Hispanics [12]. Recent declines in UADT cancers have mirrored decreases in liquor consumption which peaked in 1969 at 1.13 gallons per capita and has declined 42.5% over the past several decades to 0.65 gallons per capita in 2002. Per capita beer consumption accounted for 55.7% of the total alcohol consumed in 2002 (1.23 gallons), whereas per capita liquor consumption (0.65 gallons) and per capita wine consumption (0.33 gallons) accounted for 29.4% and 14.9%, respectively [11]. A higher percentage of whites drink than blacks and a higher percentage of males drink than females, but a decrease in alcohol use has been observed for all race-sex groups in recent years [12]. The percent of adults 18 years and over who are current smokers declined 46% between 1965 and 2002 (from 41.9% to 22.4%), but the smoking prevalence has remained consistently higher among black than among white men and among men than among women [13].

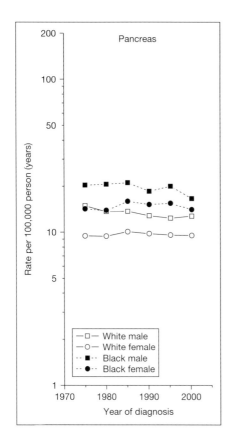

Fig. 7. Age-adjusted (2000 U.S. standard) incidence rates for pancreatic cancer by race and gender, in 9 SEER registries, 1973–2002.

Alcohol Measurement Issues in Epidemiologic Studies

Issues related to collection of alcohol data in epidemiologic studies include difficulties in quantifying prior long-term alcohol intake [14]. Alcohol data is generally obtained from structured questionnaires or interviews, but self-reports often underestimate true consumption. On the other hand, case-control and cohort studies generally assess usual past intake of alcoholic beverages which may overestimate true consumption [15]. Measurement error is also likely to occur when summarizing across different types of alcoholic beverages and studies often fail to adequately control for smoking or other confounding factors. Although there is great interest, it has been difficult to assess risk for different types of alcoholic beverages and to quantify risk for low levels of alcohol intake [16]. There is also little available data on alcohol-associated risks in U.S. minorities, especially Asian Americans, Hispanics and Latinos, and American Indians and Alaskan Natives.

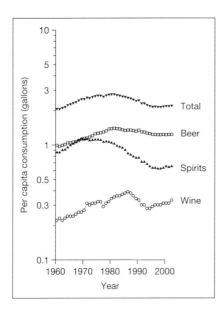

Fig. 8. Trends in per capita alcohol consumption (in gallons) in the United States by type of beverage consumed from 1960 to 2002. (Data from Lakins NE, Williams GD, Yi H-y, Smothers BA. Apparent per capita alcohol consumption: national, state, and regional trends, 1977–2002. National Institute of Alcohol Abuse and Alcoholish, Division of Epidemiology and Prevention Research, Bethesda, MD, 2004).

Alcohol-Associated Cancers: Case-Control and Cohort Studies

Upper Aerodigestive Tract Cancers

Based on various studies, 25–80% of UADT cancers of the oral cavity and pharynx, esophagus, and larynx are attributable to alcohol [17, 18]. Because heavy drinkers and heavy smokers are at greatest risk of these cancers [19–22], it is likely that the overall decline in UADT cancers can be explained in part by changes in smoking and drinking habits.

There has been much speculation in the UADT literature whether the alcohol-associated risk is due to any type of alcohol or if certain types of alcoholic beverages convey different risks. A number of studies in the U.S. and elsewhere have identified higher risks for certain types of beverages; however, the type of high-risk beverage differs in different populations and may be a combination of the most popular type used in an area and the culturally or economically determined drinking habits of the population [17, 20, 23]. Studies conducted in Italy, Switzerland, and Spain have suggested that subjects who drank alcohol out-side of meals (i.e. not with food) were at greater risk of UADT cancers than subjects who drank only with meals possibly due to the more rapid absorption of alcohol [24, 25]. Several studies have suggested that concentrated liquor is associated with a higher risk of cancer of the oral cavity, pharynx, and esophagus than diluted liquor because of local effects on tissue [24, 26–28].

Oral and Pharyngeal Cancer

It is well documented in the epidemiologic literature that alcohol is a major cause of oral and pharyngeal cancer with most studies showing increased risks with increased intensity and duration of use independent of tobacco use [19, 24, 29]. For example, a population-based case-control study in Puerto Rico found a significant dose–response trend with number of drinks of alcohol consumed for men (p < 0.001) reaching 7.7 (95% confidence interval (CI) = 3.3–17.9) for 43 or more drinks per week and an elevated risk of 9.1 (95% CI = 0.9–94.2) for the small number of women who drank 22 or more drinks per week [29]. This study also found that heavy drinkers (≥57 drinks per week) homozygous for the fast-metabolizing alcohol dehydrogenase type 3^1 allele (ADH_3^{1-1}) had substantially increased risks of alcohol-related oral cancer (odds ratio (OR) = 40.1, 95% CI = 5.4–296.0) compared with nondrinkers with the ADH_3^{1-1} genotype [30]. A role for alcohol cessation has been noted in recent studies [24, 26, 29]. Smoking is also a major risk factor for cancers of the oral cavity and pharynx [31]. Multiplicative effects of joint alcohol and tobacco use have been reported [19], with risks reaching 50-fold among heavy users of both products [24, 29, 32].

Squamous Cell Esophageal Cancer

There is strong evidence from epidemiologic studies that alcoholic beverages are a major cause of squamous cell esophageal cancer in Western populations [5, 19, 27]. Strong dose–response relationships for alcohol consumption have been demonstrated in a number of case-control studies around the world after adjustment for smoking [19, 20]. For example, in a population-based case-control study in Atlanta, Detroit, and New Jersey striking dose–response trends (p < 0.001) were seen for both white and black males, but the risks were higher for blacks at each level of consumption. Risks reached 26.9 (95% CI = 11.9–60.9) for black and 16.1 (95% CI = 6.7–38.9) for white consumers of 85 or more drinks per week compared to non or light drinkers of seven or fewer drinks per week [21]. Smoking is also a major risk factor for squamous cell esophageal cancer and heavy consumers of both alcohol and tobacco have the greatest risk of esophageal cancer in most studies [19, 33]. The dose–response gradient for alcohol consumption, however, remained strong in studies that restricted their analysis to life-long nonsmokers [20, 33–35].

Larynx Cancer

Epidemiological studies clearly indicate that consumption of alcoholic beverages of all types is causally related to laryngeal cancer [5, 19]. A dose–response trend with amount consumed has been observed for cancers of both the glottis (intrinsic) and the supraglottis (extrinsic) larynx [19, 23]. The alcohol-associated risks were stronger for the supraglottis than the glottis in some [19, 23, 36], but not all [37, 38] studies that were able to look at site-specific laryngeal cancer risks. A multiplicative effect of joint exposure to alcohol and tobacco has been documented in a number of studies with independent effects seen for both cigarette smoking and alcoholic beverage consumption [5, 19, 20, 36, 39, 40]. In a few studies, elevated laryngeal cancer risk has been noted among nonsmokers in the highest alcohol consumption category [5, 41]. In most populations, however, alcohol use appears to be a much weaker risk factor than smoking [36, 38, 40, 42]. For example, in a case-control study in northern Italy and Switzerland the odds ratios was 42.9 (95% CI = 22.8–80.9) for current smokers of at least 25 cigarettes per day and 5.9 for current drinkers of 8 or more drinks per day [36].

Other Alcohol-Associated Cancers

Liver Cancer

Primary liver cancer has been associated with chronic use of alcohol in a number of epidemiologic studies [43]. However, contrary to the UADT cancers, rates of liver cancer in the U.S. have been increasing steadily over the past 20 years, including more than an 80% increase among white and black men. This is likely a consequence of the increasing prevalence of hepatitis B virus (HBV) and hepatitis C virus (HCV) [6], but may also be related to the concomitant increase in obesity/diabetes [44, 45]. A recent population-based, case-control study found heavy alcohol use of more than 4 drinks per day (OR = 2.1, 95% CI = 0.9–4.7) or chronic use of at least 90 drink-years (OR = 2.4, 95% CI = 1.1–5.1) to be a predictor of hepatocellular carcinoma risk independent of HBV and HCV infections. Synergy (greater than additive effect) between heavy alcohol use and HBV/HCV, diabetes, and obesity were also noted [45, 46].

Colorectal Cancer

Although the IARC working group found insufficient evidence to implicate alcohol in the etiology of colorectal cancer [5], a 1996 WHO Consensus statement indicated that consumption of alcoholic beverages, even at low levels, was causally related to increased risks of colorectal adenomas and cancer [47]. A meta-analysis including six cohort and 16 case-control studies reported relative risks of 1.08 (95% CI = 1.06–1.10), 1.18 (95% CI = 1.14–1.22), and 1.38 (95% CI = 1.20–1.49) for intake of 25 (approximately 2 drinks), 50, and 100 g of alcohol per day, respectively [39]. This is in agreement with an earlier meta-analysis which reported a relative risk of 1.10 (95% CI = 1.05–1.14) for consumers of two drinks per day [48] and a recent pooled analysis of 8 cohort studies that found a relative risk of 1.4 (95% CI = 1.2–1.7) for persons who consumed at least 45 g (>3 drinks) of alcohol per day [49]. Although the alcohol-associated risk for colorectal cancer appears to be moderate, colorectal cancer is the number three cancer among both men and woman [10], thus even modest increases in risk can have important public health implications [39].

Breast Cancer

Recent epidemiologic evidence has linked moderate to heavy intake of alcohol to cancers of the breast [20, 50, 51]. Results from a recent meta-analysis show a consistent pattern of increasing risk for breast cancer starting at about 1–2 drinks/day with risk reaching 1.46 (95% CI = 1.33–1.61) for daily consumers of at last 45 g of alcohol (around four drinks) [15]. It has been estimated that there is about a 10% increase in breast cancer risk for each 10 g of alcohol consumed [15]; however, light consumers (<1 drink per day) of alcoholic beverages do not appear to be at increased breast cancer risk [15, 52, 53]. Recent data suggest that women who consume 1 or more drinks of alcoholic beverage and use postmenopausal hormones are at substantially greater risk of breast cancer than women who drink but do not take hormones [54, 55]. It has been estimated that 4% of breast cancers in developed countries may be attributable to use of alcohol [56].

Pancreas Cancer

Most epidemiologic studies have found little or no support for a causal relationship between total alcohol use and risk of pancreatic cancer [5, 20]. However, a few studies suggest that heavy alcohol consumption may be related to risk of pancreatic cancer [57–60]. For example, Silverman et al. reported risks of 2.0 or higher for black men who drank at least 57 drinks per week (OR = 2.2, 95% CI = 0.9–5.6) and for black women who drank at least 21 drinks per week (OR = 2.5, 95% CI = 1.0–5.9). Smaller risks were observed for white men who drank 57 or more drinks per week (OR = 1.4, 95% CI = 0.6–3.2); no excess risk were observed for white women who drank [59].

Conclusions

Although rates of smoking and drinking are declining, they remain the major risk factors for cancers of the oral cavity and pharynx, larynx, and squamous cell esophageal cancer. Chronic/heavy alcohol use contributes to the risk of liver cancer and moderate levels of alcohol use are associated with modest increases in the rates of cancers of the colorectum and breast. Heavy intake of alcoholic beverages may be related to a modest increase in pancreas cancer risk.

Acknowledgemants

This work was supported by the Intramural Research Program of the NIH, National Cancer Institute.

References

1   Standage T: A History of the World in Six Glasses. New York, Walker & Company, 2005.
2   McGovern PE: Ancient Wine: The Search for the Origins of Viniculture. Princeton, Princeton University Press, 2003.
3   Warren JC: Surgical Observations on Tumours with Cases and Observations. Boston, Crocker & Brewster,1837; quoted by Wynder EL: A corner of history. Malthus and population. Prev Med 4: 378–383.
4   Lamu L: Etude de statistique clinique de 131 cas de cancer de l'oesophage et du cardia. Archives Des Maladies D'Appareil Digestif Et De Maladies De La Nutrition 1910;451–456.
5   International Agency for Research on Cancer: IARC Monographs on the Evaluation of Carcinogenic Risks to Humans, Alcohol Drinking. International Agency for Research on Cancer, 1988, vol 44, pp 167–232, 258–321.
6   Poschl G, Seitz HK: Alcohol and cancer. Alcohol Alcohol 2004;39:155–165.
7   Surveillance, Epidemiology, and End Resuts (SEER) Program: SEER*Stat Database: Incidence – SEER 9 Registries Public-Use, Nov 2004 Submission (1973–2002), National Canter Institute, DCCPS, Surveillance Research Program Cancer Statistics Branch, released April 2005, based on the November 2004 submission. Available from www.seer.cancer.gov. Retrieved October 2005.
8   Surveillance Research Program: National Cancer Institute SEER*Stat software version 6.1.4. Bethesda, MD, National Cancer Institute. Available from seer.cancer.gov/seerstat. Retrieved October 2005.
9   Devesa SS, Donaldson J, Fears T: Graphical presentation of trends in rates. Am J Epidemiol 1995;141:300–304.
10 American Cancer Society, Surveillance Research: Cancer Facts & Figures 2005. Available from www.cancer.org/docroot/STT/content/STT_1x_Cancer_Facts__Figures_2005.asp.   Retrieved October 2005.
11 Latkins NE, Williams GD, Yi H, Smothers BS: Surveillance Report #66: Apparent Per Capita Alcohol Consumption: National, State, and Regional Trends 1977–2002. Bethesda, MD, National Institute on Alcohol Abuse and Alcoholism, Division of Epidemiology and Prevention Research, Alcohol Epidemiology Data System. Retrieved October 2005 from www.niaaa.nih.gov/publications/surveillance.htm
12 Caetano R, Clark CL: Trends in alcohol consumption patterns among whites, blacks and Hispanics: 1984 and 1995. J Stud Alcohol 1998;59:659–668.
13 National Center for Health Statistics: Health, United States, 2004 with Chartbook on Trends in the Health of Americans. Table 60. Hyattsville, MD. Available from http://www.cdc.gov/nchs/hus.htm. Retrieved October 2005.
14 Schottenfeld D: Alcohol as a co-factor in the etiology of cancer. Cancer 1979;43:1962–1966.
15 Collaborative Group on Hormonal Factors in Breast Cancer: Collaborative reanalysis of individual data from 53 epidemiological studies, including 58,515, women with breast cancer and 95,067 women without disease. Br J Cancer 2002;87:1234–1245.
16 Fraumeni JF Jr: Epidemiological opportunities in alcohol-related cancer. Cancer Res 1979;39: 2851–2852.
17 Franceschi S, Talamini R, Barra S, Baron AE, Negri E, Bidoli E, Serraino D, La Vecchia C: Smoking and drinking in relation to cancers of the oral cavity, pharynx, larynx, and esophagus in northern Italy. Cancer Res 1990;50:6502–6507.
18 Brown LM, Hoover R, Silverman D, Baris D, Hayes R, Swanson GM, Schoenberg J, Greenberg R, Liff J, Schwartz A, Dosemeci M, Pottern L, Fraumeni JF Jr: Excess incidence of squamous cell esophageal cancer among US Black men: role of social class and other risk factors. Am J Epidemiol 2001;153:114–122.
19 Thomas DB: Alcohol as a cause of cancer. Environ Health Perspect 1995;103(suppl 8): 153–160.
20 Jensen O, Paine SL, McMichael AJ, Ewertz M: Alcohol; in Schottenfeld D, Fraumeni JF Jr (eds): Cancer Epidemiology and Prevention. New York, Oxford University Press, 1996, pp 290–318.

21 Brown LM, Hoover RN, Greenberg RS, Schoenberg JB, Schwartz AG, Swanson GM, Liff JM, Silverman DT, Hayes RB, Pottern LM: Are racial differences in squamous cell esophageal cancer explained by alcohol and tobacco use? J Natl Cancer Inst 1994;86:1340–1345.

22 Schlecht NF, Franco EL, Pintos J, Negassa A, Kowalski LP, Oliveira BV, Curado MP: Interaction between tobacco and alcohol consumption and the risk of cancers of the upper aero-digestive tract in Brazil. Am J Epidemiol 1999;150:1129–1137.

23 Menvielle G, Luce D, Goldberg P, Bugel I, Leclerc A: Smoking, alcohol drinking and cancer risk for various sites of the larynx and hypopharynx. A case-control study in France. Eur J Cancer Prev 2004;13:165–172.

24 Castellsague X, Quintana MJ, Martinez MC, Nieto A, Sanchez MJ, Juan A, Monner A, Carrera M, Agudo A, Quer M, Munoz N, Herrero R, Franceschi S, Bosch FX: The role of type of tobacco and type of alcoholic beverage in oral carcinogenesis. Int J Cancer 2004;108:741–749.

25 Dal Maso L, La Vecchia C, Polesel J, Talamini R, Levi F, Conti E, Zambon P, Negri E, Franceschi S: Alcohol drinking outside meals and cancers of the upper aero-digestive tract. Int J Cancer 2002;102:435–437.

26 Schlecht NF, Pintos J, Kowalski LP, Franco EL: Effect of type of alcoholic beverage on the risks of upper aerodigestive tract cancers in Brazil. Cancer Causes Control 2001;12:579–587.

27 Brown LM, Hoover R, Gridley G, Schoenberg JB, Greenberg RS, Silverman DT, Schwartz AG, Swanson GM, Liff JM, Pottern LM: Drinking practices and risk of squamous-cell esophageal cancer among Black and White men in the United States. Cancer Causes Control 1997;8:605–609.

28 Huang WY, Winn DM, Brown LM, Gridley G, Bravo-Otero E, Diehl SR, Fraumeni JF Jr, Hayes RB: Alcohol concentration and risk of oral cancer in Puerto Rico. Am J Epidemiol 2003;157:881–887.

29 Hayes RB, Bravo-Otero E, Kleinman DV, Brown LM, Fraumeni JF Jr, Harty LC, Winn DM: Tobacco and alcohol use and oral cancer in Puerto Rico. Cancer Causes Control 1999;10:27–33.

30 Harty LC, Caporaso NE, Hayes RB, Winn DM, Bravo-Otero E, Blot WJ, Kleinman DV, Brown LM, Armenian HK, Fraumeni JF Jr, Shields PG: Alcohol dehydrogenase 3 genotype and risk of oral cavity and pharyngeal cancers. J Natl Cancer Inst 1997;89:1698–1705.

31 International Agency for Research on Cancer: IARC Monographs on the Evaluation of Carginogenic Risks to Humans, Tobacco Smoking. International Agency for Research on Cancer, 1986, vol 38, pp 270–275.

32 Garrote LF, Herrero R, Reyes RM, Vaccarella S, Anta JL, Ferbeye L, Munoz N, Franceschi S: Risk factors for cancer of the oral cavity and oro-pharynx in Cuba. Br J Cancer 2001;85:46–54.

33 Castellsague X, Munoz N, De Stefani E, Victora CG, Castelletto R, Rolon PA, Quintana MJ: Independent and joint effects of tobacco smoking and alcohol drinking on the risk of esophageal cancer in men and women. Int J Cancer 1999;82:657–664.

34 Tavani A, Negri E, Franceschi S, La Vecchia C: Risk factors for esophageal cancer in lifelong non-smokers. Cancer Epidemiol Biomarkers Prev 1994;3:387–392.

35 Cheng KK, Duffy SW, Day NE, Lam TH, Chung SF, Badrinath P: Stopping drinking and risk of oesophageal cancer. BMJ 1995;310:1094–1097.

36 Talamini R, Bosetti C, La Vecchia C, Dal Maso L, Levi F, Bidoli E, Negri E, Pasche C, Vaccarella S, Barzan L, Franceschi S: Combined effect of tobacco and alcohol on laryngeal cancer risk: a case-control study. Cancer Causes Control 2002;13:957–964.

37 Lopez-Abente G, Pollan M, Monge V, Martinez-Vidal A: Tobacco smoking, alcohol consumption, and laryngeal cancer in Madrid. Cancer Detect Prev 1992;16:265–271.

38 Dosemeci M, Gokmen I, Unsal M, Hayes RB, Blair A: Tobacco, alcohol use, and risks of laryngeal and lung cancer by subsite and histologic type in Turkey. Cancer Causes Control 1997;8:729–737.

39 Bagnardi V, Blangiardo M, La Vecchia C, Corrao G: A meta-analysis of alcohol drinking and cancer risk. Br J Cancer 2001;85:1700–1705.

40 Gallus S, Bosetti C, Franceschi S, Levi F, Negri E, La Vecchia C: Laryngeal cancer in women: tobacco, alcohol, nutritional, and hormonal factors. Cancer Epidemiol Biomarkers Prev 2003;12:514–517.

41 Bosetti C, Gallus S, Franceschi S, Levi F, Bertuzzi M, Negri E, Talamini R, La Vecchia C: Cancer of the larynx in non-smoking alcohol drinkers and in non-drinking tobacco smokers. Br J Cancer 2002;87:516–518.

42 Falk RT, Pickle LW, Brown LM, Mason TJ, Buffler PA, Fraumeni JF Jr: Effect of smoking and alcohol consumption on laryngeal cancer risk in coastal Texas. Cancer Res 1989;49:4024–4029.

43 Morgan TR, Mandayam S, Jamal MM: Alcohol and hepatocellular carcinoma. Gastroenterology 2004;127:S87–S96.

44 El Serag HB: Hepatocellular carcinoma: recent trends in the United States. Gastroenterology 2004;127:S27–S34.

45 Yuan JM, Govindarajan S, Arakawa K, Yu MC: Synergism of alcohol, diabetes, and viral hepatitis on the risk of hepatocellular carcinoma in blacks and whites in the U.S. Cancer 2004;101:1009–1017.

46 Marrero JA, Fontana RJ, Fu S, Conjeevaram HS, Su GL, Lok AS: Alcohol, tobacco and obesity are synergistic risk factors for hepatocellular carcinoma. J Hepatol 2005;42:218–224.

47 Scheppach W, Bingham S, Boutron-Ruault MC, Gerhardsson de Verdier M, Moreno V, Nagengast FM, Reifen R, Riboli E, Seitz HK, Wahrendorf J: WHO consensus statement on the role of nutrition in colorectal cancer. Eur J Cancer Prev 1999;8:57–62.

48 Longnecker MP, Orza MJ, Adams ME, Vioque J, Chalmers TC: A meta-analysis of alcoholic beverage consumption in relation to risk of colorectal cancer. Cancer Causes Control 1990;1:59–68.

49 Cho E, Smith-Warner SA, Ritz J, van den Brandt PA, Colditz GA, Folsom AR, Freudenheim JL, Giovannucci E, Goldbohm RA, Graham S, Holmberg L, Kim DH, Malila N, Miller AB, Pietinen P, Rohan TE, Sellers TA, Speizer FE, Willett WC, Wolk A, Hunter DJ: Alcohol intake and colorectal cancer: a pooled analysis of 8 cohort studies. Ann Intern Med 2004;140:603–613.

50 Rohan TE, Jain M, Howe GR, Miller AB: Alcohol consumption and risk of breast cancer: a cohort study. Cancer Causes Control 2000;11:239–247.

51 Feigelson HS, Calle EE, Robertson AS, Wingo PA, Thun MJ: Alcohol consumption increases the risk of fatal breast cancer (United States). Cancer Causes Control 2001;12:895–902.

52 Zhang Y, Kreger BE, Dorgan JF, Splansky GL, Cupples LA, Ellison RC: Alcohol consumption and risk of breast cancer: the Framingham Study revisited. Am J Epidemiol 1999;149:93–101.

53 Mannisto S, Virtanen M, Kataja V, Uusitupa M, Pietinen P: Lifetime alcohol consumption and breast cancer: a case-control study in Finland. Public Health Nutr 2000;3:11–18.

54 Chen WY, Colditz GA, Rosner B, Hankinson SE, Hunter DJ, Manson JE, Stampfer MJ, Willett WC, Speizer FE: Use of postmenopausal hormones, alcohol, and risk for invasive breast cancer. Ann Intern Med 2002;137:798–804.

55 Gallus S, Franceschi S, La Vecchia C: Alcohol, postmenopausal hormones, and breast cancer. Ann Intern Med 2003;139:601–602.

56 Longnecker MP: Alcoholic beverage consumption in relation to risk of breast cancer: meta-analysis and review. Cancer Causes Control 1994;5:73–82.

57 Olsen GW, Mandel JS, Gibson RW, Wattenberg LW, Schuman LM: A case-control study of pancreatic cancer and cigarettes, alcohol, coffee and diet. Am J Public Health 1989;79:1016–1019.

58 Zheng W, McLaughlin JK, Gridley G, Bjelke E, Schuman LM, Silverman DT, Wacholder S, Co-Chien HT, Blot WJ, Fraumeni JF Jr: A cohort study of smoking, alcohol consumption, and dietary factors for pancreatic cancer (United States). Cancer Causes Control 1993;4:477–482.

59 Silverman DT, Brown LM, Hoover RN, Schiffman M, Lillemoe KD, Schoenberg JB, Swanson GM, Hayes RB, Greenberg RS, Benichou J: Alcohol and pancreatic cancer in blacks and whites in the United States. Cancer Res 1995;55:4899–4905.

60 Harnack LJ, Anderson KE, Zheng W, Folsom AR, Sellers TA, Kushi LH: Smoking, alcohol, coffee, and tea intake and incidence of cancer of the exocrine pancreas: the Iowa Women's Health Study. Cancer Epidemiol Biomarkers Prev 1997;6:1081–1086.

Linda Morris Brown, Dr. P.H., CAPT USPHS
Division of Cancer Epidemiology and Genetics, National Cancer Institute
National Institutes of Health
Department of Health and Human Services
6120 Executive Blvd, Room 8026, MSC 7244
Bethesda, MD 20892–7244 (USA)
Tel. +1 301 594 7157, Fax +1 301 402 0081
E-Mail brownl@mail.nih.gov

Cho CH, Purohit V (eds): Alcohol, Tobacco and Cancer.
Basel, Karger, 2006, pp 29–47

....

Alcohol Metabolism, Tobacco and Cancer

Samir Zakhari

Division of Metabolism and Health Effects, National Institute on Alcohol
Abuse and Alcoholism, Bethesda, Md., USA

Abstract

Two modifiable life-style factors, drinking and smoking are the leading causes of
death from cancer. Excessive chronic alcohol consumption is associated with cancer of the
upper aerodigestive tract (esophagus, pharynx, larynx and oral cavity), liver and colon.
Smoking increases the risk of mouth, oropharyngeal, and esophageal cancers as well as cancer of the trachea, bronchus, lung, liver, stomach and urinary bladder. Concurrent smoking
and drinking synergistically interact to further increase the risk of carcinogenesis. The following topics are discussed: (1) Oxidative pathways of alcohol metabolism leading to the
formation of acetaldehyde and reactive oxygen species, and the induction of the microsomal
enzyme cytochrome P450 2E1 (CYP2E1); (2) Genetic variations in enzymes that metabolize alcohol (alcohol dehydrogenase and CYP2E1) and acetaldehyde (aldehyde dehydrogenase) and their relationships to risk for different types of cancer; (3) Formation of
reactive oxygen species, including superoxide, hydrogen peroxide and hydroxyl radicals, and
depletion of mitochondrial glutathione due to alcohol metabolism; (4) DNA–acetaldehyde-
adduct formation following alcohol metabolism and smoking, a critical event in carcinogenesis, and adducts involving DNA and hydroxyethyl radicals, as well as immunogenic
adducts due to acetaldehyde; (5) Tobacco smoking and cancer due to acetaldehyde and procarcinogens that are activated by CYP2E1 and CYP1A2; (6) Alcohol's effects on folate and
methionine cycles that result in hypomethylation of DNA which has been linked to an
increase in cancer risk.

Life-style factors such as diet, drinking and smoking are important
determinants of disease including several types of cancer. Excessive alcohol
consumption and smoking, two potentially modifiable risk factors, are the
leading risk factors for death from cancer worldwide. A causal association has
been reported between chronic alcohol consumption and cancer of the upper

aerodigestive tract (UADT, esophagus, pharynx, larynx, and oral cavity), liver, and colon; for pancreatic and lung cancer an association is suspected [1]. Heavy drinking increases the risk for cancer. For example, the relative risk for esophageal cancer was 5.8 following daily consumption of 72 g (about 6 drinks) of alcohol [2], and that for oral cancer, oropharyngeal cancer, and hypopharyngeal carcinoma was 13.5, 15.2, and 28.6, respectively, when more than 100 g of alcohol was consumed daily [3]. In women, chronic alcohol is associated with increased risk for breast cancer. Details of epidemiological studies of alcohol-associated cancer are discussed in chapter 2. Tobacco smoking is associated with trachea, bronchus and lung cancers, as well as mouth, oropharyngeal, esophageal, stomach, liver, and bladder cancers [4]. Concurrent smoking and drinking, which is common, further increases the risk for cancers of the UADT. For example, Tuyns [5] reported an 18-fold increase in the relative risk for esophageal cancer due to the consumption of more than 80 g/day of alcohol, a 5-fold increase due to smoking more than 20 cigarettes/day, and a 44-fold greater risk for combined heavy alcohol consumption and cigarette smoking. Another study showed more than a 35-fold increase in the odds ratio for oropharyngeal cancer risk in people who consumed two or more packs of cigarettes and more than four alcoholic drinks/day; odds ratios of 6 and 7 were observed in people who only drink or only smoke the same quantities, respectively [6]. Salaspuro and Salaspuro [7] found a 7-fold increase in salivary acetaldehyde in smokers who drank alcohol as compared to non-smokers, which may explain the synergistic effect of alcohol and smoking on UADT cancers.

A number of comprehensive reviews on alcohol and cancer have been published [8–12], and several mechanisms have been proposed for alcohol-induced cancer, including: (1) formation of acetaldehyde; (2) production of reactive oxygen species (ROS) and lipid peroxidation products; (3) changes in folate and methionine metabolism; (4) alcohol-induced increase in estrogen formation in breast cancer; (5) suppressed immune function; and (6) alcohol's solvent action enhancing the bioavailability of carcinogens from tobacco and other sources. The induction of microsomal cytochrome P450 enzymes by alcohol increases the metabolism of procarcinogens, such as nitrosamines, present in tobacco smoke, and may play an important role in the greater risk for cancer due to heavy alcohol consumption and smoking [13].

Interactions between alcohol, tobacco, and retinoid metabolism are discussed in chapter 10. This chapter focuses on ethanol metabolism and how genetic variations in the enzymes responsible for ethanol degradation increase cancer risk. It also addresses the roles of acetaldehyde and ROS in cancers due to alcohol and smoking, as well as folate and methionine metabolism, and their possible contribution to alcohol-induced cancer.

Table 1. Human ADH isozymes

Class	Gene nomenclature		Protein	K_m (mM)	V_{max} (min^{-1})	Amino acid substitution	Tissue
	new	former					
I	ADH1A	ADH1	α	4.0	30	None	Liver
	ADH1B*1	ADH2*1	β$_1$	0.05	4	Arg47, Arg369	Liver, lung
	ADH1B*2	ADH2*2	β$_2$	0.9	350	His47, Arg369	
	ADH1B*3	ADH2*3	β$_3$	40.0	300	Arg47, Cys369	
	ADH1C*1	ADH3*1	γ$_1$	1.0	90	Arg271, Ile349	Liver, stomach
	ADH1C*2	ADH3*2	γ$_2$	0.6	40	Gln271, Val349	
II	ADH4	ADH4	π	30.0	20		Liver, cornea
III	ADH5	ADH5	χ	>1,000	100		Most tissues
IV	ADH7	ADH7	σ(μ)	30.0	1,800		Stomach
V	ADH6	ADH6		?	?		Liver, stomach

K_m and V_{max} values from [99].

Alcohol Metabolism

The liver is the main organ for metabolizing ethanol. The major pathway of oxidative metabolism of ethanol to acetaldehyde in the liver involves cytosolic alcohol dehydrogenase (ADH), of which multiple isoenzymes exist (table 1). The microsomal cytochrome P450 isozymes CYP2E1, 1A2 and 3A4 also contribute to ethanol oxidation in the liver (fig. 1), particularly after chronic ethanol intake to produce acetaldehyde, a highly reactive and toxic molecule that contributes to tissue damage and cancer. CYP2E1 is induced by chronic ethanol consumption and assumes an important role in metabolizing ethanol to acetaldehyde at elevated alcohol concentration. Oxidation of ethanol by ADH is accompanied by the reduction of NAD$^+$ to NADH. Ethanol oxidation thereby generates a highly reduced cytosolic environment in cells (predominantly hepatocytes) where ADH is active. In addition, CYP2E1-dependent ethanol oxidation may occur in other tissues where ADH activity is low. It also produces highly ROS, including hydroxyethyl, superoxide anion, and hydroxyl radicals. Another enzyme, catalase, located in the peroxisomes, is capable of oxidizing ethanol in vitro in the presence of a hydrogen peroxide (H$_2$O$_2$)-generating system, such as NADPH oxidase or xanthine oxidase. Catalase seems to be the main system of oxidizing alcohol in the brain, kidney, and heart [14–17].

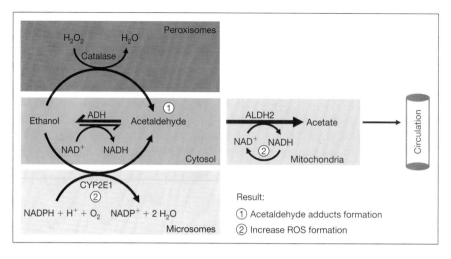

Fig. 1. Oxidative pathways of ethanol metabolism. ADH = Alcohol dehydrogenase; ALDH = aldehyde dehydrogenase; NAD = nicotinamide adenine dinucleotide; H_2O_2 = hydrogen peroxide; NADH = reduced NAD; NADP = nicotinamide adenine dinucleotide phosphate; ROS = reactive oxygen species. CYP1A2 and CYP3A4 may also oxidize ethanol to acetaldehyde [70].

Acetaldehyde, produced by ethanol oxidation through any of the above mechanisms, is rapidly metabolized to acetate, mainly by the low-K_m mitochondrial aldehyde dehydrogenase (ALDH2). However, acetaldehyde has the capacity to bind to proteins such as amino acids and enzymes, microsomal proteins, and microtubules. Acetaldehyde also form adducts with DNA and is thus implicated in cancer. Most of the acetate resulting from ethanol metabolism escapes the liver to the blood and is eventually metabolized to CO_2 by way of the tricarboxylic acid cycle in tissues such as heart, skeletal muscle, and brain.

The consequences of chronic heavy alcohol consumption, including cancer, are modulated by polymorphisms in the alcohol metabolizing enzyme ADH, mutations in ALDH, induction of CYP2E1 leading to the production of ROS and procarcinogen activation, depletion of mitochondrial glutathione (GSH), and folate and methionine metabolism. The sections that follow highlight genetic variations in ADH and ALDH, and the role of acetaldehyde; the formation of ROS and relationships to cancer; the role of alcohol and acetaldehyde adducts in autoimmunity; interactions of tobacco smoking and alcohol; and possible role of folate, methionine and DNA methylation in alcohol-induced cancers.

Genetic Variations in Ethanol Metabolism: Role of Acetaldehyde

Intra- and interindividual, and ethnic variability in ethanol pharmacokinetics have been attributed to both genetic and environmental factors. Variations in the rate of ethanol absorption, distribution and elimination contribute significantly to clinical conditions observed after chronic alcohol consumption.

Class I ADH and ALDH2 play a central role in ethanol metabolism. Allelic variations in the genes encoding ADH and ALDH produce alcohol- and acetaldehyde-metabolizing enzymes that vary in activity. These genotypes modify the susceptibility of developing alcoholism and various types of tissue damage. Cancer risk is high in individuals who accumulate acetaldehyde (due to high-activity ADH variants that increase the rate of acetaldehyde generation, or the low-activity ALDH variant which is unable to metabolize acetaldehyde). The following discussion will focus on variations in ADH and ALDH genotypes and on the activities of various isozymes in relation to cancer.

As shown in table 1, genetic polymorphism occurs at the *ADH1B* and *ADH1C* loci with different catalytic activities for ethanol metabolism. Individuals carrying the *ADH1B*2* or *ADH1B*3* alleles show higher enzymatic activity (40 and 90 times, respectively) compared to individuals with *ADH1B*1* [18, 19]. Studies on Asian populations have shown an association between the slow metabolizing *ADH1B*1* allele and increased risk of esophageal cancer [20–22]. The role of the *ADH1C*1* allele in alcohol-related head and neck cancers is controversial. In one study, pooled analysis of *ADH1C*1/1* and *ADH1C*1/2* genotypes showed no increased risk of head and neck cancers in Caucasians [13], while single studies were contradictory. Studies from Puerto Rico [23], France [24], and Germany [25] showed a significant increase in odds ratios of developing oropharyngeal cancers in patients with the *ADH1C*1/1* genotype. However, other studies could not prove such associations [26–29]. The *ADH1C*2* allele significantly increased the risk of oral squamous cell cancer in people who drink more than 15 drinks/week [30]. An analysis of ADH isomorphs in oropharyngeal tissues as correlates would yield insight into explaining these disparate findings.

Studies on ADH, alcohol consumption, and breast cancer also showed contradictory results. Thus, while a German case-control study reported a reduction in breast cancer risk associated with high consumption of alcohol in women with the fast metabolizing *ADH1B*2* allele [31], Terry et al. [32] found that average lifetime alcohol consumption of 15–30 g/day by women, especially premenopausal, with *ADH1C*1* allele increased breast cancer risk by 2–3-fold. It is possible that the genetic risk factor is not *ADH1C*1* but *ADH1B*2*, a variant of ADH which is in linkage disequilibrium with *ADH1C*1* [33]. The activities of class I ADH are much higher in cancerous than in healthy tissues, and individuals with the *ADH1C*1* allele have an increased risk to develop breast cancer from moderate amounts of ethanol [34]. Hormonal effects of alcohol in conjunction with enzyme isomorphs may greatly influence propensities to develop breast cancer.

Several isozymes of ALDH have been identified, but only the cytosolic ALDH1A1 and mainly the mitochondrial *ALDH2* metabolize acetaldehyde to acetate. There is one significant genetic polymorphism of the *ALDH2* gene, resulting in allelic variants *ALDH2*1* and *ALDH2*2* (the *ALDH2*2* [*Glu487Lys*] gene product is virtually inactive). The low activity *ALDH2*2* isozyme is a deficient phenotype, which is present in about 50% of the Taiwanese, Han Chinese, and Japanese populations [35], and shows virtually no acetaldehyde metabolizing activity in vitro. Individuals who are heterozygous or homozygous for *ALDH2*2* show increased acetaldehyde levels after ethanol consumption [36, 37]. For example, homozygous (*ALDH2*2/2*2*) and heterozygous (*ALDH2*1/2*2*) individuals who consumed alcohol at a dose of 0.1 g/kg body weight showed average peak blood acetaldehyde concentrations that were 18-fold and 5-fold, respectively, that of homozygous (*ALDH2*1/2*1*) individuals who consumed 0.8 g/kg of ethanol [38]. Japanese studies have shown increased risk of UADT cancer associated with the *ALDH2*2* allele [20]. Lymphocytes from drinkers with the inactive isozyme *ALDH2*2* have higher frequencies of sister chromatid exchanges than those from individuals with the active *ALDH2*1* [39], indicating DNA damage and chromosome instability. Although the *ALDH2*2* allele offers protection against alcoholism, it has been associated with an increased risk of various cancers including esophageal and head and neck cancers (see review by Yokoyama and Omori [20]).

Several polymorphisms of CYP2E1, an enzyme which is induced by chronic alcohol consumption, have been identified (*RsaI, DraI, and TaqI*). The *c2* allele of *RsaI* has lower enzyme activity than the *c1* allele. Significant associations between *RsaI* variant and esophageal cancer in Chinese populations [40], *RsaI* and liver cancer in smokers [41], *DraI* and oral leukoplakia (a common premalignant lesion among smokers) [42], *DraI* and oral cavity/pharyngeal cancer in heavy drinkers [27], and *DraI* and lung cancer in Mexican American smokers [43] were reported.

Formation of ROS and Decrease in Antioxidants

ROS, including superoxide ($O_2^{\bullet-}$), hydrogen peroxide (H_2O_2), hypochlorite ion (OCl^-), and hydroxyl ($^{\bullet}OH$) radicals, are generated by many reactions in multiple compartments in the cell, e.g., proteins within the plasma membrane such as NADPH oxidases; lipid metabolism within peroxisomes; and various cytosolic enzymes such as cyclooxygenases. However, in most cells the vast majority of ROS results from electron transport by the mitochondria, a process that uses the controlled oxidation of NADH to produce energy to move protons across the inner mitochondrial membrane. Mitochondria consume molecular oxygen in the cytochrome c oxidase complex of the respiratory chain to provide energy required for ATP synthesis in oxidative phosphorylation without generating ROS. However, a small percentage of oxygen is partially reduced by a single electron in the complex III of respiration to generate superoxide anions. Both acute and chronic alcohol consumption can increase the production of ROS. The major production sites of ROS along the cytochrome chain are complexes I and III. Chronic alcohol consumption decreases select iron–sulfur centers of complex I, cytochrome b content of complex III, the heme content in cytochrome oxidase, and ATP synthase activity. The ability of mitochondria to metabolize acetaldehyde depends not only on the K_m of *ALDH2*, but also on the rate of re-oxidation of the generated NADH to NAD^+ by mitochondrial complex I. Antioxidant enzymes, such as superoxide dismutase (SOD) and glutathione peroxidase (GP) scavenge ROS. Two SOD enzymes exist within cells: SOD1, a copper-containing enzyme that exists primarily in the cytosol, and SOD2, a manganese-dependent enzyme (MnSOD) that is present in the mitochondrial matrix. Both SOD1 and SOD2 convert superoxide into hydrogen peroxide, which is deactivated by GP.

Despite constant generation of ROS, mitochondria remain functional due to an antioxidant defense system that starts with MnSOD which converts $O_2^{\bullet-}$ to H_2O_2. If accumulated, H_2O_2 could generate $^{\bullet}OH$ radicals that result in protein and lipid peroxidation. However, H_2O_2 is normally disposed of by the GSH redox cycle that involves GP and the NADPH-dependant GSSG reductase. Chronic alcohol consumption depletes mitochondrial GSH, in part by decreasing GSH transport into the mitochondria from the cytosolic compartment, and by generating ROS. For more details on alcohol-induced depletion of GSH and the role this plays in liver cancer, the reader is referred to articles by Fernandez-Checa and Kaplowitz [44], and McKillop and Schrum [45].

Oxidative stress occurs when the prooxidant/antioxidant balance is changed in favor of the former. ROS are capable of extracting hydrogen atoms from a hydrogen donor, converting the donor to a free radical, or combining with stable molecules to form free radicals. ROS play an important role in ischemia-reperfusion injury, carcinogenesis, atherosclerosis, diabetes, inflammation and aging.

Ethanol-induced oxidative stress has been attributed to a decrease in the $NAD^+/NADH$ redox ratio, acetaldehyde formation, CYP2E1 induction, hypoxia, cytokine signaling, mitochondrial damage, reduction in antioxidant enzymes particularly mitochondrial and cytosolic glutathione, one electron oxidation of ethanol to 1-hydroxy ethyl radical, and the conversion of xanthine dehydrogenase to xanthine oxidase. ROS formation results in oxidative stress and peroxidation of lipids, proteins and DNA. Peroxidation of mitochondrial membranes initiates membrane permeability transition, which results in the release of cytochrome c into the cytosol. This peroxidation activates caspase 3, ultimately causing apoptosis (programmed cell death). Damage to mitochondrial membranes also causes a decrease in the mitochondrial membrane potential, which lowers cellular ATP levels, promoting necrosis. In addition, CYP2E1 induction by ethanol can activate procarcinogens to carcinogens [46], including N'-nitrosonornicotine and 4-(methylnitrosoamino)-1-butanone, constituents of tobacco smoke [47].

Adduct Formation, DNA Damage, and Cancer

Acetaldehyde Adducts

Acetaldehyde, the first metabolite of alcohol, is carcinogenic in experimental animals [48], and induces chromosomal aberrations, and microsomal and sister chromatid exchange in cultured mammalian cells [49]. It forms adducts with DNA-protein crosslinks, and with reactive lysine residues, cysteine, some aromatic amino acids, terminal α-NH_2 groups (e.g. valine in hemoglobin) and proteins (e.g. erythrocyte membranes, lipoproteins, tubulin, hemoglobin, albumin, collagens, and cytochrome enzymes). Protein adducts have been identified in blood and tissues after alcohol consumption. For example, volunteers who consumed 2 g ethanol/kg body weight over 8 h had a 2.5-fold increase in acetaldehyde–hemoglobin adducts while there was no significant increase in blood acetaldehyde levels [50]. However, the ability of acetaldehyde to form DNA adducts is likely to underlie its role in UADT cancer. These acetaldehyde–DNA adducts are elegantly discussed by Brooks and Theruvathu [51] and in chapter 6 by the same authors. Therefore, the following is a brief discussion of the genotoxicity of acetaldehyde. For a detailed discussion, the reader is referred to chapter 6.

Acetaldehyde interacts with DNA to form DNA adducts (fig. 2), a critical event in chemical carcinogenesis [52]. A stable adduct, N^2-ethyl-2'-deoxyguanosine, was detected in DNA from the liver of alcohol-treated mice [53] and in lymphocytes and granulocytes from alcoholic individuals [54], as well as in urine from healthy volunteers who consumed alcohol [55]. This DNA adduct could be a marker for alcohol-mediated DNA damage, but its biological significance in inducing cancer remains to be determined [51]. Another adduct, $1,N^2$-propanodeoxyguanosine ($1,N^2$-PdG), which is formed from acetaldehyde

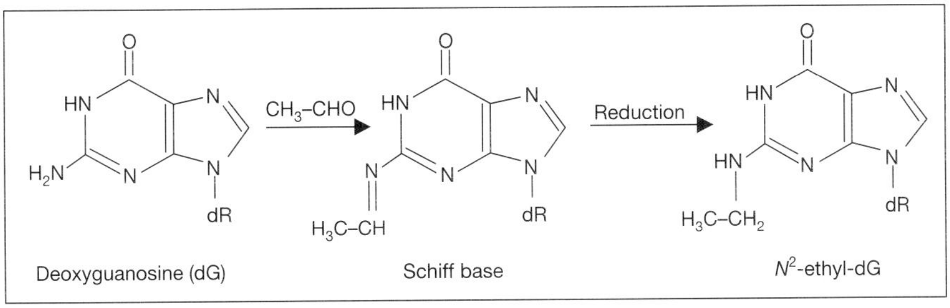

Fig. 2. Acetaldehyde (CH₃–CHO) interacts with deoxyguanosine to form a Schiff base which undergoes reduction by glutathione and ascorbate to form a stable DNA adduct (N^2-ethyl-2′-deoxyguanosine).

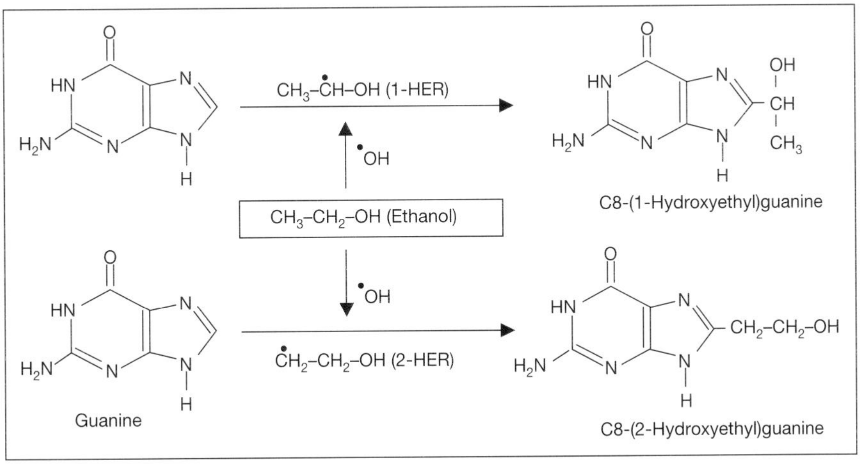

Fig. 3. 1-Hydroxyethan-1-yl, and 1-Hydroxyethan-2-yl radicals form adducts at the C8 position of the guanine base. •OH = Hydroxyl radical.

in the presence of basic molecules such as histones has been linked to the genotoxic and carcinogenic properties of crotonaldehyde [56], and may be more relevant to the carcinogenicity of acetaldehyde [51]. In addition, acetaldehyde significantly decreases the activity of enzymes involved in DNA repair [57–59].

Alcohol Adducts

Apart from acetaldehyde-induced genotoxicity, the alcohol molecule itself, in the presence of hydroxyl radicals, can form hydroxyl ethyl radicals (HER),

Table 2. Metabolites and condensation products (CP) of alcohol metabolism

Metabolites and CP	Source
Acetaldehyde	Ethanol metabolism
Malondialdehyde (MDA)	Non-enzymatic lipid peroxidation of unsaturated fatty acids, arachidonic acid catabolism in platelets
4-Hydroxynonenal (HNE)	Lipid peroxidation of long-chain PUFA
Malondialdehyde–acetaldehyde adduct (MAA)	Hybrid adducts with malondialdehyde and acetaldehyde
Hydroxyethyl radical (HER)	Ethanol oxidation in the presence of iron

which also form adducts with guanine bases in DNA (fig. 3). CYP2E1 is a major source of α-hydroxyethyl radicals [60], adducts of which are immunogenic [61]. Hydroxyethyl radicals can be generated on the external surface of hepatocytes from animals chronically fed ethanol, and, in the presence of monocytes, antihydroxyethyl radical adducts can result in antibody-dependent cell-mediated cytotoxicity [62]. In addition, the formation of ROS due to ethanol metabolism by CYP2E1 and alcohol-induced mitochondrial damage can produce reactive molecules that can bind to lipids, amino acid residues and nucleophilic molecules to produce both stable and unstable condensation products (table 2), and increase lipid peroxide adducts in DNA.

Adducts Associated with the Generation of ROS

The lipid peroxide-derived aldehydes, malondialdehyde (MDA) and 4-hydroxynonenal (HNE) produced by oxidative stress due to alcohol metabolism by CYP2E1 can also form adducts with proteins [63] and would be expected to increase lipid peroxide adducts in DNA. MDA can react with guanine bases in DNA to produce the adduct M_1G which is biologically significant [51], and can also form an intrastrand crosslink in DNA [64]. HNE is epoxidized to form 2,3-epoxy-4-hydroxynonanal (EH) which can react with DNA to generate etheno adducts. Intragastric administration of a single dose of ethanol (5 g/kg body weight) resulted in a statistically significant increase in $3,N^4$-ethenocytosine in liver DNA indicating an increased level of DNA damage produced by lipid peroxidation products [65].

In addition to its genotoxicity, acetaldehyde and MDA react with proteins in a synergistic manner generating the stable adduct, MDA–acetaldehyde–protein adduct (MAA) (fig. 4) [66]. MAA is immunogenic [67] and could contribute to carcinogenesis. While MDA reacts mainly with amino groups in proteins, HNE

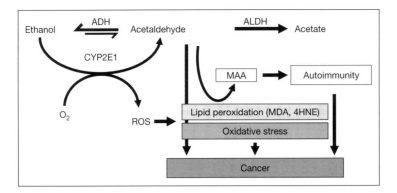

Fig. 4. ADH = Alcohol dehydrogenase; ALDH = aldehyde dehydrogenase; CYP2E1 = cytochrome P450 2E1; MAA = malonedialdehyde acetaldehyde adduct; MDA = malonedialdehyde; ROS = reactive oxygen species; 4HNE = 4-hydroxynonenal.

interacts primarily with lysine, cysteine, and histidine to form relatively stable HNE-protein adducts. While the formation of protein adducts of metabolites derived from the oxidation of ethanol and the formation of antibodies against these is well-established, the degree to which these contribute to carcinogenesis awaits further investigation.

Tobacco, Smoking, and Cancer

Cigarette smoking is causally associated with lung, laryngeal, pharyngeal, esophageal, and oral cavity cancers, the risk of which is significantly increased by concomitant alcohol consumption. Furthermore, tobacco smoking causes cancer of the stomach, liver, pancreas, kidneys, urinary bladder and ureter. Carcinogens in smoke and unburned tobacco include polyaromatic hydrocarbons, nitrosamines, aromatic amines, acetaldehyde, and phenols, among others [68]. CYP1A2 is induced by smoking and activates procarcinogens such as aromatic and heterocyclic amines [47]. In addition to CYP2E1, alcohol is also metabolized by CYP1A2 to form acetaldehyde [69, 70]. Genetic variability in CYP1A1 and CYP2E1 that lead to increased risk of tobacco related cancer were reviewed by Wu et al. [71]. Synergistic interactions between drinking and smoking in inducing cancer are depicted in figure 5.

Acetaldehyde, which is cytotoxic and genotoxic, is formed in comparatively large amounts in cigarette smoke [72] and, upon inhalation, is deposited primarily in the upper respiratory tract, including the mouth [73]. While acetaldehyde is metabolized mainly by the mitochondrial ALDH in the liver, it

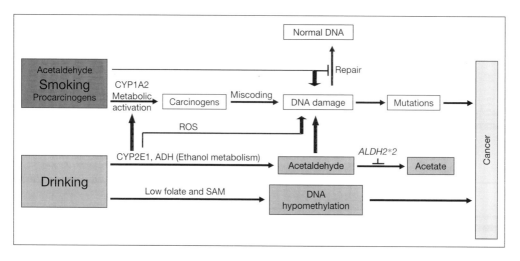

Fig. 5. Synergistic interactions between smoking and drinking resulting in cancer induction. ADH = Alcohol dehydrogenase; ALDH = aldehyde dehydrogenase; CYP1A2 = cytochrome P450 1A2; CYP2E1 = cytochrome P450 2E1; ROS = reactive oxygen species.

is also metabolized in the lungs [74]. Importantly, acrolein, another cigarette smoke aldehyde, was found to inhibit ALDH [75], which has implications for individuals who smoke and drink heavily. Acetaldehyde is metabolized rapidly and forms adducts with protein, which explains the lack of difference observed in acetaldehyde blood levels between smokers (3.7 μmol/l) and non-smokers (3.6 μmol/l) [76].

Alcohol, DNA Methylation, and Cancer Susceptibility

Several dietary factors such as vitamins (B_6, B_{12}), folate, methionine, choline, and zinc can modulate DNA methylation and cancer susceptibility. These factors are involved in one-carbon metabolism and, therefore, influence the supply of methyl groups needed for the methylation process. DNA methylation predominantly involves the addition of a methyl group to the 5′ position of cytosine that precedes a guanosine in the DNA sequence CpG dinucleotide (fig. 6). Methylation of DNA does not change the coding sequence, and therefore is referred to as an epigenetic modification. Changes in methylation patterns may contribute to the development of cancer. For example, methylation of the CpG islands in the promoter region causes the silencing of gene expression and therefore reduces the risk of cancer. On the other hand, global hypomethylation and regional DNA hypermethylation increase the risk for cancer [77]. For

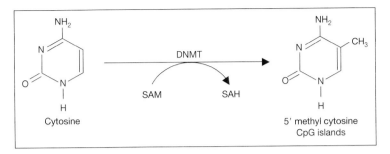

Fig. 6. DNA methylation is a crucial epigenetic modification of the genome involved in transcription, chromatin structure, genomic imprinting, and chromosome stability. DNMT = DNA methyltransferase; SAH = S-adenosylhomocysteine; SAM = S-adenosylmethionine.

example, smoking and alcohol consumption were associated with a higher risk of gastric cancer with hypermethylation of the *hMLH1* gene promoter [78].

The folate cycle plays a central role in one-carbon metabolism, and hence the methylation of DNA (fig. 7). In this cycle, the enzyme glycine hydrox-ymethyltransferase, which uses vitamin B_6 as a co-enzyme, transfers a carbon unit from glycine or serine to tetrahydrofolate (THF) to form 5,10-methylene THF [79]. This compound is then reduced to 5-methyl THF which is used to methylate homocysteine to form methionine, a reaction which is catalyzed by the vitamin B_{12}-dependent enzyme methionine synthase (MS). The enzyme methionine adenosyltransferase then transfers adenosine to methionine to form S-adenosylmethionine (SAM) [80]. Subsequently, SAM donates its labile methyl groups to DNA to form methylated DNA, a reaction catalyzed by DNA methyl-transferase, leading to the formation of S-adenosylhomocysteine (SAH). Folate deficiency leads to inhibition of homocysteine metabolism, reduction of methio-nine and SAM pools, and an increase in cellular levels of SAH (due to the fact that hydrolysis of SAH to homocysteine is a reversible reaction that favors SAH formation). The resulting reduction in methionine levels is partially compensated for by the zinc-requiring enzyme, betaine-homocysteine methyltransferase that catalyzes the transfer of a methyl group from betaine (trimethylglycine) to homocysteine resulting in the formation of methionine and dimethylglycine. However, increased levels of SAH inhibit DNA methyltransferase and conse-quently DNA methylation. Inhibition of DNA methylation due to folate defi-ciency has been associated with colorectal cancer risk [81, 82]. Animal studies showed that folate deficiency resulted in DNA hypomethylation prior to devel-oping colorectal cancer [83].

Excessive alcohol consumption interferes with the bioavailability and metab-olism of folate [84]. About 35 years ago, Halstead et al. [85] demonstrated that

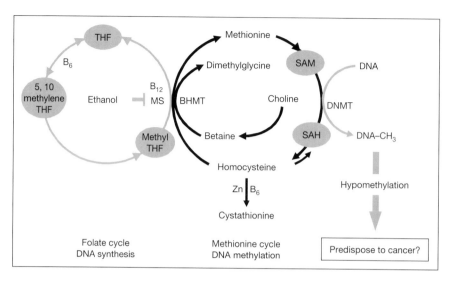

Fig. 7. Role of folate and methionine cycles in DNA methylation. Ethanol inhibits MS leading to a decreased synthesis of methionine and SAM, and increased formation of homocysteine and SAH. BHMT = betaine-homocysteine S-methyltransferase; DNMT = DNA methyltransferase; THF = tetrahydrofolate.

people who drink alcohol in excess malabsorb folate, and in 1989 McMartin et al. [86] showed that acute and chronic alcohol consumption increase urinary excretion of folate by reducing its tubular reabsorption. However, the predominant effect of alcohol is the inhibition of the enzyme MS, which catalyzes the formation of THF and methionine from the substrates 5-methyl THF and homocysteine, respectively [87, 88]. Barak [87] has attributed the inhibition of MS to the formation of adducts with acetaldehyde. As shown in figure 7, inhibition of MS leads to a decrease in SAM level and increases in homocysteine and SAH intracellular concentrations. The increase in homocysteine levels is only about 40% above control levels [88] due to SAH-homocysteine interconversion that thermodynamically favors SAH formation, and the transsulfuration of homocysteine to cystathionine. The reduction in SAM and the concurrent increase in SAH lead to a decrease in DNA methylation (fig. 6) since SAH is a potent inhibitor of numerous methyltransferases including DNA methyltransferase [89] resulting in neoplastic transformation in epithelia [90].

The inhibition of MS by alcohol results in a clinical condition that cannot be corrected by mere folate supplementation. The inefficiency of folate supplementation alone has been demonstrated in rats [91] and humans [92]. As a result, alternative approaches that are not folate- or B_{12}-dependent have been tested, including supplementation with betaine or SAM.

Apart from MS, the liver uses an alternate salvage pathway to metabolize homocysteine to methionine. This pathway is catalyzed by BHMT using betaine as a methyl donor. In addition, homocysteine is metabolized by the transsulfuration pathway which ultimately results in the formation of GSH, allowing methionine to be a precursor for GSH [93]. Thus, as shown in figure 7, betaine supplementation could result in remethylation of homocysteine and formation of methionine. Indeed, betaine was found to attenuate alcohol-induced decrease in hepatic SAM and increases in SAH and homocysteine levels in ethanol-fed rodents [94, 95].

SAM is important for transmethylation, transsulfuration and polyamine synthesis, processes that are critical for intracellular functions including nucleic acid synthesis and DNA methylation [96]. Chronic depletion of hepatic SAM could lead to malignant transformation [97]. Thus supplementation with SAM could be preventative for liver cancer through increased DNA methylation and inhibition of growth-related genes, including H-*ras* and c-*myc* [98]. The role of SAM in alcohol-associated cancer is discussed in chapter 11.

References

1 Boffetta P, Hashibe M: Alcohol and cancer. Lancet Oncol 2006;7:149–156.
2 Boffetta P, Garfinkel L: Alcohol drinking and mortality among men enrolled in an American Cancer Society prospective study. Epidemiology 1990;1:342–348.
3 Bruguere J, Guenel P, Leclerc A, Rodriguez J: Differential effects of tobacco and alcohol in cancer of the larynx, pharynx and mouth. Cancer 1986;57:391–395.
4 Danaei G, Hoorn SV, Lopez AD, Murray CJL, Ezzati M, The Comparative Assessment Collaborating Group (Cancers): Causes of cancer in the world: comparative risk assessment of nine behavioural and environmental risk factors. Lancet 2005;366:1784–1793.
5 Tuyns A: Alcohol and cancer. Alcohol Health Res World 1978;2:20–31.
6 Blot WJ, McLaughlin JK, Winn DM, Austin DF, Greenberg RS, Preston-Martin S, Bernstein L, Schoenberg JB, Stemhagen A, Fraumeni JF Jr: Smoking and drinking in relation to oral and pharyngeal cancer. Cancer Res 1988;48:3282–3287.
7 Salaspuro V, Salaspuro M: Synergistic effect of alcohol drinking and smoking on in vivo acetaldehyde concentration in saliva. Int J Cancer 2004;111:480–483.
8 Poschl G, Seitz HK: Alcohol and cancer. Alcohol Alcohol 2004;39:155–165.
9 Seitz HK, Poschl G, Stickel F: Alcohol and colorectal cancer; in Scheppach W, Scheuerle M (eds): Exogenous Factors in Colonic Carcinogenesis. Boston, London, Kluwer Dordrecht 2003, pp 128–124.
10 Seitz HK, Stickel F, Homann N: Pathogenetic mechanisms of upper aerodigestive tract cancer in alcoholics. Int J Cancer 2004;108:483–487.
11 Stickel F, Schuppan D, Hahn EG, Seitz HK: Cocarcinogenic effects of alcohol in hepatocarcinogenesis. Gut 2002;51:132–139.
12 Salaspuro MP: Alcohol consumption and cancer of the gastrointestinal tract. Best Pract Res Clin Gestroenterol 2003;174:679–694.
13 Brennan P, Lewis S, Hashibe M, Bell DA, Boffetta P, Bouchardy C, Caporaso N, Chen C, Coutelle C, Diehl SR, Hayes RB, Olshan AF, Schwartz SM, Sturgis EM, Wei Q, Zavras AI, Benhamou S: Pooled analysis of alcohol dehydrogenase genotypes and head and neck cancer: a huge review. Am J Epidemiol 2004;159:1–16.
14 Vasiliou V, Ziegler TL, Bludeau P, Peterson DR, Gonzalez FJ, Deitrich RA: CYP2E1 and catalase influence ethanol sensitivity in the central nervous system. Pharmacogen Genomics 2006;16:51–58.

15   Soffia F, Penna M: Ethanol metabolism by rat heart homogenates. Alcohol 1987;4:45–48.

16   Aragon CM, Rogan F, Amit Z: Ethanol metabolism in rat brain homogenates by a catalase-H_2O_2 system. Biochem Pharmacol 1992;44:93–98.

17   Hamby-Mason R, Chen JJ, Schenker S, Perez A, Henderson GI: Catalase mediates acetaldehyde formation from ethanol in fetal and neonatal rat brain. Alcohol Clin Exp Res 1997;21:1063–1072.

18   Bosron WF, Crabb DW, Li T-K: Relationship between kinetics of liver alcohol dehydrogenase and alcohol metabolism. Pharmacol Biochem Behav 1983;18(suppl 1):223–227.

19   Neumark YD, Friedlander S, O'Connor S, Ranchandani V, Carr L, Li TK: The influence of alcohol dehydrogenase polymorphisms on alcohol metabolism among Jewish males in Israel. Alcohol Clin Exp Res 2001;25:126A.

20   Yokoyama A, Omori T: Genetic polymorphisms of alcohol and aldehyde dehydrogenases and risk for esophageal and head and neck cancers. Jpn J Clin 2003;33:111–121.

21   Yokoyama A, Omori T: Genetic polymorphisms of alcohol and aldehyde dehydrogenases and risk for esophageal and head and neck cancers. Alcohol 2005;35:175–185.

22   Higuchi S, Matsushita S, Masaki T, Yokoyama A, Kimura M, Suzuki G, Mochizuki H: Influence of genetic variations of ethanol-metabolizing enzymes on phenotypes of alcohol-related disorders. Ann NY Acad Sci 2004;1025:472–480.

23   Harty LC, Caporaso NE, Hayes RB, Winn DM, Bravo-Otero E, Blot WJ, Kleinman DV, Brown LM, Armenian HK, Fraumeni JF, Shields PG: Alcohol dehydrogenase 3 genotype and risk of oral cavity and pharyngeal cancers. J Natl Cancer Inst 1997;89:1698–1705.

24   Coutelle C, Ward PJ, Fleury B, Quattrocchi P, Chambrin H, Iron A, Couzigou P, Cassaigne A: Laryngeal and oropharyngeal cancer, and alcohol dehydrogenase 3 and glutathione S-tranferase M1 polymorphisms. Hum Genet 1997;99:319–325.

25   Homann N, Stickel F, Konig IR, Jacobs A, Junghanns K, Benesova M, Schuppan D, Himsel S, Zuber-Jerger I, Hellerbrand C, Ludwig D, Caselmann WH, Seitz HK: Alcohol dehydrogenase 1C*1 allele is a genetic marker for alcohol-associated cancer in heavy drinkers. Int J Cancer 2006;118: 1998–2002.

26   Olshan AF, Weissler MC, Watson MA, Bell DA: Risk of head and neck cancer and the alcohol degydrogenase 3 genotype. Carcinogenesis 2001;22:57–61.

27   Bouchardy C, Hirvonen A, Coutelle C, Ward RJ, Dayer P, Benhamou S: Role of alcohol dehydrogenase 3 and cytochrome P-4502E1 genotypes in susceptibility to cancers of the upper aerodigestive tract. Int J Cancer 2000;87:734–740.

28   Sturgis EM, Dahlstrom KR, Guan Y, Eicher SA, Strom SS, Spitz MR, Wei Q: Alcohol dehydrogenase 3 genotype is not associated with risk of squamous cell carcinoma of the oral cavity and pharynx. Cancer Epidemiol Biomarkers Prev 2001;10:273–275.

29   Risch A, Ramroth H, Raedts V, Rajaee-Behbahani N, Schmezer P, Bartsch H, Becher H, Dietz A: Laryngeal cancer risk in Caucasians is associated with alcohol and tobacco consumption but not modified by genetic polymorphisms in class I alcohol-dehydrogenases ADH1B and ADH1C and glutathione-S-transferases GSTM1 and GSTT. Pharmacogenetics 2003;13:225–230.

30   Schwartz SM, Doody DR, Fitzgibbons ED, Ricks S, Porter PL, Chen C: Oral Squamous cell cancer risk in relation to alcohol consumption and alcohol dehydrogenase-3 genotypes. Cancer Epidemiol Biomarkers Prev 2001;10:1137–1144.

31   Lilla C, Koehler T, Kropp S, Wang-Gohrke S, Chang-Claude J: Alcohol dehydrogenase 1B (ADH1B) genotype, alcohol consumption and breast cancer risk by age 50 years in a German case-control study. Br J Cancer 2005;1–3.

32   Terry MB, Gammon MD, Zhang FF, Knight JA, Wang Q, Britton JA, Teitelbaum SL, Neugut AI, Santella RM: ADH3 genotype, alcohol intake and breast cancer risk. Carcinogenesis 2006;27: 840–847.

33   Osier MV, Pakstis AJ, Soodyall H, Comas D, Goldman D, Odunsi A, Okonofua F, Parnas J, Schulz LO, Bertranpetit J, Bonne-Tamir B, Lu R-B, Kidd JR, Kidd KK: A global perspective on genetic variation at the ADH genes reveals unusual patterns of linkage disequilibrium and diversity. Am J Hum Genet 2002;71:84–99.

34   Coutelle C, Hohn B, Benesova M, Oneta CM, Quattrochi P, Roth HJ, Schmidt-Gayk H, Schneeweiss A, Bastert G, Seitz HK: Risk factors in alcohol associated breast cancer: alcohol dehydrogenase polymorphism and estrogens. Int J Oncol 2004;25:1127–1132.

35   Shen YC, Fan JH, Edenberg Hj, Li TK, Cui YH, Wang YF, Tian CH, Zhou CF, Zhou RL, Wang J, Zhao ZL, Xia GY: Polymorphism of ADH and ALDH genes among four ethnic groups in China and effects upon the risk for alcoholism. Alcohol Clin Exp Res 1997;21:1272–1277.

36   Luu SU, Wang MF, Lin DL, Kao MH, Chen ML, Chiang CH, Pai L, Yin SJ: Ethanol and acetaldehyde metabolism in Chinese with different aldehyde dehydrogenase-2 genotypes. Proc Natl Sci Counc Repub China B 1995;19:129–136.

37   Wall TL, Peterson CM, Peterson KP, Johnson ML, Thomasson HR, Cole M, Ehlers CL: Alcohol metabolism in Asian-American men with genetic polymorphisms of aldehyde dehydrogenase. Ann Intern Med 1997;127:376–379.

38   Enomoto N, Takase S, Yasuhara M, Takada A: Acetaldehyde metabolism in different aldehyde dehydrogenase-2 genotypes. Alcohol Clin Exp Res 1991;15:141–144.

39   Morimoto K, Takeshita T: Low Km aldehyde dehydrogenase (ALDH2) polymorphism, alcohol-drinking behavior, and chromosome alterations in peripheral lymphocytes. Environ Health Perspect 1996:104(suppl 3):563–567.

40   Lin D-X, Tang Y-M, Peng Q, Lu S-X, Ambrosone B, Kadlubar FF: Susceptibility to esophageal cancer and genetic polymorphisms in glutathione S-transferases T1, P1, and M1 and cytochrome P450 2E1. Cancer Epidemiol Biomarkers Prev 1998;7:1013–1018.

41   Munaka M, Kohshi K, Kawamoto T, Takasawa S, Nagata N, Itoh H, Oda S, Katoh T: Genetic polymorphisms of tobacco and alcohol-related metabolizing enzymes and the risk of hepatocellular carcinoma. J Cancer Res Clin Oncol 2003;129:355–360.

42   Sikdar N, Mahmud Sk A, Paul RR, Roy B: Polymorphisms in CYP1A1 and CYP2E1 genes and susceptibility to leukoplakia in Indian tobacco users. Cancer Lett 2003;195:33–42.

43   Wu X, Amos CI, Kemp BL, Shi H, Jiang H, Wan Y, Spitz MR: Cytochrome P450 2E1 DraI polymorphisms in lung cancer in minority populations. Cancer Epidemiol Biomarkers Prev 1998;7:13–18.

44   Fernandez-Checa JC, Kaplowitz N: Hepatic mitochondrial glutathione: transport and role in disease and toxicity. Toxicol Appl Pharmacol 2005;204:263–273.

45   McKillop IM, Schrum LW: Alcohol and liver cancer. Alcohol 2005;35:195–203.

46   Ma XL, Baraona E, Lasker JM, Lieber CS: Effects of ethanol consumption on bioactivation and hepatotoxicity of N-nitrosodimethylamine in rats. Biochem Pharmacol 1991;42:585–591.

47   Nishikawa A, Mori Y, Lee I-S, Tanaka T, Hirose M: Cigarette smoking, metabolic activation and carcinogenesis. Curr Drug Metab 2004;5:363–373.

48   International Agency for Research on Cancer: Re-evaluation of some organic chemicals, hydrazine and hydrogen peroxide; in IARC Monographs on the Evaluation on the Carcinogenic Risks to Human. Lyon, IARC, 1999, vol 71, pp 319–335.

49   Dellarco VL: A mutagenicity assessment of acetaldehyde. Mutat Res 1988;195:1–20.

50   Niemela O, Israel Y, Mizoi Y, Fukunaga T, Eriksson CJ: Hemoglobin-acetaldehyde adducts in human volunteers following acute ethanol ingestion. Alcohol Clin Exp Res 1990;14:838–841.

51   Brooks PJ, Theruvathu JA: DNA adducts from acetaldehyde: implications for alcohol-related carcinogenesis. Alcohol 2005;35:187–193.

52   Vaca CE, Fang JL, Schweda EKH: Studies of the reaction of acetaldehyde with deoxynucleosides. Chem Biol Interact 1995;98:51–67.

53   Fang JL, Vaca CE: Development of a 32P-postlabelling method for the analysis of adducts arising through the reaction of acetaldehyde with 2′-deoxygeanosine-3′-monophysphate and DNA. Carcinogenesis 1995;16:2177–2185.

54   Fang JL, Vaca CE: Detection of DNA adducts of acetaldehyde in peripheral white blood cells of alcohol abusers. Carcinogenesis 1997;18:627–632.

55   Matsuda T, Terashima I, Matsumoto Y, Yabushita H, Matsui S, Shibutani S: Effective utilization of N2-ethyl-2′-deoxyguano-sine triphosphate during DNA synthesis catalyzed by mammalian replicative DNA polymerase. Biochemistry 1999;38:929–935.

56   Eder E, Budiawan T: Cancer risk assessment for the environmental mutagen and carcinogen crotonaldehyde on the basis of TD50 and comparison with 1,N2-propanodeoxyguanosine adduct levels. Cancer Epidemiol Biomark Prev 2001;10:883–888.

57   Garro AJ, Espina N, Farinati F, Salvagnin M: The effects of chronic ethanol consumption on carcinogen metabolism and on O6-methylguanine transferase-mediated repair of alkylated DNA. Alcohol Clin Exp Res 1986;10(suppl):73S–77S.

58   Grafstrom RC, Dypbukt JM, Sundqvist K, Atzori L, Neilsen I, Curren RD, Harris CC: Pathobiological effects of acetaldehyde in cultured human epithelial cells and fibroblasts. Carcinogenesis 1994;15:985–990.

59   Brooks PJ, Theruvathu JA: DNA damage resulting from alcohol abuse with special reference to the brian; in Preedy VR, Watson RR (eds): Comprehensive Handbook of Alcohol Related Pathology. Elsevier Academic Press, 2005, pp 1049–1067.

60   Reinke LA, Moore DR, Hange CM, McCay PB: Metabolism of ethanol to 1-hydroxyethyl radicals in rat liver microsomes: comparative studies with three spin trapping agents. Free Radic Res 1994;21:213–222.

61   Moncada C, Torres V, Varghese G, Albano E, Israel Y: Ethanol-derived immunoreactive species formed by free radical mechanisms. Mol Pharmacol 1994;46:786–791.

62   Clot P, Parola M, Bellomo G, Dianzani U, Carini R, Tabone M, Arico S, Ingelman-Sundberg M, Albano E: Plasma membrane hydroxyethyl radical adducts cause antibody-dependent cytotoxicity in rat hepatocytes exposed to alcohol. Gastroenterology 1997;113:265–276.

63   Worall S, Thiele GM: Protein modification in ethanol toxicity. Adverse Drug React Toxicol Rev 2001;20:133–159.

64   Niedernhofer LJ, Daniels JS, Rouzer CA, Greene RE, Marnett LJ: Malondialdehyde, a product of lipid peroxidation, is mutagenic in human cells. J Biol Chem 2003;278:31426–31433.

65   Navasumrit P, Ward TH, O'Conner PJ, Nair J, Frank N, Bartsch H: Ethanol enhances the formation of endogenously and exogenously derived adducts in rat hepatic DNA. Mutat Res 2001;479: 81–94.

66   Tuma DJ: Role of malondialdehyde-acetaldehyde adducts in liver injury. Free Radic Biol Med 2002;32:303–308.

67   Thiele GM, Tuma DJ, Willis MS, Miller JA, McDonald TL, Sorrell MF, Klassen LW: Soluble proteins modified with acetaldehyde and malondialdehyde are immunogenic in the absence of adjuvant. Alcohol Clin Exp Res 1998;22:1731–1739.

68   Wogan GN, Hecht SS, Felton JS, Conney AH, Loeb LA: Environment and chemical carcinogenesis. Semin Cancer Biol 2004;14:473–486.

69   Asai H, Imaoka S, Kuroki T, Monna T, Funae Y: Microsomal ethanol oxidizing system activity by human hepatic cytochrome P450s. J Pharmacol Exp Ther 1996;277:1004–1009.

70   Lieber CS: The discovery of the microsomal ethanol oxidizing system and its physiologic and pathologic role. Drug Metabol Rev 2004;36:511–529.

71   Wu X, Zhao H, Suk R, Christiani D: Genetic susceptibility to tobacco-related cancer. Oncogene 2004;23:6500–6523.

72   Smith CJ, Hansch C: The relative toxicity of compounds in mainstream cigarette smoke condensate. Food Chem Toxicol 2000;38:636–646.

73   Seeman JI, Dixon M, Jaussmann S-J: Acetaldehyde in mainstream tobacco smoke: formation and occurrence in smoke and bioavailability in the smoker. Chem Res Toxicol 2002;1511:1331–1350.

74   Von Wartburg JP: Acetaldehyde; in Sandler M (ed): Psychopharmacology of Alcohol. New York, Raven Press, pp 137–147.

75   Ren S, Kalhorn TF, Slattery JT: Inhibiton of human acetaldehyde dehydrogenase 1 by the 4-hydroxycyclo-phosphamide degradation product acrolein. Drug Metab Dispos 1999;27:133–137.

76   Helander A, Curvall M: Comparison of blood aldehyde dehydrogenase activities in moist snuff users, cigarette smokers and nontobacco users. Clin Exp Res 1991;15:1–6.

77   Davis CD, Uthus EO: DNA methylation, cancer susceptibility, and nutrient interactions. Exp Biol Med (Maywood) 2004;229:988–995.

78   Nan H-M, Song YJ, Yun HY, Park JS, Kim H: Effects of dietary intake and genetic factors on hypermethylation of the hMLH1 gene promoter in gastric cancer. World J Gastroenterl 2005;1125: 3834–3841.

79   Scott JM, Weir DG: Folic acid, homocysteine and one-carbon metabolism: a review of the essential biochemistry. J Cardiovasc Risk 1998;5:223–227.

80   Ross SA: Diet and DNA methylation interactions in cancer prevention. Ann NY Acad Sci 2003;983:197–207.

81   La Vecchia C, Negri E, Pelucchi C, Franceschi S: Dietary folate and coloredtal cancer. Int J Cancer 2002;102:545–547.

82   Kim Y-I: Folate and DNA methylation: a mechanistic link between folate deficiency and colorectal cancer? Cancer Epidemiol Biomarkers Prev 2004;13:511–519.

83   Jacob RA, Gretz DM, Taylor PC, James SJ, Pogribny IP, Miller BJ, Henning SM, Swendseid ME: Moderate folate depletion increases plasma homocysteine and decreases lymphocyte DNA methylation in postmenopausal women. J Nutr 1998;128:1204–1212.

84   Mason JB, Choi S-W: Effects of alcohol on folate metabolism: implications for carcinogenesis. Alcohol 2005;35:235–241.

85   Halsted CH, Robles EA, Mezey E: Decreased jejunal uptake of labeled folic acid (3H-PGA) in alcoholic patients: roles of alcohol and nutrition. New Engl J Med 1971;285:701–706.

86   McMartin KE, Collins TD, Eisenga BH, Fortney T, Bates WR, Bainsfather L: Effects of chronic ethanol and diet treatment on urinary folate excretion and development of folate deficiency in rat. J Nutr 1989;119:1490–1497.

87   Barak A: Ralationship of blood alcohol levels to methionine metabolism in liver. Med Sci Res 1989;17:265–266.

88   Halsted CH, Villanuva JA, Devlin AM, Niemela O, Parkkila S, Garrow TA, Wallock LM, Shigenaga MK, Melnyk S, James SJ: Folate deficiency disturbs hepatic methionine metabolism and promotes liver injury in the ethanol-fed micropig. Proc Natl Acad Sci USA 2002;99: 10072–10077.

89   Yi P, Melnyk S, Pogribna M, Pogribny IP, Hine RJ, James SJ: Increases in plasma S-adenosylhomocysteine and lymphocyte DNA hypomethylation. J Biol Chem 2000;275: 29318–29323.

90   Jones PA, Baylin SB: The fundamental role of epigenetic events in cancer. Nat Rev Genet 2002;3:415–428.

91   Stickel F, Choi SW, Kim YI, Bagley PJ, Seitz HK, Russell RM, Selhub J, Mason JB: Effect of chronic alcohol consumption on total plasma homocysteine level rats. Alcohol Clin Exp Res 2000;24:259–264.

92   Cravo ML, Gloria LM, Selhub J, Nadeau MR, Camilo ME, Resende MP, Cardoso JN, Leitao C, Mira FC: Hyperhomocysteinemia in chronic alcoholism: correlation with folate, vitamin B12, and vitamin B6 status. Am J Clin Nutr 1996;63:220–224.

93   Lu SC: Regulation of hepatic glutathione synthesis: current concepts and controversies. FASEB J 1999;13:1169–1183.

94   Barak AJ, Beckenhauer HC, Mailiard ME, Kharbanda KK, Tuma DJ: Betaine lovers elevated S-adenosylhomocysteine levels in hepatocytes from ethanol-fed rats. J Nutr 2003;133:2845–2848.

95   Ji C, Kaplowitz N: Betaine decreases hyperhomocysteinemia, endoplasmic reticulum stress, and liver injury in alcohol-fed mice. Gastroenterology 2003;124:1488–1499.

96   Lu SC: Regulation of glutathione synthesis. Curr Top Cell Regul 2000;36:95–116.

97   Lu SC, Mato JM: Role of menthionine adenosyltransferase and S-adenosylmethionine in alcohol-associated liver cancer. Alcohol 2005;35:227–234.

98   Pascale RM, Simile MM, Seddaiu MA, Daino L, Vinci MA, Pinna G, Bennati S, Gaspa L, Feo F: Chemoprevention of rat liver carcinogenesis by S-adenosyl-L-methionine: is DNA methylation involved? Basic life Sci 1993;61:219–237.

99   Hurley TD, Edenberg HJ, Li T-K: Pharmacogenomics of Alcoholism; in Licinio J, Wong M-L (eds): Pharmacogenomics – The Search for Individualized Therapies. Weiheim (Germany), Wiley-VCH, 2002.

Samir Zakhari, PhD, Director
Division of Metabolism and Health Effects
National Institute on Alcohol Abuse and Alcoholism
5635 Fishers Lane, Room 2031
Bethesda, MD 20892–9304 (USA)
Tel. +1 301 443 0799, E-mail szakhari@mail.nih.gov

Cho CH, Purohit V (eds): Alcohol, Tobacco and Cancer.
Basel, Karger, 2006, pp 48–62

....

Interaction of Alcohol and Tobacco in Upper Aerodigestive Tract and Stomach Cancer

Mikko Salaspuro[a], *Ville Salaspuro*[a], *Helmut K. Seitz*[b]

[a]Research Unit of Substance Abuse Medicine, Biomedicum Helsinki, University of
Helsinki, Helsinki, Finland; [b]Department of Medicine, Salem Medical Center
Heidelberg and Laboratory of Alcohol Research, Liver Disease and Nutrition,
Heidelberg, Germany

Abstract

There is increasing epidemiological, biochemical, and genetic evidence supporting the
role of the first metabolite of alcohol oxidation – acetaldehyde – as a common denominator
in the pathogenesis of upper aerodigestive tract cancers. Accordingly, acetaldehyde derived
either from ethanol or tobacco appear to act in the digestive tract as a local carcinogen in a
dose-dependent and synergistic way. Alcohol is metabolized to acetaldehyde locally in the
oral cavity mainly by microbes representing normal oral flora. Also tobacco smoke contains
high levels of acetaldehyde, which during smoking is dissolved in the saliva. Via swallowing
salivary acetaldehyde of either origin will be distributed further to the pharynx, esophagus,
and stomach. However, strongest evidence for the local carcinogenic action of acetaldehyde
provide biochemical studies with individuals who have either a decreased ability to detoxify
acetaldehyde or an enhanced ability to produce it. ALDH2-deficient Asian heavy drinkers
and Caucasian alcoholics homozygous for the fast ADH1C*1 allele have markedly increased
risk for aerodigestive tract cancers. Both of these groups also have markedly higher salivary
acetaldehyde levels after drinking of alcohol than in those with the normal enzymes. In
affected individuals the cells of the salivary glands and/or mucous membranes, obviously
contribute in addition to the microbes to the local production of acetaldehyde in the saliva.
On this basis, these genetic variants form an exceptional human model for long-term
acetaldehyde exposure. With regard to cancer prevention, it is important to characterize all
those factors that regulate the concentration of acetaldehyde in the upper aerodigestive tract.
In addition to the hereditary factors, these include smoking and drinking habits, individual
differences in the gut microflora, salivary ethanol levels, and some acetaldehyde-binding
agents.

Heavy drinking, cigarette smoking, and some gene polymorphisms are the main known independent causes of upper aerodigestive tract cancers. In tobacco smoke there are many potentially carcinogenic agents. With regard to alcoholic beverages, there is no evidence that ethanol itself or its additives are carcinogenic. However, there is strong evidence for the local carcinogenic action of acetaldehyde in the digestive tract [1]. Acetaldehyde is the first metabolite of ethanol oxidation and is produced locally in the mouth by some microbes representing the normal oral flora and also by the alcohol dehydrogenase (ADH) enzyme of the salivary glands and mucous membranes. Also tobacco smoke contains high concentrations of acetaldehyde that during smoking becomes dissolved in the saliva. Via swallowing salivary acetaldehyde is distributed from the mouth to the pharynx, esophagus, and stomach.

Other possible mechanisms for alcohol related upper aerodigestive tract cancers include enhanced activation of precarcinogens, alterations in the metabolism of retinol and retinoic acid, enhanced formation of reactive oxygen species via cytochrome P-4502E1 and nutritional deficiencies [2]. The major focus of this chapter will be in the increasing epidemiological, biochemical, and genetic evidence supporting the role of acetaldehyde as a common denominator in the pathogenesis of upper aerodigestive tract cancers. Accordingly, interactions of alcohol and tobacco in the pathogenesis of cancers at following anatomical sites will be discussed in this chapter: oral cavity, tongue, pharynx, larynx, esophagus, and stomach.

Epidemiological Interactions

Independent and Synergistic Effects of Alcohol and Tobacco on Cancer Risk

It has been estimated that up to 80% of upper digestive tract cancers can be avoided by abstaining from alcohol drinking and smoking [3–5]. In a meta-analysis concerning alcohol related cancers and including 235 studies (over 117,000 cases) strong trends in risk were observed for cancers of oral cavity, pharynx, esophagus, and larynx [6]. Pooled relative risks (RR) for alcohol 25, 50 or 100 g per day were for oral cavity and pharynx cancer 1.76, 2.87, and 6.10 for larynx cancer 1.38, 1.94, and 3.95 and for esophageal cancer 1.51, 2.21, and 4.231, respectively. A bottle of wine corresponds approximately to 80 g of alcohol. Less strong direct relation was observed for the cancer of the stomach (RR 1.32/100 g alcohol daily). For all these cancers a significantly increased risk was found also for ethanol intake of 25 g (about two drinks) per day. There was no evidence of a threshold effect for most alcohol-related cancers. This meta-analysis confirms by and large the conclusions of the systematic review of International Agency for Research on Cancer from 1988 [7].

The overall risk of oral cancer among smokers has been estimated to be 7–10 times higher than for never smokers [8]. Furthermore, there is a strong dose–response relationship that has been observed in several studies. In a classical study by Brugere et al. [9] (n = 2,540) it was shown that the RR of oral, pharyngeal, and laryngeal cancers was 3.9 for those smoking 10–19 g and 15.4 for those smoking over 30 g daily. In another study, the RR of esophageal cancer for those smoking 0–10 g daily was 1.0 and for those smoking over 30 g daily it was 7.8 (n = 200) [10].

In USA it has been estimated that 28% of stomach cancer deaths in men and 14% among women are attributable to tobacco use [11]. In a recent European Prospective Investigation into Cancer and Nutrition including 521,468 individuals and 10 European countries a significant association between cigarette smoking and gastric cancer risk was also demonstrated [12]. The hazard ratio for ever smokers was 1.45 and for current smokers 1.73 in males and 1.87 in females. Hazard ratios increased with intensity and duration of cigarette smoked. In both of these studies it was concluded that stomach cancer should be added to the burden of diseases caused by smoking.

The synergistic effect of alcohol and tobacco on aerodigestive tract cancer risk is clearly demonstrated in a recent meta-regression analysis [13]. In this analysis fourteen studies (4,585 cases) met the final selection criteria. The carcinogenic effects of alcohol and tobacco were found to be multiplicative on the RR scale. Tobacco appeared to have a much stronger effect on the larynx than on any other aerodigestive sites, while alcohol's effect was strongest on the pharynx. The effect of alcohol on the esophagus appeared to depend strongly on cell type. Squamous cell carcinoma had a slope nearly 4 times that of adenocarcinoma. The RR for individuals consuming over 30 cigarettes and four or more drinks daily were for different cancer sites as follows: 21.2 oropharynx, 35.6 pharynx, 34.6 larynx, and 12.7 esophagus. Also this meta-analysis confirms the conclusions achieved in several earlier studies [3, 14, 15].

There is an increasing evidence that tobacco, alcohol, and *Helicobacter pylori* may act together in increasing the risk of gastric cancer [11, 16, 17]. This association between three independent carcinogenic factors may explain the international variations in the incidence of gastric cancer – so-called African and Asian enigmas [18].

Effect of Gene Polymorhism on Cancer Risk
Enzymes regulating the degradation and formation of the first metabolite of ethanol oxidation, acetaldehyde, appear to play an important role in the

incidence of alcohol related aerodigestive tract cancers especially among regular or heavy alcohol consumers.

Aldehyde Dehydrogenase Deficiency

Aldehyde dehydrogenase (ALDH2) is polymorphic, located in the mitochondria, and is responsible for most of acetaldehyde oxidation [19]. The ALDH2*2 is caused by a point mutation in chromosome six coding the normal ALDH2*1 allele. Homozygotes for this mutation have very low ALDH2 activity. If they drink alcohol, acetaldehyde accumulates in their blood and other tissues. As a consequence, they develop unpleasant symptoms such as flushing, tachycardia, drop in blood pressure, headache, and nausea. Due to these adverse reactions homozygous ALDH2-deficient subjects are by and large protected against alcoholism [20].

Heterozygotes with ALDH2*1/*2 genotype have 30–50% of the normal ALDH2 activity [19]. Some Asian populations show high frequencies of the ALDH2*2 allele, e.g. about 50% of the Japanese are ALDH2-deficient, while they are extremely rare in Caucasian populations [21]. Acetaldehyde accumulation and flushing reaction are slighter among heterozygotes and therefore they may become heavy drinkers and alcoholics [20].

Many epidemiological studies have uniformly shown that the risk of alcohol related aerodigestive tract cancers is markedly increased in Asians with low-activity ALDH2 enzyme. After adjustment for confounders the RR to those with the normal enzyme were 11.1 for oropharyngolaryngeal, 12.5 esophageal, 3.5 stomach, 3.4 colon, and 8.2 for lung cancer [22]. In a recent meta-analysis including seven studies and 905 cases carried out in Japan, Taiwan, and Thailand these findings with regard to the risk of esophageal cancer were confirmed [23]. The review provided additional evidence for the important role of alcohol intake in the risk of esophageal cancer. Individuals whose genotype results in markedly lower alcohol intake (homozygotic flushers) appear to be protected. However, the most important message of this meta-analysis was that acetaldehyde may play an important carcinogenic role in the pathogenesis of esophageal cancer.

High Active Alcohol Dehydrogenases

The main enzyme for alcohol oxidation is ADH. ADH1C gene is polymorphic and the ADH1C*1 allele metabolizes ethanol to acetaldehyde about 2.5 times faster than the ADH1C*2 allele. Already in 1997 it was reported that French alcoholics with the ADH1C*1/1 (homozygotes) had an almost 4-fold risk of head and neck associated cancer than the individuals without this genotype [24]. This positive finding has later on been confirmed in three other studies [25–27]. Conflicting results – no association – have been reported in four studies [28–31]. In a recent pooled analysis including seven studies (1,325 cases and 1,760 controls) but missing the positive studies of Visapää et al. [26] and Homann et al. [27] an increased risk of head and neck cancer was not observed for the ADH1C*1/2 (OR 1.00) or ADH1C*1/1 genotype (OR 1.14) [32]. However, it should be emphasized that in all studies showing a positive association with the fast metabolizing ADH1C*1/1 (homozygotes) the majority of the cases have been alcoholics or heavy drinkers consuming over 60 drinks per week. Whereas in the negative studies the number of alcoholics and heavy drinkers has been markedly lower.

Acetaldehyde as a Carcinogen

There is sufficient evidence to identify acetaldehyde as a carcinogen in animals [33]. An acetaldehyde inhalation experiment in rats showed an increased incidence of carcinomas in the nasal mucosa [34]. Another inhalation study with hamsters resulted in enhanced number of laryngeal carcinomas [35]. Long-term administration of acetaldehyde in drinking water to rats causes hyperplastic and hyperproliferating changes in the tongue, epiglottis, and the forestomach [36].

Acetaldehyde is able to form several types of DNA adducts [37–39]. However, most of these adducts are formed only at unphysiological acetaldehyde concentrations. Most recently, it was demonstrated that polyamines are able to facilitate the formation of mutagenic α-methyl-γ-hydroxy-1, N2-propano-2′-deoxyguanosine adducts from biologically relevant acetaldehyde concentrations (50–100 μM) [40]. Furthermore, spermidine appeared to be able to directly react with acetaldehyde and generate crotonaldehyde. These findings suggest that acetaldehyde derived from ethanol metabolism in saliva may be converted to crotonaldehyde by polyamines in dividing cells, forming crotonaldehyde adducts, which may be responsible for the carcinogenicity of acetaldehyde [40].

Biochemical Interactions Mediated by Microbes

Accumulation of Acetaldehyde in the Saliva after Alcohol Drinking

Alcohol is absorbed from the stomach and upper duodenum. It is rapidly transported by blood circulation to other organs including mucous membranes and salivary glands. Due to its water solubility ethanol levels in the saliva and other parts of the digestive tract are equal to those of the blood for as long as alcohol is available [41].

Table 1. The effect of alcohol drinking with and without chlorhexidine rinsing, smoking and gene polymorphism on salivary acetaldehyde concentration in relation to acetaldehyde level (50–100 μM) known to be able to form mutagenic DNA-adducts in vitro [40]. Note that after ethanol challenge acetaldehyde levels in the saliva are markedly higher than those in the blood.

Condition	Salivary acetaldehyde, range of peak concentrations or mean (μM)	Blood acetaldehyde (μM) when measured
Caucasian volunteers with normal ALDH after a small dose of alcohol (0.5 g/kg); no ADH-genotyping [42, 71]		Under the limit of detection (<1 μM) in both cases
· without chlorhexidine rinsing	19–143	
· after chlorhexidine rinsing	9–49	
Caucasians with normal ALDH after a moderate dose of alcohol (0.8 g/kg); no ADH-genotyping [50]		
· nonsmokers	11–41	
· smokers	44–90	
Asians after a small dose of alcohol (0.5 g/kg) [43, 71]		
· ALDH2-deficients	33–94	Mean 6.6 μM
· those with normal ALDH2	12–34	<1 μM
Caucasians during smoking [50]	207–614	
Caucasians, alcohol ad libitum, mean blood alcohol concentration 40 mM [26]		
· ADH1C1*1 (fast ADH)	80 (mean)	
· ADH1C2*2 (normal ADH)	38 (mean)	

In the saliva ethanol is effectively oxidized to acetaldehyde by many microbes representing normal oral flora [42]. As compared to the liver the capacity of oral microbes and mucous membranes to detoxify acetaldehyde is limited. Therefore, acetaldehyde concentrations found in the saliva and intestinal contents during and after drinking of alcoholic beverages are 10–100 times higher than in the blood (table 1) [1, 42, 43]. From the mouth salivary acetaldehyde is via swallowing transported to the mucous membranes of the pharynx, esophagus, and stomach.

Microbial Acetaldehyde Production in the Saliva
Many microbes representing normal oral flora show ADH activity with a variety of individual kinetic characteristics. Under anaerobic conditions

prevailing in some parts of the mouth and gingiva these microbes produce energy from glucose through alcoholic fermentation. The reaction runs as follows:

Alcoholic fermentation
Glucose → Acetaldehyde $\xleftarrow{\text{ADH}}$ Ethanol

Under aerobic or microaerobic conditions prevailing in many parts of the mouth and close to the mucous membranes ADH-reaction is reversed with acetaldehyde being the end-product [44]. After drinking of alcohol salivary ethanol is metabolized to acetaldehyde via this reversed reaction. Consequently, mutagenic amounts of acetaldehyde (50–150 μM) can be detected in the saliva of healthy volunteers after ingestion of a moderate dose of ethanol (0.5 g/kg body weight) (table 1) [42]. Salivary acetaldehyde production is reduced by 30–50%, if the volunteers rinse their mouth with an antiseptic mouthwash [42]. This emphasizes the essential role played by the oral microflora in the production of salivary acetaldehyde.

The kinetic characteristics of microbial ADH may vary to a great extent. Some microbial ADHs are not saturated with ethanol. Therefore, in increasing salivary ethanol concentrations the production of acetaldehyde is enhanced [42, 26]. Accordingly, at higher blood and salivary ethanol concentrations, salivary acetaldehyde levels are also higher (table 1). This may explain the well-established epidemiological finding of increased cancer risk associated with more heavier and intoxicating drinking.

Acute Effects of Smoking on Salivary Acetaldehyde

Acetaldehyde is a known constituent of tobacco smoke [45], and may be one of the most toxic compounds in cigarette smoke condensates [46]. It is formed during burning process and inhaled to the aerodigestive tract during smoking [47]. Smokers have been shown to have elevated breath acetaldehyde levels both after acute cigarette smoking and also endogenously after several weeks of smoking cessation [48, 49]. During the period of cigarette smoking, the mean in vivo salivary acetaldehyde concentration is close to 400 μM (fig. 1) [50]. This finding proves that acetaldehyde of tobacco smoke becomes at least in part dissolved in the saliva, and that smoking is one of the most important factors in the regulation of salivary acetaldehyde levels (table 1).

Effect of Chronic Smoking on Microbial Acetaldehyde Production in the Saliva

During ethanol challenge smokers have about 2 times higher salivary acetaldehyde concentration in vivo than nonsmokers [50]. This is most probably

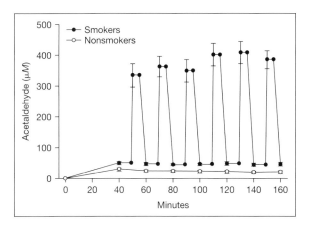

Fig. 1. The synergistic effect of alcohol drinking (0.8 g/kg body weight) and smoking on acetaldehyde levels in the saliva. Note the 7-fold increase in the salivary acetaldehyde concentration during each 5 min period of active smoking. Adopted from Salaspuro and Salaspuro [50].

due to the changes in the capacity of oral microflora to produce acetaldehyde from ethanol in the saliva. Smoking about 20 cigarettes daily increases in vitro salivary acetaldehyde production by about 50% [51]. An increased incidence of oral yeast infections and appearance of oral Gram-positive bacteria has been reported among chronic smokers [52, 53]. On the other hand, in smokers these same groups of microbes also associate with a particularly high in vitro capacity to produce salivary acetaldehyde from ethanol [51].

Effect of Heavy Alcohol Consumption on Microbial
Acetaldehyde Production in the Saliva

Heavy drinking, exceeding 40 g/day increases the in vitro acetaldehyde production in the saliva dose-dependently by about 50% [51]. The induction of ethanol metabolizing microbial enzymes might explain this finding, since high salivary acetaldehyde production is observed only among heavy drinkers.

The effect of chronic smoking and heavy drinking on salivary acetaldehyde production appears to be additive. Together, they increase salivary acetaldehyde production by about 100% as compared to nonsmokers and moderate drinkers [50, 51]. However, due to the high peak level of acetaldehyde in the saliva during active smoking the main risk factors for upper digestive tract cancer – smoking and drinking – appear to increase the salivary acetaldehyde concentration independently, jointly, and synergistically [50].

Effect of Poor Oral Hygiene on Microbial Acetaldehyde
Production in the Saliva

Poor dental hygiene, tooth loss, and insufficient oral hygiene habits are additional but weak risk factors for oral cancer [54, 55]. After adjustment for smoking, alcohol consumption, age and gender, poor oral hygiene associated with about 2-fold increase in salivary acetaldehyde production from ethanol in a study with 132 volunteers [56]. Mainly aerobic bacteria and yeasts appeared to be associated with high acetaldehyde production in these in vitro studies [56, 57].

Biochemical Interactions Mediated by Genes

Strongest evidence for the local carcinogenic effect of acetaldehyde in the aerodigestive tract provide biochemical studies with individuals who have either a decreased ability to detoxify acetaldehyde or an enhanced ability to produce it. As mentioned before, ALDH2-deficient Asian heavy drinkers and Caucasian alcoholics homozygous for the fast ADH1C*1 allele have markedly increased risk for aerodigestive tract cancers. Both of these groups also appeared to have higher salivary acetaldehyde levels after drinking of alcohol than in those with the normal enzymes. These findings indicate that in affected individuals the cells of the salivary glands and/or mucous membranes markedly contribute to the regulation of acetaldehyde concentration in the saliva. On this basis, these individuals form an exceptional human model for long-term acetaldehyde exposure.

ALDH2-Deficient Asians

After ingestion of a moderate dose of alcohol (0.5 g/kg body weight) in vivo salivary acetaldehyde levels are 2–3 times higher in flushing ALDH2-deficient Asians (heterozygotes) than in those with the normal enzyme (fig. 2) (table 1) [43]. In the flushers acetaldehyde concentration of the saliva appeared to be 9 times higher than that in the blood [43]. This indicates that in ALDH2-deficient subjects the main sources of additional salivary acetaldehyde are the salivary glands and mucous membranes – not the blood. In accordance with this, only in flushers the sterile parotid gland saliva contained acetaldehyde [43]. All these results indicate that the ALDH2-enzyme most probably is deficient also in the salivary glands and therefore acetaldehyde derived from ethanol oxidation in these glands is in part secreted to the saliva. This conclusion is further supported by the finding that ADH-inhibitor (4-methylpyrazole) reduces salivary acetaldehyde levels only in ALDH2-deficient individuals [58]. Four-methylpyrazole is a poor inhibitor of microbial ADHs [42] and a null effect in subjects with the normal ALDH2-enzyme suggest that the main source

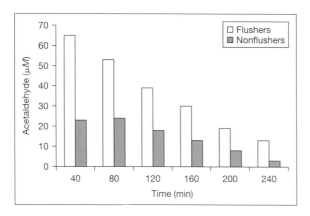

Fig. 2. The effect of ALDH2-deficiency on the concentration of acetaldehyde in the saliva after a dose of alcohol (0.5 g/kg body weight) in flushing (flushers) ALDH2 deficient Chinese volunteers as compared to those with the normal ALDH2-enzyme (nonflushers). Adopted from Väkeväinen et al. [43].

of salivary acetaldehyde in these individuals are the microbes representing the normal oral flora [58].

High ADH-Activity Caucasians

In addition to ALDH2-deficient Asians, salivary acetaldehyde levels are significantly increased after drinking of different doses of alcohol also among Caucasian volunteers, who are homozygous for the fast ADH1C*1 allele [26]. In these individuals, the acetaldehyde production of the cells of the mucous membranes and/or salivary glands is most probably enhanced due to the fast ADH-enzyme.

Acetaldehyde in the Stomach

The possible effect of tobacco smoke on the acetaldehyde concentration of the gastric juice is so far not known. However, during active smoking considerable amounts of salivary acetaldehyde can be expected to reach the stomach via swallowing.

H. pylori is an established risk factor for gastric cancer. This association has been confirmed in several meta analyses. The reported summary odds ratios in different populations range from 1.92 to 3.0 [59–61]. Many *H. pylori* strains possess significant ADH-activity and are able to produce acetaldehyde from ethanol under microaerophilic conditions [62, 63]. So far it is not known whether local acetaldehyde production by *H. pylori* could contribute to the pathogenesis of alcohol related gastric cancer.

Achlorhydric atrophic gastritis is considered to be the major premalignant condition of gastric cancer [64]. The pathogenetic mechanisms behind increased gastric cancer risk in these patients, however, is still without final explanation. Enhanced local microbial production of endogenous ethanol and acetaldehyde could be one of the explaining factors for the increased gastric cancer risk among patients with achlorhydria. Normal human stomach is free of microbes, because of its low pH. However, microbes may survive and even proliferate in exceeding intragastric pH [65]. Accordingly, in patients with achlorhydric atrophic gastritis the stomach is colonized with microbes representing the normal oral flora. In these individuals, bacterial overgrowth results in the presence of glucose, formation of minor concentrations of endogenous ethanol, and acetaldehyde in the gastric juice [66]. Furthermore, in achlorhydria acetaldehyde production increases to 6.5-fold (mean 44.5 μM) after administration of a small amount of alcohol (0.3 g/kg body weight) by gastric infusion as compared to the healthy controls [66].

Possible Future Preventive Actions

The capacity of the microbes representing normal oral flora to produce or to detoxify acetaldehyde varies to a great extent. It remains to be examined whether some of these microbes with particularly high acetaldehyde producing capacity could be over presented in the patients with upper aerodigestive tract cancers, especially among certain risk groups such as smokers, heavy drinkers, ALDH2-deficients and those with high active ADH. This hypothesis is at least in part supported by our earlier finding of increased in vitro acetaldehyde production of the mouthwashings of patients with oral cavity, pharyngeal and laryngeal cancers [67].

Another possible approach could be the trapping of salivary acetaldehyde with l-cysteine. Cysteine is a nonessential amino acid that reacts covalently with acetaldehyde and forms an inactive derivative 2-methylthiazolidine-4-carboxylic acid [68]. By this means cysteine could prevent acetaldehyde to form carcinogenic DNA-adducts. Indeed, up to two-thirds of carcinogenic acetaldehyde formed during ethanol challenge can be bound in human saliva with a tablet releasing slowly and continuously l-cysteine [69]. In this tablet, formula hydroxypropyl methylcellulose and carbomer molecules were used as tablet binders in order to retard the release of l-cysteine from the tablet matrices. In accordance with the above, carcinogenic acetaldehyde could be totally inactivated in the saliva during smoking by sucking a tablet containing only 5 mg of l-cysteine [70]. It remains to be established whether these types of tablets could be used for cancer prevention among above mentioned high-risk groups.

References

1   Salaspuro M: Acetaldehyde, microbes, and cancer of the digestive tract. Crit Rev Clin Lab Med 2003;40:183–208.
2   Pöschl G, Seitz HK: Alcohol and cancer. Alcohol Alcohol 2004;39:155–165.
3   Tuyns AJ, Esteve J, Raymond L, Berrino F, Benhamou E, Blanchet F, Boffetta P, Crosignani P, Del Moral A, Lehman W, Merletti F, Pequignot G, Riboli E, Sancho-Garnier H, Terracini B, Zubiri A, Zubiri L: Cancer of the larynx/hypopharynx, tobacco and alcohol: IARC international case-control study in Turin and Varese (Italy), Zaragoza and Navarra (Spain), Geneva (Switzerland) and Calvados (France). Int J Cancer 1988;41:483–491.
4   Franceschi S, Talamini R, Barra S, Baron AE, Negri E, Bidoli E, Serraino D, La Vecchia C: Smoking and drinking in relation to cancers of the oral cavity, pharynx, larynx, and esophagus in northern Italy. Cancer Res 1990;50:6502–6507.
5   Blot WJ: Alcohol and cancer. Cancer Res 1992;52(suppl):2119–2123.
6   BagnardiV, Blangiardo M, La Vecchia C, Corrao G: A meta-analysis of alcohol drinking and cancer risk. Br J Cancer 2001;85:1700–1705.
7   IARC: Alcohol drinking. IARC monographs on the evaluation of the carcinogenic risk to humans. Lyon, International Agency for Research on Cancer, 1988, vol 44, pp 1–416.
8   Warnakulasuriya S, Sutherland G, Scully C: Tobacco, oral cancer, and treatment of dependence. Oral Oncol 2005;41:244–260.
9   Brugere J, Guenel P, Leclerc A, Rodriguez J: Differential effects of tobacco and alcohol in cancer of the larynx, pharynx, and mouth. Cancer 1986;57:391–395.
10   Tuyns AJ, Pequignot G, Jensen OM: Les cancers del'oesophage an Ille-et-Villaine en fonction de niveaux de consommation d'alcool et de tabac. Des risques qui se multiplient. Bull Cancer 1977:64:45–60.
11   Chao A, Thun MJ, Henley J, Jacobs EJ, McCullough ML, Calle EE: Cigarrette smoking, use of other tobacco products and stomach cancer mortality in US adults: the cancer prevention study II. Int J Cancer 2002;101:380–389.
12   Conzales CA, Pera G, Agudo A, Palli D, Krogh V, Vineis P, Tumino R, Panico S, Berglund G, Siman H, Nyren O, Agren A, Martinez C, Dorronsoro M, Barricarte A, Tormo MJ, Quiros JR, Allen N, Bingham S, Day N, Miller A, Nagel G, Boeing H, Overvad K, Tjonneland A, Bueno-de-Mesquita HB, Boshuizen HC, Peeters P, Numans M, Clavel-Chaplion F, Helen I, Agapitos E, Lund E, Fahey M, Saracci R, Kaaks R, Riboli E: Smoking and the risk of gastric cancer in the European prospective investigation into cancer and nutrition (EPIC). Int J Cancer 2003;107:629–634.
13   Zeka A, Gore R, Kriebel D: Effects of alcohol and tobacco on aerodigestive tract cancer risk: a meta-regression analysis. Cancer Causes Control 2003;14:897–906.
14   Choi SY, Kahyo H: Effect of cigarette smoking and alcohol consumption in the aetiology of cancer of the oral cavity, pharynx and larynx. Int J Epidemiol 1991;20:878–885.
15   Castellague X, Quintana MJ, Martinez MC, Nieto A, Sanchez MJ, Juan A, Monner A, Carrera M, Agudo A, Quer M, Munoz N, Herrero R, Franceschi S, Bosch FX: The role of type of tobacco and type of alcoholic beverage in oral carcinogenesis. Int J Cancer 2004;108:741–749.
16   Chen M-J, Chiou Y-Y, Wu D-C, Wu S-L: Lifestyle habits and gastric cancer in a hospital-based case-control study in Taiwan. Am J Gastroenterol 2000;95:3242–3249.
17   Siman JH, Forsgren A, Berglund G, Floren CH: Tobacco smoking increases the risk for gastric adenocarcinoma among Helicobacter pylori-infected individuals. Scand J Gastroenterol 2001;36: 208–213.
18   Lunet N, Barros H: Helicobacter pylori infection and gastric cancer: facing the enigmas. Int J Cancer 2003;106:953–960.

19   Crabb DW, Edenberg HJ, Bosron WF, Li TK: Genotypes for aldehyde dehydrogenase deficiency and alcohol sensitivity. The inactive ALDH2*2 allele is dominant. J Clin Invest 1989;83:314–316.

20   Peng GS, Wang, MF, Chen CY, Luu SU, Chou HC, Li TK, Yin SJ: Involvement of acetaldehyde for full protection against alcoholism by homozygosity of the variant allele of mitochondrial aldehyde dehydrogenase gene in Asians. Pharmacogenetics 1999;9:463–476.

21   Agarwal DP, Goedde HW: Pharmocogenetics of alcohol metabolism and alcoholism. Pharmacogenetics 1992;2:48–62.

22   Yokoyama A, Muramatsu T, Ohmori T, Yokoyama T, Okoyama K, Takahashi H, Hasegawa Y, Higuchi S, Maruyama K, Shirakura K, Ishii H: Alcohol-related cancers and aldehyde dehydrogenase-2 in Japanese alcoholics. Carcinogenesis 1998;19:1383–1387.

23   Lewis SJ, Smith GD: Alcohol, ALDH2, and esophageal cancer: a meta-analysis which illustrates the potentials and limitations of a Mendelian randomization approach. Cancer Epidemiol Biomarkers Prev 2005;14:1967–1971.

24   Coutelle C, Ward PJ, Fleury B, Quattrocchi P, Chambrin H, Iron A, Couzigou P, Cassaigne A: Laryngeal and oropharyngeal cancer, and alcohol dehydrogenase 3 and glutathione S-transferase M1 polymorphisms. Hum Genet 1997;99:319–325.

25   Harty LC, Caporaso NE, Hayes RB, Winn DM, Bravo-Otero E, Blott WJ, Kleinmen DV, Brown LM, Armenian HK, Fraumeni JF Jr, Shields PG: Alcohol dehydrogenase 3 genotype and risk of oral cavity and pharyngeal cancers. J Natl Cancer Inst 1997;89:1689–1705.

26   Visapää J-P, Götte K, Benesova M, Li J, Homann N, Conradt C, Inoue H, Tisch M, Hörrmann K, Väkeväinen S, Salaspuro M, Seitz HK: Increased cancer risk in heavy drinkers with the alcohol dehydrogenase 1C*1 allele, possibly due to salivary acetaldehyde. Gut 2004;53:871–876.

27   Homann N, Stickel F, König IR, Jacobs A, Junghanns K, Benesova M, Schuppan D, Himsel S, Zuber-Jerger I, Hellerbrand C, Ludwig D, Caselmann WH, Seitz HK: Alcohol dehydrogenase 1C*1 allele is a genetic marker for alcohol-associated cancer in heavy drinkers. Int J Cancer 2005;118:1998–2002.

28   Bouchardy C, Hirvonen A, Coutelle C, Ward PJ, Dayer P, Benhamou S: Role of alcohol dehydrogenase 3 and cytochrome P-4502E1 genotypes in susceptibility to cancers of upper aerodigestive tract. Int J Cancer 2000;87:734–740.

29   Schwartz SM, Doody DR, Fitzgibbons ED, Ricks S, Potter PL, Chen C: Oral squamous cell cancer risk in relation to alcohol consumption and alcohol dehydrogenase-3 genotypes. Cancer Epidemiol Biomarkers Prev 2001;10:1137–1144.

30   Olshan AF, Weissler MC, Watson AM, Bell DA: Risk of head and neck cancer and the alcohol dehydrogenase 3 genotype. Carcinogenesis 2001;22:57–61.

31   Peters ES, McClean MD, Liu M, Eisen EA, Mueller N, Kelsey KT: The ADH1C polymorphism modifies the risk of squamous cell carcinoma of the head and neck associated with alcohol and tobacco use. Cancer Epidemiol Biomarkers Prev 2005;14:476–482.

32   Brennan P, Lewis S, Hahibe M, Bell DA, Bofetta P, Bouchardy C, Caporaso N, Chen C, Coutelle C, Diehl SR, Hayes RB, Olshan AF, Schwartz SM, Sturgis EM, Wei Q, Zavras AI, Benhamou S: Pooled analysis of alcohol dehydrogenase genotypes and head and neck cancer: a HuGe review. Am J Epidemiol 2004;159:1–16.

33   IARC: Re-evaluation of some organic chemicals, hydrazine and hydrogen peroxide; in Monographs on the Evaluation of the Carcinogenic Risk of Chemicals to Humans. Acetaldehyde. Lyon, International Agency for Research on Cancer, 1999, vol 77, pp 319–335.

34   Woutersen RA, Appelman LM, van Garderen-Hoetmer A, Feron VJ: Inhalation toxicity of acetaldehyde in rats. III. Carcinogenicity study. Toxicology 1986;41:213–231.

35   Feron VJ, Kruysse A, Woutersen RA: Respiratory tract tumours in hamsters exposed to acetaldehyde vapour alone or simultaneously to benzo(a)pyrene or diethylnitrosamine. Eur J Cancer Clin Oncol 1982;18:13–31.

36   Homann N, Kärkkäinen P, Koivisto T, Nosova T, Jokelainen K, Salaspuro M: Effects of acetaldehyde on cell regeneration and differentiation of the upper gastrointestinal tract mucosa. J Natl Cancer Inst 1997;89:1692–1697.

37   Fang JL, Vaca CE: Detection of DNA adducts of acetaldehyde in peripheral white blood cells of alcohol abusers. Carcinogenesis 1997;18:627–632.

38   Wang M, McIntee EJ, Cheng G, Shi Y, Villalta PW, Hecht SS: Identification of DNA adducts of acetaldehyde. Chem Res Toxicol 2000;13:1149–1157.

39   Inagaki S, Esaka Y, Goto M, Deyashiki Y, Sako M: LC-MS study on the formation of cyclic 1, N2-propane guanine adduct in the reactions of DNA with acetaldehyde in the presence of histone. Biol Pharm Bull 2004;27:273–276.

40   Theravathu JA, Jaruga P, Nath RG, Dizdaroglu M, Brooks PJ: Polyamines stimulate the formation of mutagenic 1, N2-propanodeoxyguanosine adducts from acetaldehyde. Nucleic Acids Res 2005;33:3513–3520.

41   Jones AW: Distribution of ethanol between saliva and blood in man. Clin Exp Pharmacol Physiol 1979;6:53–59.

42   Homann N, Jousimies-Somer H, Jokelainen K, Heine R, Salaspuro M: High acetaldehyde levels in saliva after ethanol consumption: methodological aspects and pathogenetic implications. Carcinogenesis 1997;18:1739–1743.

43   Väkeväinen S, Tillonen J, Agarwal DP, Svirastava N, Salaspuro M: High salivary acetaldehyde after a moderate dose of alcohol in ALDH2-deficient subjects: strong evidence for the local carcinogenic action of acetaldehyde. Alcohol Clin Exp Res 2000;24:873–877.

44   Salaspuro V, Nyfors S, Heine R, Siitonen A, Salaspuro M, Jousimies-Somer H: Ethanol oxidation and acetaldehyde production in vitro by human intestinal strains of Escherichia coli under aerobic, microaerobic, and anaerobic conditions. Scand J Gastroenterol 1999;34:967–973.

45   IARC: Monographs on the evaluation of the carcinogenic risk of chemicals to humans; in Tobacco Smoking. Lyon, International Agency for Research on Cancer, 1986, vol 38.

46   Smith CJ, Hansch C: The relative toxicity of compounds in mainstream cigarette smoke condensate. Food Chem Toxicol 2000;38:637–646.

47   Seeman JI, Dixon M, Haussman H-J: Acetaldehyde in mainstream tobacco smoke: formation and occurrence in smoke and bioavailability in the smoker. Chem Res Toxicol 2002;15:1332–1350.

48   Shaskan EG, Dolinsky ZS: Elevated endogenous breath acetaldehyde levels among abusers of alcohol and cigarettes. Prog Neuropsychopharmacol 1984;9:267–272.

49   McLaughlin SD, Scott BK, Peterson SM: The effect of cigarette smoking on breath and whole blood-associated acetaldehyde. Alcohol 1990;7:285–287.

50   Salaspuro V, Salaspuro M: Synergistic effect of alcohol drinking and smoking on in vivo acetaldehyde concentration in saliva. Int J Cancer 2004;111:480–483.

51   Homann N, Tillonen J, Meurman JH, Rintamäki H, Lindqvist C, Rautio M, Jousimies-Somer H, Salaspuro M: Increased salivary acetaldehyde levels in heavy drinkers and smokers: a microbiological approach to oral cavity cancer. Carcinogenesis 2000;22:663–668.

52   Holmstrup P, Besserman M: Clinical, therapeutic and pathogenetic aspects of chronic oral multifocal candiasis. Oral Surg Oral Med Oral Pathol 1983;56:388–395.

53   Colman G, Beighton D, Chalk AJ, Wake S: Cigarrette smoking and the microbial flora of the mouth. Aust Dent J 1976;21:111–118.

54   Zheng TZ, Boyle P, Hu HF, Duan J, Jian PJ, Ma DQ, Shui LP, Niu SR, Scully C, MacMahon B: Dentition, oral hygiene, and risk of oral cancer: a case-control study in Beijing, People's Republic of China. Cancer Causes Control 1990;1:235–241.

55   Maier H, Zöller J, Herrmann A, Kreiss M, Heller WD: Dental status and oral hygiene in patients with head and neck cancer. Otolaryngol Head Neck Surg 1993;1088:655–661.

56   Homann N, Tillonen J, Rintamäki H, Salaspuro M, Lindqvist C, Meurman JH: Poor dental status increases the acetaldehyde production from ethanol in saliva: a possible link to the higher risk of oral cancer among alcohol-consumers. Oral Oncol 2001;37:153–158.

57   Tillonen J, Homann N, Rautio M, Jousimies-Somer H, Salaspuro M: Role of yeasts in the salivary acetaldehyde production from ethanol among risk groups for ethanol-associated oral cavity cancer. Alcohol Clin Exp Res 1999;23:1409–1415.

58   Väkeväinen S, Tillonen J, Salaspuro M: 4-Methylpyrazole decreases salivary acetaldehyde levels in ALDH2-deficient subjects but not in subjects with normal ALDH2. Alcohol Clin Exp Res 2001;25:829–834.

59   Huang J-Q, Subbaramiah S, Chen Y, Hunt RH: Meta-analysis of the relationship between Helicobacter pylori seropositivity and gastric cancer. Gastroenterology 1998;114:1169–1179.

60   Eslick GD, Lim LL, Byles JE, Xia HH, Talley NJ: Association of Helicobacter pylori infection with gastric carcinoma: a meta-analysis. Am J Gastroenterol 1999;94:2373–2379.

61 Xue F-B, Xu Y-Y, Pan B-R, Fan D-M: Association of H. pylori infection with gastric carcinoma: a meta analysis. World J Gastroenterol 2003;7:801–804.
62 Salmela KS, Roine RP, Koivisto T, Höök-Nikanne J, Kosunen TU, Salaspuro M: Characteristics of Helicobacter pylori alcohol dehydrogenase. Gastroenterology 1993;105:325–330.
63 Salmela KS, Roine RP, Höök-Nikanne J, Kosunen TU, Salaspuro M: Acetaldehyde and ethanol production by Helicobacter pylori. Scand J Gastroenterol 1994;29:309–312.
64 Morson BC, Sobin LH, Grundmann E, Johansen A, Nagayo T, Serck-Hanssen A: Precancerous conditions and epithelial dysplasia in the stomach. J Clin Pathol 1980;33:711–721.
65 Stockbruegger RW, Cotton PB, Menon GG, Beilby JO, Bartholomew BA, Hill MJ, Walters CL: Pernicious anaemia, intragastric bacterial overgrowth and possible consequences. Scand J Gastroenterol 1984;19:355–364.
66 Väkeväinen S, Mentula S, Nuutinen H, Salmela KS, Jousimies-Somer H, Färkkilä M, Salaspuro M: Ethanol-derived microbial production of carcinogenic acetaldehyde in achlorhydric atrophic gastritis. Scand J Gastroenterol 2002;37:648–655.
67 Jokelainen K, Heikkonen E, Roine R, Lehtonen H, Salaspuro M: Increased acetaldehyde production by mouthwashings from patients with oral cavity, laryngeal, or pharyngeal cancer. Alcohol Clin Exp Res 1996;20:1206–1210.
68 Sprince H, Parker CM, Smith GG, Gonzales LJ: Protection against acetaldehyde toxicity in the rat by l-cysteine, thiamin and l-2-methylthiazolidine-4-carboxylic acid. Agents Actions 1974;4:125–130.
69 Salaspuro V, Hietala J, Kaihovaara P, Pihlajarinne H, Marvola M, Salaspuro M: Removal of acetaldehyde from saliva by a slow-release buccal tablet of l-cysteine. Int J Cancer 2002;97:361–364.
70 Salaspuro V, Hietala JM, Marvola MK, Salaspuro MP: Eliminating carcinogenic acetaldehyde by cysteine from saliva during smoking. Cancer Epidemiol Biomarkers Prev 2006;15:146–149.
71 Sarkola T, Iles MR, Kohlenberg-Mueller K, Eriksson CJ: Ethanol, acetaldehyde, acetate, and lactate levels after alcohol intake in white men and women: effect of 4-methylpyrazole. Alcohol Clin Exp Res 2002;26:239–245.

Prof. Mikko Salaspuro, MD, PhD
Research Unit of Substance Abuse Medicine
University Central Hospital of Helsinki
Biomedicum Helsinki, PL 700
FIN–00029 HUS Helsinki (Finland)
Tel. +358 9 47171860, Fax +358 9 47171862
E-Mail mikko.salaspuro@hus.fi

Cho CH, Purohit V (eds): Alcohol, Tobacco and Cancer.
Basel, Karger, 2006, pp 63–77

....

Alcohol and Cancer of the Large Intestine

Helmut K. Seitz[a]*, Gudrun Pöschl*[a]*, Mikko Salaspuro*[b]

[a]Department of Medicine, Salem Medical Centre and Laboratory of Alcohol
Research, Liver Disease and Nutrition, Heidelberg, Germany;
[b]Research Unit of Substance Abuse Medicine, Biomedicum Helsinki,
University of Helsinki, Helsinki, Finland

Abstract

Epidemiological data from humans and biochemical data from animal experiments
have identified chronic alcohol consumption as a significant risk factor for polyps and cancer of the colorectum. Mechanisms involved in ethanol associated carcinogenesis emphasize
the action of acetaldehyde (AA), the first and most toxic metabolite of ethanol oxidation.
Chronic alcohol consumption enhances chemically induced colorectal carcinogenesis, and
when AA degradation is inhibited by an aldehyde dehydrogenase (ALDH) inhibitor, cancer
development is further enhanced. AA can be produced from ethanol by gastrointestinal
mucosal enzymes and by microbes representing normal gut flora. During chronic alcohol
administration high concentrations of AA can be detected in the mucosa of the large intestine
and this associates with increased mucosal crypt cell production rates and hyperregeneration.
Genetic linkage studies in Japan have identified individuals with a deficient ALDH2-
enzyme, and elevated AA concentrations to have a special risk for colorectal cancer. In
Caucasians, a similar observation has been made for the alcohol dehydrogenase (ADH)1C*1
allele. This allele encodes for an ADH enzyme with 2.5 times higher AA production as compared to the ADHC1*2 allele. Thus, individuals homozygous for ADH1C*1 seem also to
have an increased risk for large intestinal cancer. In addition to AA, reactive oxygen species
produced by Cytochrome P-4502E1 may also contribute to alcohol associated carcinogenesis. There is a striking induction of Cytochrome P-4502E1 in the colorectal mucosa of alcoholics, which disappears within 3–4 days after withdrawal. The administration of radical
scavengers such as alpha-tocopherol to rats almost normalizes colorectal cellular hyperproliferation, supporting the idea of involvement of reactive oxygen species in carcinogenesis.
Nutritional deficiencies such as those of folate, vitamin B6 and retinoids may also be risk
factors for colorectal cancer.

Multiple mechanisms are involved in alcohol associated carcinogenesis of the gastrointestinal tract including acetaldehyde (AA), the first metabolite of ethanol oxidation, the induction of Cytochrome P-4502E1 (CYP2E1), reactive oxygen species (ROS) and an enhanced procarcinogen activation, and as well as the modulation of cellular regeneration and nutritional deficiency. In this review, major emphasis is laid on the effect of AA and its role in cancer of the colorectum (CR). In addition, some aspects of the role of CYP2E1 in carcinogenesis will be discussed. The role of nutritional factors, such as retinoids and folate as well as alcohol associated effects on transmethylation will be discussed elsewhere within this book. Therefore, it is referred to a recent review article [1].

Epidemiology

Breslow and Enstrom [2] were the first to consider the possibility of an association between beer drinking and the occurrence of rectal cancer. In 1992, a review on this issue was conducted by Kune and Vitetta [3] gathering the results of more than 50 major epidemiologic studies from 1957–1991. This review included seven correlation studies, more than 40 case control studies and 17 prospective cohort studies on the role of alcohol on the development of colorectal cancer (CRC). An association was found in five of the seven correlation studies and in more than half of the case control studies. In the majority of case control studies in which community-based individuals were used as controls, a positive correlation between alcohol consumption and CRC was detected. However, this was not the case when hospital-based controls were used, possibly due to the high prevalence of alcohol consumption and alcohol related diseases in hospital controls. Eleven of the seventeen cohort studies also demonstrated a positive association with alcohol. A positive trend with respect to dose–response was found in five of the ten case control studies and in all prospective cohort studies in which a dose–response analysis has been taken into consideration.

Between 1992 and 1997 another twelve epidemiological studies have been published [4–15] and the data on alcohol consumption and the risk for CRC have not been consistent. A prospective cohort study in Japan reported a positive dose–response relationship between alcohol intake and colon cancer risk in both sexes [13]. On the other hand, a Danish population based cohort study showed no association [14].

Most recently, Cho et al. [15] pooled eight cohort studies from North America and Europe and showed a significant trend between an increased amount of alcohol intake and the risk for CRC. Consumption of more than 45 g alcohol per day increased the risk by 45%.

Five out of six studies of the effect of alcohol on the occurrence of adeno-matous polyps in the large intestine showed a positive association with alcohol [3, 16]. The same was true for hyperplastic polyps. When more than 30 g alcohol per day was consumed the relative risk for men was 1.8 and for women 2.5 [17].

Finally, alcohol may influence the adenoma–carcinoma sequence at differ-ent early steps as reported recently by Boutron et al. [10]. High alcohol intake favours high risk polyps or CRC occurrence among patients with adenoma [18]. The authors reported that a reduction in ethanol intake for individuals with genetic predisposition for CRC had a large beneficial effect on tumour inci-dence [19].

In 1999, epidemiological data on alcohol and CRC were reviewed by a panel of experts at the WHO Consensus Conference on Nutrition and CRC [20]. It was concluded that although the data are still somewhat controversial, chronic alcohol ingestion even at low daily intake (10–40 g), especially when consumed as beer, results in a 1.5–3.5-fold risk for rectal cancer and to a lesser extent for colonic cancer in both sexes, but predominantly in males. Epidemiologic studies also underline the importance of nutritional deficiencies such as methionine, folate and vitamin B6, which may modulate the ethanol associated CRC risk [4, 21].

Subsequently, a recent prospective follow-up study of more than 10,000 U.S. citizens concluded that the consumption of one or more alcoholic bever-ages per day at baseline is associated with an approximately 70% greater risk of colon cancer with a strong positive dose–response relationship [22]. The most important factor for CRC appeared to be liquor consumption.

Animal Experiments

The results of animal experiments on alcohol and CRC depend on the experimental design, the type of carcinogen used, its time, duration of exposure and dosage, as well as the route of alcohol administration. Chronic alcohol administration alone, without the application of a primary or secondary car-cinogen dose does not produce tumours [23]. Table 1 summarizes the effect of chronic ethanol ingestion on CRC [24–31]. In two of the eight studies ethanol was given in the drinking water, and the results of these experiments therefore have to be questioned [24, 25].

When the two procarcinogens, dimethylhydrazine (DMH) and azoxymethane (AOM) were used to induce colorectal tumours, different results were reported, depending on the experimental conditions [26–29]. In these studies it is important to note that both compounds need metabolic activation by cytochrome P-450 dependent microsomal enzymes to become carcinogenic. The results of these

Table 1. Effect of ethanol on chemically induced colorectal carcinogenesis in rats

Carcinogen	Ethanol administration	Ethanol effect	Ref.
DMH, s.c.	6% LD (36% total calories), preinduction	Increased rectal but not colonic tumours	26
DMH, s.c.	5% DW, induction	No effect	24
DMH, s.c.	5% DW, preinduction/induction	No effect	25
DMH, s.c.	6% LD (36% total calories), preinduction	No effect	27
AMMN, i.r.	6% LD (36% total calories), preinduction/induction	Increased rectal tumours	30
AOM, s.c.	LD (11%, 22%, 33% total calories), preinduction/ induction, postinduction	Inhibition of tumour development in the left but less than in the right colon. Higher ethanol intake has a stronger inhibitory effect. No effect when ethanol is given in the postinduction phase	28
AOM, s.c.	LD (9% total calories ethanol) (12%, 23% total calories beer), preinduction/induction	High ethanol inhibits tumours in the right colon, but not in the left colon, while low ethanol enhances tumours in the left colon, but not in the right colon. No effect of beer	29
AMMN, i.r.	i.g. (4.8 g/kg body weight per day), preinduction/ induction	Increased rectal tumours. Carcinogenesis was further stimulated when cyanamide, an acetaldehyde dehydrogenase inhibitor, was administered additionally	31

AMMN = Azoxymethylmethylnitrosamine, AOM = azoxymethane, DMH = dimethylhydrazine, DW = drinking water, i.g. = intragastrically, i.r. = intrarectally, LD = liquid diet, s.c. = subcutaneously.

studies depend on the ethanol dose used and on the timing of ethanol administration. The conclusion derived from these experiments are as follows:

(1) Carcinogenesis in the right and left CR is affected differently by alcohol and may depend on the levels of alcohol consumption. Thus, high alcohol intake (18–33% of total calories) inhibits carcinogenesis in the right colon and has no effect on the left colon, while lower ethanol consumption (9–12% of total calories) enhances tumour development in the left colon without effect on the right colon. One explanation could be that the presence of alcohol

(e.g. at higher alcohol intake) at the time of procarcinogen activation may inhibit this activation due to a competitive inhibition at the CYP2E1 binding site. On the other hand, chronic ethanol ingestion enhances CYP2E1 induction and this may lead to an enhanced procarcinogen activation at a time when ethanol levels are low.

(2) Ethanol affects carcinogenesis during the preinduction and/or induction phase, including carcinogen metabolism, but not in the postinduction phase (promotion).

(3) An interaction between ethanol and procarcinogen metabolism does occur, and this may influence tumour incidences.

When the primary carcinogen acetoxymethylmethylnitrosamine (AMMN), a compound which does not require metabolic activation was used to induce rectal tumours, acceleration of carcinogenesis was noted [30, 31]. Thus, it seems most likely that alcohol enhances carcinogenesis, at least in part, by local mechanisms in the rectal mucosa. In these experiments AA concentrations were experimentally increased by the administration of cyanamide, an AA-dehydrogenase (ALDH) inhibitor, and this led to a stimulation of colorectal carcinogenesis induced by AMMN, emphasizing the pathogenetic role of AA [32].

Mechanisms by Which Alcohol Stimulates Colorectal Carcinogenesis

Acetaldehyde
Acetaldehyde Toxicity
There is increasing evidence that AA rather than alcohol itself is responsible for the cocarcinogenic effect of alcohol [33]. In the gastrointestinal tract, AA can be generated from ethanol through mucosal and/or bacterial ADH [34]. AA is highly toxic, mutagenic and carcinogenic. It interferes at many sites with DNA synthesis and repair and can, consequently, result in tumour development [33].

Numerous in vitro and in vivo experiments in prokaryotic and eukaryotic cell cultures as well as in animal models have shown that AA has direct mutagenic and carcinogenic effects [35]. It causes point mutations in the hypoxanthine-guanine-phosphoribosyl transferase locus in human lymphocytes, induces sister chromatide exchanges, and gross chromosomal aberrations [32, 35–37]. It induces inflammation and metaplasia of tracheal epithelium, delays cell cycle progression and enhances cell injury associated with hyperregeneration [38, 39]. Thus, when AA was administered in drinking water to rodents [40], the mucosal lesions of the upper aerodigestive tract observed resembled those following chronic alcohol ingestion [41].

It has also been shown that AA interferes with the DNA repair machinery. AA directly inhibits O6 methyl-guanyltransferase, an enzyme important for the repair of adducts caused by alkylating agents [42]. Moreover, when inhaled, AA causes nasopahryngeal and laryngeal carcinomas [34, 43]. AA also binds rapidly to cellular proteins and DNA, which results in morphological and functional impairment of the cell and may lead to an immunologic cascade reaction. The binding to DNA and the formation of stable adducts represent one mechanism by which AA could trigger the occurrence of replication errors and/or mutations in oncogenes or tumour suppressor genes [44]. The occurrence of stable DNA adducts has been shown in different organs of alcohol fed rodents and in leucocytes of alcoholics [45]. It has been shown that the major stable DNA adduct N^2-ethyl-desoxyguanosine (N^2-Et-dG) can be used efficiently by eukaryotic DNA polymerase [46]. Most recently, another DNA adduct of AA namely 1,N^2-propano-desoxyguanosine (PdG) has been identified [47]. Its generation occurs in the presence of basic amino acids, histones or polyamines. While N^2-Et-dG is not mutagenic and may be used as a marker, PdG has mutagenic properties. The action of AA in carcinogenesis is summarized in figure 1. According to the International Agency for Research on Cancer there is sufficient evidence to classify AA as a carcinogen in experimental animals [32, 34].

Acetaldehyde Metabolism in Colorectal Mucosa

In the mucosa of the human CR, ADH1 and ADH3 are present, whereas ADH4 is expressed only in the mucosa of the upper gastrointestinal tract [33]. In addition, ALDH in relatively low activities is also detectable. Since the net amount of AA present in a tissue may determine its toxic or carcinogenic action, it is understandable that individuals with an increased production or decreased detoxification of AA are especially on risk for cancer development. In addition, the relationship between the total producing capacity of AA or its degradation seems also to be of importance. It has been proposed that the ALDH activity of colonic mucosa may be sufficient for the removal of the AA produced by colonic mucosal ADH during ethanol oxidation [48]. It may be, however, insufficient for the removal of the AA produced by intracolonic bacteria [48].

The most striking evidence of the causal role of AA in ethanol-associated carcinogenesis derives from genetic linkage studies in alcoholics and/or heavy drinkers. Individuals who accumulate AA due to polymorphism and/or mutation in the gene coding for enzymes responsible for AA generation and detoxification have been shown to have an increased cancer risk [49]. In Japan as well as in other Asian countries, a high percentage of individuals carry a mutation of the ALDH2 gene, which codes for an enzyme with low activity leading to elevated AA levels after alcohol consumption. While homozygotes are completely protected against

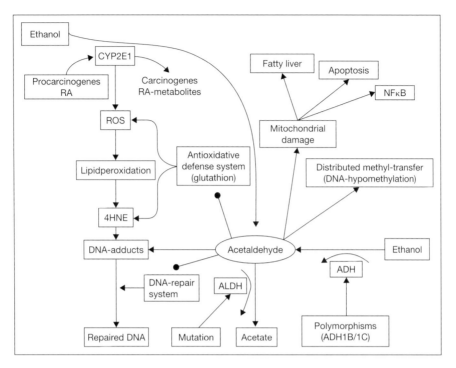

Fig. 1. Acetaldehyde and its role in ethanol associated carcinogenesis. AA is produced through ADH which are polymorphic. The ADH1C*1 and the ADH1B*2 allele code for enzymes with an increased AA production. In addition, AA is produced via cytochrome P-4502E1, which is induced following chronic alcohol ingestion. AA is oxidized primarily by ALDH2. Which reveals a mutation in approximately 50% of Asians (40 heterozygotes, 10 homozygotes) with low enzyme activity and AA accumulation following ethanol ingestion. AA binds to protein resulting in structural and functional alterations including microtubular and mitochondrial damage leading to fatty liver, apoptosis and activation of nuclear factor kappa B. In addition AA inhibits various enzymes involved in methyl transfer resulting among others in DNA hypomethylation. AA also affects the antioxidative defense system, mainly through binding to glutathione and the DNA repair system.

alcoholism and alcohol associated diseases due to the fact that they cannot tolerate alcohol even at very small doses, heterozygotes (ALDH2*1,2) have an increased risk for alimentary tract cancer [49].

In addition to the mutation of the ALDH2 gene, polymorphisms of the ADH1B and ADH1C gene may also modulate AA levels. While the ADH1B*2 allele encodes for an enzyme which is approximately 40 times more active than the enzyme encoded by the ADH1B*1 allele, ADH1C*1 transcription leads to an ADH isoenzyme 2.5 times more active than that from ADH1C*2

[50]. However, the ADH1B*2 allele frequency is high in Asians but low in Caucasians. It protects Asians from alcoholism, because of the high amount of AA produced [51, 52].

We have investigated more than 800 individuals with various alcohol associated cancers including oropharynx, larynx, oesophagus and liver and have compared them to individuals with other alcohol associated diseases such as cirrhosis of the liver and pancreatitis but a similar intake of alcohol. Multivariate analysis has identified ADH1C*1 homozygosity as a strong risk factor for these cancers with OR between 2.2 and 3.6 [53, 54]. More recently, preliminary results for the CR underline the role of AQDH1C*1 allele. Also here a significant risk for tumour development was observed, which however started already at lower level of chronic alcohol consumption [55].

Clinically, it has been found that alcoholics reveal a disturbed morphology in rectal biopsies with crypt destruction and inflammation, which completely returns to normal after 3 weeks of withdrawal [56].

It has been shown that after alcohol administration, the amount of AA per gram of tissue is highest for the colonic mucosa compared to all other tissues in the body [57]. This is primarily due to the production of AA by the faecal bacteria [48]. AA has toxic effects on the colon mucosa resulting in a decreased number of cells in the functional compartment of the colonic crypt [58]. This AA mediated toxicity induces a secondary compensatory hyperregeneration, with increased crypt cell production rates, and an extension of the proliferative compartment towards the lumen of the crypt [39, 58, 59]. This parallels by a significant increase in rectal mucosal ornithine decarboxylase activity [31]. Such a change in crypt cell dynamics represents a condition associating with increased risk for CRC.

The alcohol associated hyperregeneration of the colonic mucosa is especially pronounced with increasing age [39]. This may have practical implications since age by itself is a risk factor for CRC. Furthermore, when we extended our findings to humans, we again found that chronic alcoholics had an increased proliferation in their crypts with extension of their proliferative compartment, a condition associated with an increased cancer risk [59]. In contrast, cell differentiation regarding cytokeratin expression pattern was found to be unchanged, as well as regulatory factors involved in carcinogenesis and/or apoptosis [59].

Bacterial Acetaldehyde Metabolism

As pointed out above, high AA levels have been found after alcohol administration in the colon of rats and these concentrations were significantly lower in germ free animals as compared to the conventional rats suggesting that faecal bacteria are capable to produce AA [31]. Indeed, the reversed microbial ADH reaction produces under aerobic or microaerobic conditions striking amounts of AA when human colonic contents or some microbes representing normal colonic flora are incubated in vitro at 37°C with increasing ethanol concentrations [60–62]. This reaction is active already at comparatively low ethanol concentrations (10–100 mg%), which exist in the colon following social drinking [63]. AA formation catalysed by microbial ADH takes place at a pH normally found in the colon and is rapidly reduced with decreasing pH [60].

Antibiotics have been used to study the role of colonic bacteria in the oxidation of ethanol to AA. The reduction of aerobic gastrointestinal flora with ciprofloxacin decreases total ethanol elimination by a rate of about 10% both in rats and in man [64, 65]. This is associated with a significant decrease in the mean ADH activity of faecal samples, in almost total abolishment of the formation of endogenous ethanol in the colon, and in a remarkable reduction of the intracolonic AA production from ethanol [48, 65].

Taking all these data together, there is multiple evidence for the involvement of AA in alcohol associated colorectal carcinogenesis:

(1) High AA levels occur in the colon due to bacterial and mucosal ethanol oxidation.
(2) Animal experiments show an increased occurrence of CR tumours induced by the specific locally acting carcinogen AMMN, when cyanamide, an ALDH inhibitor is applied and AA levels are elevated.
(3) Crypt cell production rate correlates significantly with AA levels in the colonic mucosa.
(4) Colonic AA levels show a significant inverse relationship with mucosal folate concentrations, which supports in vitro data showing a destruction of folate by AA.
(5) Individuals with the inactive form of ALDH2 resulting in elevated AA concentrations exhibit an increased risk for CRC when they consume alcohol.
(6) Preliminary data also show that individuals homozygous for the ADH1C*1 allele coding for an enzyme with a 2.5 times higher AA production have also an increased risk for CRC.

Induction of Cytochrome P-4502E1

Chronic alcohol consumption leads to an induction of CYP2E1, which metabolizes ethanol to AA. Such an induction also takes place in the CR mucosa of rodents [66] and man [67]. CYP2E1 is also involved in the activation of various xenobiotics, including procarcinogens such as nitrosamines, aflatoxin, vinylchloride, polycyclic hydrocarbons, and hydrazines to their ultimative carcinogenic metabolites [68].

It has been shown in the liver that the concentration of CYP2E1 can be correlated with the generation of ROS and hydroxyethyl radicals and, thus, with lipid peroxidation. Induction of CYP2E1 resulted in enhanced hepatic injury, and inhibition of CYP2E1 was associated with an improvement of these lesions [69]. It was concluded that this is mainly due to a stimulation and inhibition, respectively, of free radical formation. ROS generated by CYP2E1 results in lipid peroxidation with lipid peroxidation products such as malonedialdehyde and 4-hydroxynonenal. These products can bind to DNA to form adducts, e.g. exocyclic DNA etheno adducts which are carcinogenic [70]. The role of CYP2E1 induction and cell injury has been studied in detail in the liver. However, the CR data on the role of CYP2E1 induction and alcohol-related cancer is limited.

In humans, the extent of CYP2E1 induction is individually determined, but may already be significant following the ingestion of 40 g of alcohol per day (corresponding to 400 ml of 12.5 vol% wine) over 1 week [71].

The effect of chronic ethanol consumption on the induction of CYP2E1 and the activation of procarcinogens has been elegantly demonstrated in various animal experiments using AOM as an inducing agent for CRC. Sohn et al. [72] demonstrated that the metabolism of AOM is inhibited in the presence of ethanol, but significantly enhanced when ethanol is withdrawn in a condition where CYP2E1 is induced and is completely available for the activation of AOM. Thus, the induction of CYP2E1 in the colorectal mucosa can result in an enhanced activation of nitrosamines or polycyclic hydrocarbons ubiquitously present in diets, and thus also in the faeces, and therefore be one mechanism by which chronic ethanol ingestion stimulates CRC development.

Interestingly, colorectal hyperregeneration observed after chronic alcohol administration to rats – most likely due to AA – was also attenuated by the concomitant administration of alpha-tocopherol, a radical scavanger [73].

Nutritional Factors

Since the effect of alcohol on the role of various nutritional deficiencies and their interaction in alcohol associated carcinogenesis is discussed in detail by others in this book, this topic will only be briefly mentioned. Epidemiologic studies have shown that low methionine and low folate levels may be associated with an increased risk for alcohol associated colorectal carcinogenesis [4]. This is due to a lack of methyl groups and possibly a disturbed methyl transfer which has been shown to occur in the liver [74]. Chronic ethanol consumption increases the requirement for methyl groups. Methionine obtained from the diet and synthesized by several reactions is the main precursor of S-adenosylmethionine (SAMe), the primary methyl donor in the body. Disruption in methionine metabolism and methylation reaction may be involved in the cancer process. SAMe is involved in the methylation of a small percentage of cytosine bases of the DNA. Enzymatic DNA methylation is an important component of gene control and may serve as a silencing mechanism for gene function. DNA hypomethylation has been observed in many cancers and tumours. Chronic ethanol consumption decreases intake of methionine and its conversion to SAMe. Folate deficiency which is common in the alcoholics may additionally contribute to an inhibition of transmethylation which is an important factor in one carbon transport. Most recently, it has been demonstrated that vitamin B6 deficiency, which also frequently occurs in the alcoholics may also lead to a decrease of SAMe and to CRC [21].

Another factor for DNA hypomethylation is AA-mediated inhibition of methyl adenosyl transferase (MAT) activity. In the colon of chronically ethanol fed rats, a significant reduction of folate associating with increased AA levels has been reported [75]. This may explain, at least in part, the genomic DNA hypomethylation observed in the colonic mucosa following alcohol ingestion [76]. However, no effect of alcohol consumption was found when the region of the p53 gene, which is most closely linked to colonic carcinogenesis was examined [76].

In addition, alcohol also leads to a decrease in retinoic acid (RA), at least in the liver. In this context it is referred to the article by Dr. Wang in this book. For the colon, the data with respect to alcohol and RA are very limited. Thus, it is interesting that ADH4, in contrast to ADH1, is not expressed in the human colorectal mucosa. However, it was found recently that in a number of biopsies from colorectal polyps of alcoholics ADH4 was expressed [77]. One explanation for such a denovo expression of ADH4 could be RA deficiency in a critical premalignant condition to guarantee increased generation of RA to suppress mucosal hyperregeneration. However, this hypothesis needs to be confirmed. The observed mucosal hyperregeneration following chronic ethanol ingestion may be not only due to the direct toxic action of AA but also to RA deficiency, and AA may contribute by preventing its generation.

References

1   Pöschl G, Seitz HK: Alcohol and cancer. Alcohol Alcohol 2004;39:155–165.
2   Breslow NE, Enstrom JE: Geographic correlations between mortality rates and alcohol, tobacco consumption in the United States. J Natl Cancer Inst 1974;53:631–639.
3   Kune GA, Vitetta L: Alcohol consumption and the etiology of colorectal cancer: a review of the scientific evidence from 1957 to 1991. Nutr Cancer 1992;18:97–111.
4   Giovannucci E, Rimm EB, Ascherio A, Stampfer MJ, Colditz GA, Willett WC: Alcohol, low-methionine – low folate diets, and risk of colon cancer in men. J Natl Cancer Inst 1995;87:265–273.
5   Goldbohm RA, van den Brandt PA, van't Veer P, Dorant E, Sturmans F, Hermus RJ: Prospective study on alcohol consumption and the risk of cancer of the colon and rectum in the Netherlands. Cancer Causes Control 1994;5:95–104.

6   Gapstur SM, Potter JD, Folsom AR: Alcohol consumption and colon and rectal cancer in post-menopausal women. Int J Epidemiol 1994;23:50–57.

7   Gerhardsson de Verdier M, Romelsjo A, Lundberg M: Alcohol and cancer of the colon and rectum. Eur J Cancer Prev 1993;2:401–408.

8   Meyer F, White E: Alcohol and nutrients in relation to colon cancer in middle-aged adults. Am J Epidemiol 1993;138:225–236.

9   Franceschi S, La Vecchia C: Alcohol and the risk of cancers of the stomach and colon-rectum. Dig Dis 1994;12:276–289.

10   Boutron MC, Faivre J, Dop MC, Quipourt V, Senesse P: Tobacco, alcohol, and colorectal tumors: a multistep process. Am J Epidemiol 1995;141:1038–1046.

11   Glynn SA, Albanes D, Pietinen P, Brown CC, Rautalahti M, Tangrea JA, Taylor PR, Virtamo J: Alcohol consumption and risk of colorectal cancer in a cohort of Finnish men. Cancer Causes Control 1996;7:214–223.

12   Yamada K, Araki S, Tamura M, Sakai I, Takahashi Y, Kashihara H, Kono S: Case-control study of colorectal carcinoma in situ and cancer in relation to cigarette smoking and alcohol use (Japan). Cancer Causes Control 1997;8:780–785.

13   Shimizu N, Nagata C, Shimizu H, Kametani M, Takeyama N, Ohnuma T, Matsushita S: Height, weight, and alcohol consumption in relation to the risk of colorectal cancer in Japan: a prospective study. Br J Cancer 2003;88:1038–1043.

14   Pedersen A, Johansen C, Gronbaek M: Relations between amount and type of alcohol and colon and rectal cancer in a Danish population based cohort study. Gut 2003;52:861–867.

15   Cho E, Smith-Warner SA, Ritz J, van den Brandt PA, Colditz GA, Folsom AR, Freudenheim JL, Giovannucci E, Goldbohm RA, Graham S, Holmberg L, Kim DH, Malila N, Miller AB, Pietinen P, Rohan TE, Sellers TA, Speizer FE, Willett WC, Wolk A, Hunter DJ: Alcohol intake and colorectal cancer: a pooled analysis of 8 cohort studies. Ann Intern Med 2004;140:603–613.

16   Seitz HK, Pöschl G, Simanowski UA: Alcohol and cancer. Recent Dev Alcohol 1998;14:67–95.

17   Kearney J, Giovannucci E, Rimm EB, Stampfer MJ, Colditz GA, Ascherio A, Bleday R, Willett WC: Diet, alcohol, and smoking and the occurrence of hyperplastic polyps of the colon and rectum (United States). Cancer Causes Control 1995;6:45–56.

18   Bardou M, Montembault S, Giraud V, Balian A, Borotto E, Houdayer C, Capron F, Chaput JC, Naveau S: Excessive alcohol consumption favours high risk polyp or colorectal cancer occurrence among patients with adenomas: a case control study. Gut 2002;50:38–42.

19   Le Marchand L, Wilkens LR, Hankin JH, Kolonel LN, Lyu LC: Independent and joint effects of family history and lifestyle on colorectal cancer risk: implications for prevention. Cancer Epidemiol Biomarkers Prev 1999;8:45–851.

20   Scheppach W, Bingham S, Boutron-Ruault MC, Gerhardsson de Verdier M, Moreno V, Nagengast FM, Reifen R, Riboli E, Seitz HK, Wahrendorf J: WHO consensus statement on the role of nutrition in colorectal cancer. Eur J Cancer Prev 1999;8:57–62.

21   Larsson SC, Giovannucci E, Wolk A: Vitamin B6 intake, alcohol consumption, and colorectal cancer: a longitudinal population-based cohort of women. Gastroenterology 2005;128:1830–1837.

22   Su LJ, Arab L: Alcohol consumption and risk of colon cancer: evidence from the national health and nutrition examination survey I epidemiologic follow-up study. Nutr Cancer 2004;50:111–119.

23   Ketcham AS, Wexler H, Mantel N: Effects of alcohol in mouse neoplasia. Cancer Res 23:703–709.

24   Howarth AE, Pihl E: High-fat diet promotes and causes distal shift of experimental rat colonic cancer – beer and alcohol do not. Nutr Cancer 1984;6:229–235.

25   Nelson RL, Samelson SL: Neither dietary ethanol nor beer augments experimental colon carcinogenesis in rats. Dis Colon Rectum 1985;28:460–462.

26   Seitz HK, Czygan P, Waldherr R, Veith S, Raedsch R, Kassmodel H, Kommerell B: Enhancement of 1,2-dimethylhydrazine-induced rectal carcinogenesis following chronic ethanol consumption in the rat. Gastroenterology 1984;86:886–891.

27   McGarrity TJ, Via EA, Colony PC: Qualitative and quantitative changes in sialomucins during 1,2-dimethylhydrazine-induced colon carcinogenesis in the rat. J Natl Cancer Inst 1987;79: 1375–1382.

28   Hamilton SR, Sohn OS, Fiala ES: Effects of timing and quantity of chronic dietary ethanol consumption on azoxymethane-induced colonic carcinogenesis and azoxymethane metabolism in Fischer 344 rats. Cancer Res 1987;47:4305–4311.

29 Hamilton SR, Hyland J, McAvinchey D, Chaudhry Y, Hartka L, Kim HT, Cichon P, Floyd J, Turjman N, Kessie G, Nair PP, Dick J: Effects of chronic dietary beer and ethanol consumption on experimental colonic carcinogenesis by azoxymethane in rats. Cancer Res 1987;47: 1551–1559.
30 Garzon FT, Simanowski UA, Berger MR, Schmahl D, Kommerell B, Seitz HK: Acetoxymethyl-methylnitrosamine (AMMN) induced colorectal carcinogenesis is stimulated by chronic alcohol consumption. Alcohol Alcohol Suppl 1987;1:501–502.
31 Seitz HK, Simanowski UA, Garzon FT, Rideout JM, Peters TJ, Koch A, Berger MR, Einecke H, Maiwald M: Possible role of acetaldehyde in ethanol-related rectal cocarcinogenesis in the rat. Gastroenterology 1990;98:406–413.
32 Helander A, Lindahl-Kiessling K: Increased frequency of acetaldehyde-induced sister-chromatide exchanges in human lymphocytes treated with an aldehyde dehydrogenase inhibitor. Mutat Res 1991;264:103–107.
33 Anonymous: Acetaldehyde. IARC Monogr Eval Carcinog Risk Chem Hum 1985;36.
34 Seitz HK, Oneta CM: Gastrointestinal alcohol dehydrogenases. Nutr Rev 1998;56:52–60.
35 IARC, Acetaldehyde IARC Monographs on the evaluation of the carcinogenic risk to humans. Reevaluation of some organic chemicals, hydrazine and hydrogen peroxides. Lyon, International Agency for Research on Cancer 1999, vol 71, pp 319–335.
36 Dellarco VL: A mutagenicity assessment of acetaldehyde. Mutat Res 1988;195:1–20.
37 Obe G, Jonas R, Schmidt S: Metabolism of ethanol in vitro produces a compound which induces sister-chromatid exchanges in human peripheral lymphocytes in vitro: acetaldehyde not ethanol is mutagenic. Mutat Res 1986;174:47–51.
38 Seitz HK, Matsuzaki S, Yokoyama A, Homann N, Väkeväinen S, Wang XD: Alcohol and cancer. Alcohol Clin Exp Res 2001;25:137–143.
39 Simanowski UA, Suter P, Russel RM, Heller M, Waldherr R, Ward R, Peters TJ, Smith D, Seitz HK: Enhancement of ethanol induced rectal mucosal hyperregeneration with age in F244 rats. Gut 1994;35:1102–1106.
40 Homann N, Kärkkäinen P, Koivisto T, Nosova T, Jokelainen K, Salaspuro M: Effects of acetaldehyde on cell regeneration and differentiation of the upper gastrointestinal tract mucosa. J Natl Cancer Inst 1997;89:1692–1697.
41 Simanowski UA, Suter PM, Stickel F, Maier H, Waldherr R, Smith D, Russell RM, Seitz HK: Esophageal epithelial hyperregeneration following chronic ethanol ingestion: effect of age and salivary gland function. J Natl Cancer Inst 1993;85:2030–2033.
42 Espina N, Lima V, Lieber CS, Garro AJ: In vitro and in vivo inhibitory effect of ethanol and acetaldehyde on O6-methylguanine transferase. Carcionogenesis 1988;9:761–766.
43 Feron VJ, Kruysee A, Wouterson RA: Respiratory tract tumours in hamsters exposed to acetaldehyde vapour alone or simultaneously with benzo(a)pyrene or diethylnitrosamine. Eur J Cancer Clin Oncol 1982;18:13–31.
44 Fang JL, Vaca CE: Development of a 32P-postlabeling method for the analysis of adducts arising through the reaction of acetaldehyde with 2′-deoxyguanosine-3′-monophosphate and DNA. Carcinogenesis 1995;16:2177–2185.
45 Fang JL, Vaca CE: Detection of DNA adducts of acetaldehyde in peripheral white blood cells of alcohol abusers. Carcinogenesis 1997;18:627–632.
46 Matsuda T, Terashima I, Matsumoto Y, Yabushita H, Matsui S, Shibutani S: Effective utilization of N2-ethyl-2′-deoxyguanosine triphosphate during DNA synthesis catalyzed by mammalian replicative DNA polymerases. Biochemistry 1999;38:929–935.
47 Brooks PJ, Thiruvathu JA: DNA adducts from acetaldehyde: implications for alcohol-related carcinogenesis. Alcohol 2005;35:187–193.
48 Salaspuro M: Acetaldehyde, microbes, and cancer of the digestive tract. Crit Rev Clin Lab Sci 2003;40:183–208.
49 Yokoyama A, Muramatsu T, Ohmori T, Yokoyama T, Okuyama K, Takahashi H, Hasegawa Y, Higuchi S, Maruyama K, Shirakura K, Ishii H: Alcohol-related cancers and aldehydrogenase-2 in Japanese alcoholics. Carcinogenesis 1998;19:1383–1387.
50 Bosron WF, Li TK: Genetic polymorphism of human liver alcohol and aldehyde dehydrogenase, and their relationship to alcohol metabolism and alcoholism. Hepatology 1986;6:502–510.

51   Borras E, Coutelle C, Rosell A, Fernandez-Muixi F, Broch M, Crosas B, Hjelmqvist L, Lorenzo A, Gutierrez C, Santos M, Szczepanek M, Heilig, M. Quattrocchi P, Farres J, Vidal F, Richart C, Mach T, Bogdal J, Jornvall H, Seitz HK, Couzigou P, Parex X: Genetic polymorphism of alcohol dehydrogenase in Europeans: the ADH1B*2 allele decreases the risk for alcoholism and is associated with ADH1C*1. Hepatology 2000;31:984–989.

52   Chao YC, Young TH, Thang HS, Hsu CT: Alcoholism and alcoholic organ damage and genetic polymorphisms of alcohol metabolizing enzymes in Chinese patients. Hepatology 1997;25: 112–117.

53   Visapää JP, Götte K, Benesova, Li JJ, Homann N, Conradt C, Inoue H, Tisch M, Hörmann K, Väkeväinen S, Salaspuro M, Seitz HK: Increased cancer risk in heavy drinkers with the alcohol dehydrogenasse 3*1-Allel possibly due to salivary acetaldehyde. Gut 2004;53:871–876.

54   Homann N, Stickel F, König IR, Jacobs A, Junghanns K, Benesova M, Schuppan D, Himsel S, Zuber-Jerger I, Hellerbrand C, Ludwig D, Caselmann WH, Seitz HK: Alcohol dehydrogenase 1C*1 allele is a genetic marker for alcohol-associated cancer in heavy drinkers. Int J Cancer 2005;118:1998–2002.

55   Marks M, Benesova M, Stickel F, Büchler MW, Seitz HK: Alcohol intake, ADH1C genotype and colorectal cancer. Alcohol Alcohol 2005;40:i44–i45.

56   Brozinski S, Fami K, Grosberg JJ: Alcohol ingestion-induced changes in the human rectal mucosa: light and electronmicroscopic studies. Dis Colon Rectum 1979;21:329–335.

57   Seitz HK, Simanowski UA, Garzon FT, Peters TJ: Alcohol and cancer (Letter to the Editor). Hepatology 1987;7:616.

58   Simanowski UA, Seitz HK, Baier B, Kommerell B, Schmidt-Gayk H, Wright NA: Chronic ethanol consumption selectively stimulates rectal cell proliferation in the rat. Gut 1986;27:278–282.

59   Simanowski UA, Homann N, Knühl M, Arce C, Waldherr R, Conradt C, Bosch FX, Seitz HK: Increased rectal cell proliferation following alcohol abuse. Gut 2001;49:418–422.

60   Jokelainen K, Roine R, Väänänen H, Salaspuro M: In vitro acetaldehyde formation by human colonic bacteria. Gut 1994;35:1271–1274.

61   Jokelainen K, Siitonen A, Jousimies-Somer H: In vitro alcohol dehydrogenase-mediated acetaldehyde production by aerobic bacteria representing the normal colonic flora in man. Alcohol Clin Exp Res 1996;20:967–972.

62   Salaspuro V, Nyfors S, Heine R, Siitonen A. Salaspuro M, Jousimies-Somer H: Ethanol oxidation and acetaldehyde production in vitro by human intestinal strains of *Eschericia coli* under aerobic, microaerobic, and anaerobic conditions. Scand J Gastroenterol 1999;34:967–973.

63   Jokelainen K, Matysiak-Budnik T, Mäkisalo H, Höckerstedt K, Salaspuro M: High intracolonic acetaldehyde values produced by a bacteriocolonic pathway for ethanol oxidation in piglets. Gut 1996;39:100–104.

64   Jokelainen K, Nosova T, Koivisto T, Väkeväinen S, Jousimies-Somer H, Heine R, Salaspuro M: Inhibition of bacteriocolonic pathway for ethanol oxidation by ciprofloxacin in rats. Life Sci 1997;61:1755–1762.

65   Tillonen J, Homann N, Rautio M, Jousimies-Somer H, Salaspuro M: Ciprofloxacin decreases the rate of ethanol elimination in humans. Gut 1999b;44:347–352.

66   Shimizu M, Lasker JM, Tsutsumi M, Lieber CS: Immunohistochemical localization of ethanol inducible cytochrome P4502E1 in rat alimentary tract. Gastroenterology 1990;93:1044–1050.

67   Baumgarten G, Waldherr R, Stickel F, Simanowski UA, Ingelmann-Sundberg M, Seitz HK: Enhanced expression of cytochrome P450 2E1 in the oropharyngeal mucosa in alcoholics with cancer. Annual meeting international society of biomedical researchers, alcoholism. Washington DC, June 22–27, 1996.

68   Seitz HK, Osswald B: Effect of ethanol on procarcinogen activation; in Watson RR (ed): Alcohol and Cancer. Boca Raton, Ann Arbor, London, Tokyo, CRC Press 1992, pp 55–72.

69   Gouillon Z, Lucas D, Li JJ, Hagbjork AL, French BA, Fu P, Fang C, Ingelmann-Sundberg M, Donohue TM, French SW: Inhibition of ethanol-induced liver disease in the intragastric feeding rat model by chlormethiazole. Proc Soc Exp Biol Med 2000;224:302–308.

70   Frank A, Seitz HK, Bartsch H, Frank N, Nair J: Immunohistochemical detection of 1,N[6]-ethenodeoxyadenosine in nuclei of human liver affected by diseases predisposing to hepatocarcinogenesis. Carcinogenesis 2004;25:1027–1031.

71   Oneta CM, Lieber CS, Li JJ, Ruttimann S, Schmid B, Lattmann J, Rosman AS, Seitz HK: Dynamics of cytochrome P4502E1 activity in man: induction by ethanol and disappearance during withdrawal phase. J Hepatol 2002;36:47–52.

72   Sohn OA, Fiala ES, Puz C: Enhancement of rat liver microsomal metabolism of azoxymethane to methylazoxymethanol by chronic ethanol administration: similarity to the microsomal metabolism of N-nitrosomethylamine. Cancer Res 1987;47:3123–3129.

73   Vincon P, Wunderer J, Simanowski UA, Koll M, Preedy VR, Peters TJ, Werner J, Waldherr R, Seitz HK: Effect of ethanol and vitamin E on cell regeneration and BCL-2 expression in the colorectal mucosa of rats. Alcohol Clin Exp Res 2003;27:100–106.

74   Stickel F, Herold C, Seitz HK, Schuppan D: Alcohol and methyl transfer: implications for alcohol-related hepatocarcinogenesis; in Ali S, Friedman SL, Mann DA (eds): Liver Diseases: Biochemical Mechanisms and New Therapeutic Insights. New Delhi, Oxford and IBH Publ Co PRT.Ltd., 2005, pp 57–70.

75   Homann N, Tillonen J, Salaspuro M: Heavy alcohol intake leads to local colonic folate deficiency in rats: evidence of microbal acetaldehyde production from ethanol is the pathogenic substance. Int J Cancer 2000;86:169–173.

76   Choi SW, Stickel F, Baik W, Kim YI, Seitz HK, Mason JC: Chronic alcohol consumption induces genomic, but not p53-specific DNA hypomethylation in the colon of the rat. J Nutr 1999;129: 1945–1950.

77   Seitz HK: Alcohol and retinoid metabolism (Editorial). Gut 2000;47:748–750.

Helmut K. Seitz, MD
Salem Medical Centre
Zeppelinstrasse 11–33
DE–69121 Heidelberg (Germany)
Tel. +49 6221 483 200, Fax +49 6221 483 494
E-Mail helmut_karl.seitz@urz.uni-heidelberg.de

Cho CH, Purohit V (eds): Alcohol, Tobacco and Cancer.
Basel, Karger, 2006, pp 78–94

....

Acetaldehyde-DNA Adducts: Implications for the Molecular Mechanism of Alcohol-Related Carcinogenesis

P.J. Brooks, Jacob A. Theruvathu

Section on Molecular Neurobiology, Laboratory of Neurogenetics,
National Institute on Alcohol Abuse and Alcoholism, National Institutes of Health,
Bethesda, Md., USA

Abstract

Alcoholic beverage consumption is classified as a known human carcinogen, causally related to an increased risk of cancer of the upper aerodigestive tract. The formation of acetaldehyde from ethanol metabolism appears to be the major mechanism underlying this effect. Acetaldehyde has been shown to cause cancer in animals, and is classified as a suspected human carcinogen. In human cells, acetaldehyde causes sister chromatid exchanges and chromosomal aberrations. In this chapter, we focus on the mechanisms by which a specific DNA adduct, α-methyl-γ-hydroxy-1, N^2-propano-2′-deoxyguanosine (Cr-PdG), is formed from the reaction of acetaldehyde and DNA. We compare the mechanism of formation and biological properties of this adduct with another, well known acetaldehyde-DNA adduct, N^2-ethyl-dG, and describe how the genotoxic properties of the Cr-PdG adduct can explain many of the known genotoxic effects of acetaldehyde. We propose that while N^2-ethyl-dG is a good marker for acetaldehyde reaction with DNA, the Cr-PdG adduct is likely to play a central role in the mechanism of alcoholic beverage related carcinogenesis. We also discuss the DNA repair pathways responsible for protecting the human genome from this adduct, with the goal of identifying candidate DNA repair genes in which variation may affect the risk of cancer.

According to the 11th Report on Carcinogens, issued by the US National Toxicology Program [1], alcoholic beverage consumption is classified as a known human carcinogen causally related to an increased risk of cancer of the upper aerodigestive tract (UADT). This is the strongest classification on this list. The role of alcohol (ethanol) in UADT cancer is quite substantial. It has been estimated that between 20 and 80% of head and neck cancers may be attributable to alcohol consumption [2]. In light of these considerations, it is of obvious importance to understand the mechanisms by which alcohol consumption increases cancer risk.

In principle, there are several ways in which alcohol consumption could increase cancer risk (for review see [3, 4]). For several reasons, the formation of acetaldehyde appears to be the most important mechanism. The genotoxicity of acetaldehyde has been well documented [5]. According to the Report on Carcinogens [1], acetaldehyde is currently classified as 'reasonably anticipated to be a human carcinogen', and is classified as 'possibly carcinogenic to humans' by the IARC [6]. Moreover, individuals who carry the *ALDH2*2* genotype and lack the capacity to metabolize acetaldehyde to acetate, have been shown in several studies to have an increased risk of UADT cancer from alcoholic beverage consumption [7, 8]. This topic is also addressed elsewhere in this volume.

In this chapter, we will focus on the molecular basis of the carcinogenicity of acetaldehyde. In particular, we will focus on a specific DNA adduct, α-methyl-γ-hydroxy-1, N^2-propano-2'-deoxyguanosine (Cr-PdG), and describe how this adduct is formed from biologically relevant acetaldehyde concentrations. We will also describe how the biological properties of this adduct may explain many of the known genotoxic effects of acetaldehyde, and by extension, how this adduct may play a central role in the mechanism of alcoholic beverage related carcinogenesis. In the last part of the chapter, we will also discuss the role of DNA repair in protecting the genome against the Cr-PdG adduct and its derivatives, with the goal of identifying rational candidate DNA repair genes in which polymorphisms could affect susceptibility to alcoholic beverage related cancer.

Genotoxicity of Acetaldehyde

The evidence that acetaldehyde is an animal carcinogen comes from studies by Woutersen and colleagues [9–12]. These investigators found that chronic acetaldehyde exposure significantly increased the incidence of nasal tumors, notably including squamous cell carcinomas, the same type of tumor resulting from alcohol consumption. Thus it is clear that acetaldehyde is a carcinogen, at least in the rodent nasal epithelium.

Regarding the mechanism of acetaldehyde carcinogenicity, an important series of studies by Obe and colleagues, as well as work from other investigators (reviewed in [13]) showed that exposure of cells to acetaldehyde increased the frequency of sister chromatid exchanges (SCEs) and chromosomal aberrations in mammalian cells. Notably, Obe and Ristow [14] showed that acetaldehyde increased SCEs in CHO cells at concentrations as low as 3.9 µg/ml, which corresponds to approximately 88 μM acetaldehyde. This concentration is likely to be relevant to the situation in alcoholic beverage consumption, as salivary acetaldehyde concentrations as high as 143 μM have been measured during ethanol drinking in laboratory settings [15]. As pointed out by Homann et al. [16], by extrapolation of their data on human volunteers, local salivary acetaldehyde concentrations could reach up to 450 μM at higher blood alcohol concentrations. These high levels appear to be due in large part to the generation of acetaldehyde by microorganisms in the human oral cavity [16] (see also contribution by Salaspuro in this volume). However, the possibility that intracellular formation of acetaldehyde by alcohol dehydrogenases in the gastrointestinal tract [17, 18] may also play a role in alcohol-related carcinogenesis cannot be excluded.

DNA Adducts from Acetaldehyde

Formation of N^2-ethyl-dG: A Marker, But Not a Mechanism?
The most well-known and best studied DNA lesion resulting from acetaldehyde is N^2-ethyl-2′-deoxyguanosine (N^2-ethyl-dG) (fig. 1). This lesion was originally identified by Vaca et al. [19] in studies to identify DNA adducts resulting from the reaction of acetaldehyde and deoxynucleosides. It is important to note that, in addition to the reaction of acetaldehyde with DNA, the formation of the stable N^2-ethyl-dG adduct requires a subsequent reduction step [19]. Without this reduction step, the product is an unstable imine that, at least under in vitro conditions, is lost upon removal of the acetaldehyde. In vitro, the reduction step requires the use of high concentrations of either sodium borohydride or cyanoborohydride. For the lesion to form in vivo, as has been documented, this reduction step must be accomplished by some intracellular molecule or enzyme. Vitamin C and glutathione (GSH) can carry out the reduction step in vitro [20], but whether these or other compounds (such as NADH or NADPH) carry out the reduction step in vivo is not known.

Using a very sensitive ^{32}P-postlabeling assay, Fang and Vaca [20] detected low levels of lesion (1 lesion/10^8 nucleotides) in liver DNA obtained from rats given 10% EtOH in the drinking water. N^2-ethyl-dG was undetectable in DNA from control animals. In a later study, the same group used the postlabeling method to assay the lesion in white blood cell DNA from human alcoholics and controls [21]. They found that the levels of lesion in white blood cell DNA from the alcoholic subjects were on the order of 2–3 lesions/10^7 nucleotides, significantly higher than the levels in non-drinking controls. Interestingly, the lesion levels in the human alcoholics were about 1 order of magnitude higher than the levels measured in alcohol treated rat liver. Also, in this human study, detectable basal levels of lesion were observed in samples from some non-alcoholic controls.

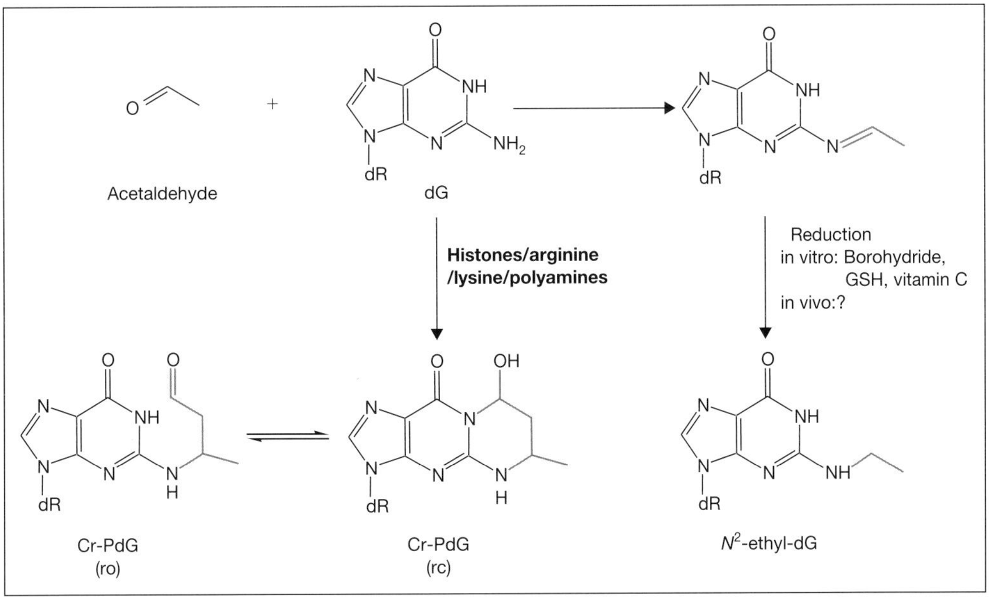

Fig. 1. N^2-ethyl-dG and Cr-PdG adduct formation from the reaction of acetaldehyde with 2'-deoxyguanosine. When AA reacts with dG, a Schiff base intermediate is formed, which can be reduced to N^2-ethyl-dG using borohydride, cyanoborohydride, GSH or vitamin C. When the AA reaction occurs in the presence of histones, basic amino acids (arginine or lysine), or polyamines, the Cr-PdG adduct, not the Schiff base, is formed. Cr-PdG adduct can exist in the rc form or in the ro form.

While the measurement of DNA adducts in blood or tissue samples has been very useful, the practical difficulties in getting appropriate samples, especially in humans, is leading many investigators in the DNA adduct field to investigate the feasibility of measuring DNA adducts in urine as a surrogate biomarker of DNA damage. Using an LC-MS method, Matsuda et al. [22] were able to detect the N^2-ethyl-dG lesion in urine samples from healthy volunteers. However, detectable levels of N^2-ethyl-dG were found in urine samples from individuals who had abstained from drinking alcohol for 1 week prior to testing. We have made similar observations (Theruvathu and Brooks, 2004, unpublished observations). These observations suggest that, in contrast to the situation in cellular DNA, the lesion in urine results either from acetaldehyde formed endogenously, or from some other endogenous ethylating compound. Therefore, assay of this lesion in urine does not appear to be useful as a biomarker for acetaldehyde related DNA damage from alcohol consumption.

One of the difficulties in relating the N^2-ethyl-dG lesion to alcoholic beverage related carcinogenesis is that there is little evidence that N^2-ethyl-dG is a mutagenic lesion. While the *E. coli* DNA polymerase incorporates both deoxy-cytosine (dC) and dG opposite N^2-ethyl-dG by in vitro [23], the situation in mammalian cells is different. Terashima et al. [22] found that the major replicative DNA polymerase in mammalian cells, polymerase delta, incorporated N^2-ethyl-dGTP opposite template dC, the correct base. This observation indicates that bypass of the adduct during DNA replication by polymerase delta in mammalian cells is unlikely to be mutagenic. Perrino et al. [24] showed that N^2-ethyl-dG strongly blocks polymerization by DNA polymerase alpha. However, polymerase alpha is not the major replicative DNA polymerase; it is responsible for the initiation of DNA replication, as well as having a limited role in lagging strand synthesis [25]. The same authors [24] also found that DNA polymerase eta can bypass the lesion, incorporating a dC residue opposite the lesion. In contrast to the replicative DNA polymerases, which are highly accurate during DNA synthesis, DNA polymerase eta is one of a growing number of specialized DNA polymerases, sometimes referred to as translesion synthesis polymerases, which are very inaccurate when copying undamaged DNA, but can bypass DNA lesions that are strong blocks to replicative DNA polymerases [25–29]. Taken together, the results of these studies indicate that mammalian cells have at least two ways of accurately replicating past N^2-ethyl-dG lesions. However, the possibility that this lesion could be mutagenic in a particular sequence context has not been ruled out, and the effect of the lesion during replication by DNA polymerase epsilon, the other major replicative DNA polymerase in eukaryotic cells [25], has not yet been examined.

Another difficulty in relating N^2-ethyl-dG to alcohol related carcinogenesis is that, in contrast to the Cr-PdG adduct, it is not clear how this lesion could produce the major cytogenetic effects that have been demonstrated in mammalian cells exposed to acetaldehyde, i.e. SCEs and chromosomal aberrations. Therefore, for several reasons, while the presence of N^2-ethyl-dG in cellular DNA appears to be a good biomarker for DNA damage from acetaldehyde, including acetaldehyde derived from alcohol metabolism, the role of this lesion in the carcinogenic effects of acetaldehyde, if any, is unclear at present.

Cr-PdG: A Better Candidate DNA Lesion for Acetaldehyde Related Carcinogenesis

The original studies of Vaca et al. [19] investigated the reaction of acetaldehyde with individual deoxynucleosides. More recently, studies carried out in the Hecht laboratory [30] utilized a variety of analytical techniques to investigate the reaction of acetaldehyde with double-stranded DNA. They found that exposure of acetaldehyde to DNA can result in several DNA lesions in addition to N^2-ethyl-dG. Amongst these lesions, one that is of particular significance is Cr-PdG. Previous studies have shown that Cr-PdG is an endogenous DNA lesion

[31], perhaps resulting from lipid peroxidation [32]. We refer to this lesion as a Cr-PdG adduct to distinguish it from other propano-dG adducts (i.e. those resulting from malondialdehyde and acrolein [33, 34]) because it has previously been identified as being responsible for the mutagenic, genotoxic and carcinogenic properties of crotonaldehyde (CrA), an environmental pollutant [35].

In the studies by Wang et al. [30], formation of the Cr-PdG lesion was only detectable using acetaldehyde concentrations of 40 mM. Such levels would not occur in vivo as a result of ethanol metabolism. However, experiments by Sako et al. [36] showed that the basic amino acids arginine and lysine, which are present at high levels in histones, could facilitate the formation of Cr-PdG from acetaldehyde and DNA. Indeed, other studies by the same group have directly shown that histones can facilitate the formation of Cr-PdG from acetaldehyde and DNA [37]. Importantly, the formation of Cr-PdG from acetaldehyde and dG does not require any reduction step, in contrast to the formation of N^2-ethyl-dG. As in the studies of Wang et al. [30] however, the concentrations of acetaldehyde used by Sako and colleagues were far above the levels that could reasonably be expected to occur in the human body as a result of alcohol consumption.

Based on the studies described above, we considered the possibility that polyamines might stimulate the formation of the Cr-PdG adduct from acetaldehyde and DNA. Polyamines, such as spermine, spermidine, and putrescine, are small, highly basic molecules that are present at millimolar concentrations in cells [38, 39]. Several different roles have been proposed for cellular polyamines, including protective roles against oxidative stress [40, 41] and DNA damage from ionizing radiation [42] though a complete understanding of all of their functions has not yet been achieved [43–45]. Importantly, however, many studies have shown that polyamine synthesis is highest in rapidly dividing cells [46].

Using mass spectrometric techniques, we investigated a possible role of polyamines in Cr-PdG formation. We found that biologically relevant concentrations of polyamines could in fact stimulate Cr-PdG formation from acetaldehyde and deoxyguanosine, or from acetaldehyde and DNA [47]. Importantly, we showed that in the presence of physiologically relevant polyamines concentrations, acetaldehyde concentrations as low as 100 μM, could stimulate Cr-PdG formation in DNA [47]. This concentration is below acetaldehyde concentrations that have been measured in human saliva during alcohol drinking in laboratory conditions, and is therefore likely to be biologically relevant.

Polyamines Directly Stimulate the Conversion of Acetaldehyde to Crotonaldehyde

From the observation of the relative amounts of the two diastereomers of Cr-PdG formed in our experiments, we hypothesized that the formation of Cr-PdG adduct from acetaldehyde and DNA in the presence of polyamines is a

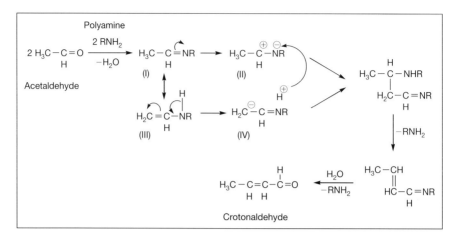

Fig. 2. The chemical mechanism for the formation of CrA from acetaldehyde and polyamine. In this mechanism, one of the amino groups of the polyamine reacts with carbonyl group of acetaldehyde to form a Schiff base intermediate (I), which can polarize the electron density on the C–N double bond and become a carbocation (II). The Schiff base can also tautomerize to give (III) and subsequently become a carbanion (IV). The proton eliminated from III can bind to the negatively charged nitrogen. Then the oppositely charged molecules can combine together to form a dimer molecule, which can eliminate a molecule of polyamine to form an unsaturated imine. The unsaturated imine can be further hydrolyzed to give CrA and another polyamine molecule.

two-step process, in which acetaldehyde first reacts with polyamine to form CrA, and the CrA reacts with DNA to give the Cr-PdG adduct. In support of this hypothesis, we found that CrA is generated by incubating acetaldehyde with polyamine in the absence of DNA.

The chemical mechanism for the formation of CrA from acetaldehyde is essentially an enamine aldol condensation in which two molecules of acetaldehyde are condensed to from one molecule of CrA. The mechanism is shown and described in more detail in figure 2.

Based on our work, as well as the previous studies Sako and colleagues [37, 48], it is clear that basic molecules containing free amino groups can stimulate the formation of CrA from acetaldehyde. It seems likely that similar reactions might occur with other aldehydes, and perhaps with other basic molecules as well. An interesting molecule in this regard is Tris (tris(hydroxymethyl)aminomethane), commonly used as a buffer in DNA solutions. It is possible that Tris could act like a polyamine to stimulate CrA formation from acetaldehyde, and we have preliminary evidence to support this possibility (unpublished observations). On the other hand, Tris has been shown to react with the PdG adduct formed from malondialdehyde [49], and might do the same with other PdG adducts, including Cr-PdG. For this reason, the report that high concentrations of acetaldehyde produced DNA lesions leading to GG-TT mutations in plasmid DNAs [50] is intriguing but difficult to interpret, because the acetaldehyde reactions were carried out in the presence of Tris buffer. However, it is possible that Cr-PdG adducts may be involved in these mutations.

What is so Special About the Cr-PdG Adduct?

An important aspect of the genotoxicity of the Cr-PdG adduct is that, like other PdG adducts [33, 34], the Cr-PdG adduct can exist in either of two forms, ring-opened (ro) or ring-closed (rc) (fig. 1), depending on the state of the DNA. The rc forms of Cr-PdG adducts are favored in single-stranded DNA, whereas double-stranded DNA favors the ro forms. The rc form is expected to strongly interfere with Watson–Crick base pairing, as has been directly shown for the highly related acrolein-PdG adduct [34]. This is in contrast to the N^2-ethyl-dG adduct, which can accommodate normal base paring by rotation of the ethyl group.

The ro form of Cr-PdG can also accommodate Watson–Crick base pairing. However, the existence of the free aldehyde group in the open chain form of Cr-PdG allows the possibility of forming additional, more complex DNA lesions, including DNA-protein crosslinks and DNA interstrand crosslinks (ICLs). These may be thought of as derivatives of the Cr-PdG adduct. Figure 3 shows the chemical structures of ICLs derived from the ro form of the Cr-PdG adduct that have been reported in the literature [30, 51, 52].

The Genotoxic Effects of Acetaldehyde May be Explained by the Genotoxic Effects of the Cr-PdG Adduct and its Derivatives

Based on the discussion above, in figure 4 we present a schematic model of how the Cr-PdG adducts could play a central role in the genotoxicity of acetaldehyde, and by extension in alcohol-related carcinogenesis. It should be emphasized that this is a working model, which like all models is useful to the extent that it stimulates experiments to test or disprove it.

According to this model, acetaldehyde derived from ethanol reacts with intracellular polyamines to form CrA, according to the mechanisms described above. In the UADT, the acetaldehyde is most likely formed from alcohol metabolism by oral microorganisms [16]. The CrA formed in the cell reacts with deoxyguanosine residues in DNA to form the Cr-PdG adduct. In the rc form, the Cr-PdG adduct may be bypassed to a limited extent during DNA replication, resulting in mutagenesis [53, 54]. Indeed, direct evidence that the Cr-PdG adduct can stimulate single-base mutations in mammalian cells has been provided by Fernandes et al. [55].

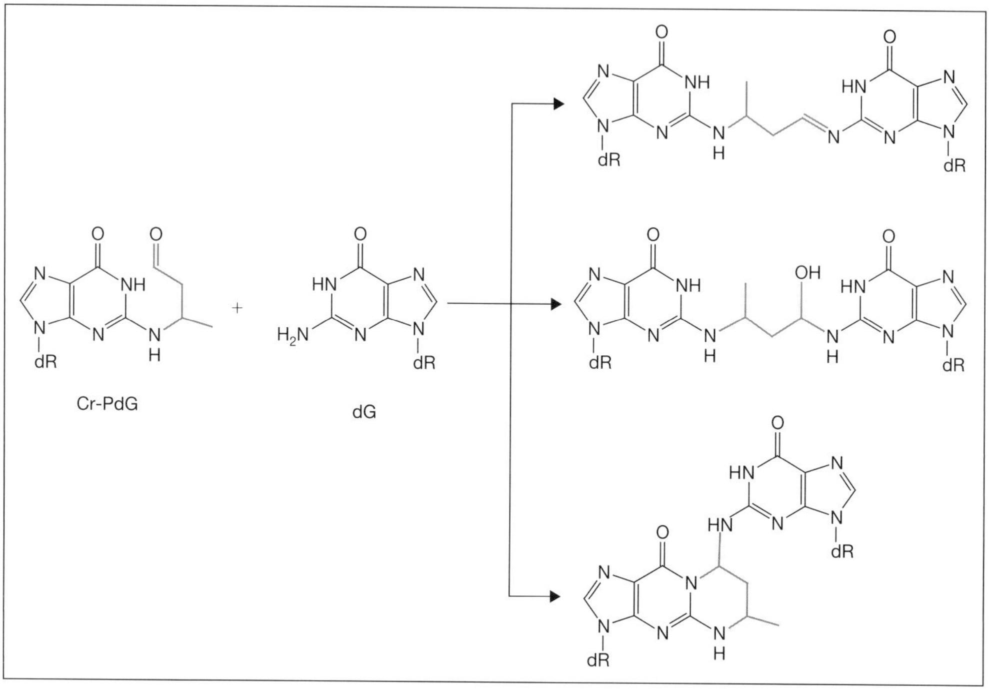

Fig. 3. Different types of DNA ICL from the ro form of the Cr-PdG [30, 51, 52].

In double-stranded DNA, however, the Cr-PdG adduct is most likely to undergo a ro reaction, as described above. The ro form can undergo subsequent reactions to generate either DNA-protein crosslinks or ICLs (fig. 3). In regard to DNA-protein crosslinks, Kurtz and Lloyd [56] demonstrated cross linking of the Cr-PdG adduct with peptides containing lysine, and shown that the formation of these crosslinks is more efficient in double-stranded than single-stranded DNA, consistent with the expectation that double-stranded DNA favors the open-chain form of the adduct. Acetaldehyde has been shown to produce DNA-protein crosslinks [57], and DNA-protein crosslinks can be precursors to SCEs [58]. Therefore this mechanism provides a possible explanation for the increase in SCEs that has been consistently observed in mammalian cells after acetaldehyde exposure [13]. As indicated in the figure, however, the relationship of SCEs to carcinogenesis is unclear [59].

With regard to ICLs, chemicals that produce ICLs, such as mitomycin C or diepoxybutane, generate chromosomal aberrations, indicating that ICLs can lead to chromosomal aberrations. In contrast to SCEs, chromosomal aberrations are generally accepted as causal events in cancer development [59]. The

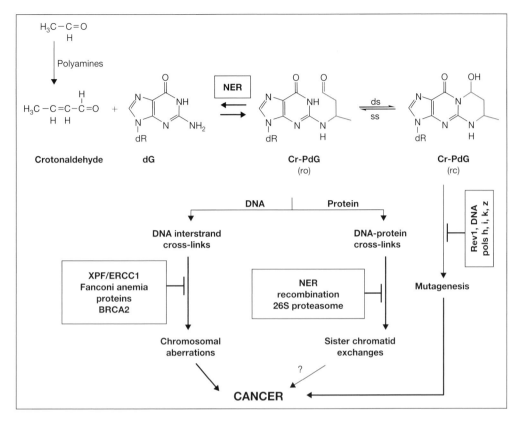

Fig. 4. Summary diagram to illustrate the formation of the central role of the Cr-PdG adduct and its derivatives in alcohol-related carcinogenesis. The solid boxes denote proteins or pathways that have been proposed to act to protect the cell from mutagenesis, ICLs, or DNA-protein crosslinks as indicated. The question mark above the arrow between SCEs and cancer reflects the uncertainty regarding the relationship between these events (see [59]). NER – nucleotide excision repair.

observation by Obe et al. [60] that acetaldehyde (approximately 178 or 356 μM) can increase chromosomal aberrations in human cells from a patient with the genetic disease Fanconi anemia (FA) indicates that acetaldehyde at concentrations that can form in vivo from alcohol drinking produces ICLs in vivo. As suggested in figure 3, the formation of ICLs from the ro form of the Cr-PdG could be the mechanism of this effect. Cells from FA patients have an elevated basal level of chromosomal aberrations, and are very sensitive to chromosomal aberrations when treated with drugs that cause ICLs. Indeed, this sensitivity forms the basis of a clinical test for FA that is in use today [61].

In summary, the genotoxic effects of the Cr-PdG adduct, along with its derivatives, DNA-protein crosslinks and ICLs, can in principle account for the known genotoxic effects of acetaldehyde: mutagenesis, DNA-protein crosslinks, SCEs, and chromosomal aberrations. By extension, therefore, to the extent that the genotoxic effects of acetaldehyde are involved in the increased risk of cancer from alcoholic beverage consumption, the formation of the Cr-PdG may explain this process as well.

DNA Repair Gene Polymorphism as Risk Factors for Alcohol-Related Cancer

One might reasonably ask what the value is of trying to determine which DNA lesion is responsible for alcoholic beverage related carcinogenesis. An answer to this question is that determining which lesion is involved can give clues as to which DNA repair pathways are the most relevant in protecting cells against the lesion. In turn, the genes encoding the relevant DNA repair proteins are rational 'candidate' genes in which genetic variation may influence the risk of developing cancer from alcohol drinking. Such DNA repair genotype information would be most useful when combined with genotyping of polymorphisms in ADH and ALDH genotypes that have already been shown to modulate cancer risk from alcoholic beverage consumption [7].

The multiple DNA repair pathways that are involved in protecting cells against the Cr-PdG adduct and its derivatives (DNA-protein crosslinks and ICLs) are also illustrated in figure 4. Beginning with the possibility that bypass of the rc form of the lesion can be mutagenic, the genes encoding the relevant translesion synthesis DNA polymerases that allow incorporation of dC opposite the Cr-PdG lesion might be useful to consider. Based on in vitro studies, the bypass of related DNA lesions, [62–64] the relevant enzymes would include Rev1, and DNA polymerases pol eta, zeta, iota, and kappa.

The nucleotide excision repair (NER) pathway plays a role in protecting against the Cr-PdG adduct in several ways. In addition to a role in repairing the lesion itself [65, 66], NER has been implicated in the repair of DNA-protein crosslinks [67]. The *XPF* (or *ERCC4*) and *ERCC1* gene products form a nuclease complex that is essential for NER, and is also involved in homologous recombination [68] and in the repair of ICLs [69]. Therefore, polymorphisms in either of these genes could affect the ability of the cell to repair the Cr-PdG lesion itself, as well as both the DNA-protein crosslinks and ICLs resulting from the ro form of the lesion. Interestingly, based on the data available, there is an approximately 3-fold higher frequency of non-synonomous amino acid substitution variants in *XPF* than *ERCC1*, even after correcting for the different sizes of the proteins (905 for *XPF* versus 297 for *ERCC1*).

The observation of Obe et al. [60], that cells from a FA patient showed a significant increase in chromosomal aberrations following exposure to acetaldehyde under conditions where normal cells showed little or no increase, clearly indicates an important role for FA proteins in protecting human cells from acetaldehyde-induced chromosomal aberrations. FA patients are known to have a higher rate of hematopoietic and epithelial cancers than the normal population [70]. In view of the strong relationship between alcoholic beverage consumption and UADT cancer, it is of particular relevance that the risk of developing head and neck (including esophageal) cancers in FA patients is especially high [70].

In regard to studying the significance of genetic variation in FA genes in relation to alcohol related cancers, there are at least 11 FA complementation groups known at the present time, and the identities of the mutated genes have been determined for 9 of them [71]. The complementation group of the FA patient whose cells were studied by Obe et al. [60] was not indicated in the report, but the majority (approximately 66%) of FA patients that have been characterized to date have mutations in the *FANCA* gene [72]. It would be useful to repeat and extend the work of Obe et al. [60] using cells from patients in different FA complementation groups, which might give some clues to which of the various FA genes to focus on for susceptibility of alcoholic beverage related cancers. In view of the complexity of the FA pathway [71–73], it is possible that mutations in different genes may have different effects on chromosomal aberrations in response to acetaldehyde, and therefore that some FA genes will be better candidate genes for susceptibility to alcohol related carcinogenesis than others. Finally, it is particularly noteworthy that the FANCD1 protein has been shown to be identical to BRCA2, one of the well-known breast cancer susceptibility genes [74]. This may be of mechanistic significance in understanding the possible relationship between alcohol consumption and the development of breast cancer [75].

The Human DNA Repair Genes Database (described in [76]) is an important resource for studies of genetic variation in DNA repair genes. The database lists over 150 DNA repair genes, including the FA genes, and includes functional information as well as links to other genome resources and a list of SNPs for each gene. In addition, a recent commentary [77] contains a very useful description of the problems and considerations involved in selecting DNA repair gene polymorphisms for analysis of cancer risk.

Conclusions and Future Directions

In this chapter, we have proposed an important role for a particular acetaldehyde-derived DNA adduct, Cr-PdG, in alcoholic beverage related carcinogenesis. This proposal can be tested in various ways. One approach would be to demonstrate the formation of the lesion in cells or the relevant target tissues for acetaldehyde related carcinogens. We are currently working on the development of antibodies against the lesion to address this topic at the cellular level. One prediction of our model is that varying polyamine levels in cells will affect the level of Cr-PdG adduct formation. This can be tested by engineering cells to produce different polyamine levels. Finally, an important future direction would be to develop an animal model for alcohol or acetaldehyde related UADT cancer. In light of the strong evidence that alcoholic beverage related UADT cancer results from relatively high concentrations of acetaldehyde formed from local ethanol metabolism by microorganisms in the mouth, a rational model would provide acetaldehyde to animals in the drinking water. A prediction of the model proposed in this chapter is that mice lacking various DNA repair pathways, in particular the NER and FA pathways, would be more susceptible to carcinogenesis from acetaldehyde in such a model than wild-type mice. The success of such an experiment would directly show that acetaldehyde could be a complete carcinogen in the UADT, and also provide a compelling case for the role of specific DNA repair pathways in protecting against acetaldehyde, and by extension, alcoholic beverage related, carcinogenesis.

Acknowledgements

We thank Anoop Patel and Cheryl Marietta for helpful comments on the manuscript. Work in our laboratory is supported by the DICBR, NIAAA, and NIH.

References

1   U.S. Department of Health and Human Services PHS, National Toxicology Program, Alcoholic Beverage Consumption; in Report on Carcinogens, ed 11. 2005.
2   Brown LM: Epidemiology of alcohol-associated cancers. Alcohol 2005;35:161–168.
3   Brooks PJ: DNA damage, DNA repair, and alcohol toxicity – a review. Alcohol Clin Exp Res 1997;21:1073–1082.
4   Poschl G, Seitz HK: Alcohol and cancer. Alcohol Alcohol 2004;39:155–165.
5   Dellarco VL: A mutagenicity assessment of acetaldehyde. Mutat Res 1988;195:1–20.
6   IARC: Acetaldehyde; in IARC Monographs on the Evaluation of Carcinogenic Risks to Humans: Re-Evaluation of Some Organic Chemicals, Hydrazine and Hydrogen Peroxide. Lyon, France, IARC, 1999, vol 71, p 319.
7   Yokoyama A, Omori T: Genetic polymorphisms of alcohol and aldehyde dehydrogenases and risk for esophageal and head and neck cancers. Jpn J Clin Oncol 2003;33:111–121.
8   Eriksson CJ: The role of acetaldehyde in the actions of alcohol (update 2000). Alcohol Clin Exp Res 2001;25:15S–32S.
9   Feron VJ, Kruysse A, Woutersen RA: Respiratory tract tumours in hamsters exposed to acetaldehyde vapour alone or simultaneously to benzo(a)pyrene or diethylnitrosamine. Eur J Cancer Clin Oncol 1982;18:13–31.

10 Woutersen RA, Appelman LM, Feron VJ, Van der Heijden CA: Inhalation toxicity of acetaldehyde in rats. II. Carcinogenicity study: interim results after 15 months. Toxicology 1984;31:123–133.

11 Woutersen RA, Appelman LM, Van Garderen-Hoetmer A, Feron VJ: Inhalation toxicity of acetaldehyde in rats. III. Carcinogenicity study. Toxicology 1986;41:213–231.

12 Woutersen RA, Feron VJ: Inhalation toxicity of acetaldehyde in rats. IV. Progression and regression of nasal lesions after discontinuation of exposure. Toxicology 1987;47:295–305.

13 Obe G, Anderson D: International commission for protection against environmental mutagens and carcinogens. ICPEMC working paper no. 15/1. Genetic effects of ethanol. Mutat Res 1987;186: 177–200.

14 Obe G, Ristow H: Acetaldehyde but not ethanol induced sister chromatid exchanges in Chinese hamster cells. Mutat Res 1977;56:211–213.

15 Visapaa JP, Gotte K, Benesova M, Li J, Homann N, Conradt C, Inoue H, Tisch M, Horrmann K, Vakevainen S, Salaspuro M, Seitz HK: Increased cancer risk in heavy drinkers with the alcohol dehydrogenase 1C*1 allele, possibly due to salivary acetaldehyde. Gut 2004;53:871–876.

16 Homann N, Jousimies-Somer H, Jokelainen K, Heine R, Salaspuro M: High acetaldehyde levels in saliva after ethanol consumption: methodological aspects and pathogenetic implications. Carcinogenesis 1997;18:1739–1743.

17 Yin SJ, Chou FJ, Chao SF, Tsai SF, Liao CS, Wang SL, Wu CW, Lee SC: Alcohol and aldehyde dehydrogenases in human esophagus: comparison with the stomach enzyme activities. Alcohol Clin Exp Res 1993;17:376–381.

18 Haselbeck RJ, Duester G: Regional restriction of alcohol/retinol dehydrogenases along the mouse gastrointestinal epithelium. Alcohol Clin Exp Res 1997;21:1484–1490.

19 Vaca CE, Fang JL, Schweda EK: Studies of the reaction of acetaldehyde with deoxynucleosides. Chem Biol Interact 1995;98:51–67.

20 Fang JL, Vaca CE: Development of a 32P-postlabelling method for the analysis of adducts arising through the reaction of acetaldehyde with 2′-deoxyguanosine-3′-monophosphate and DNA. Carcinogenesis 1995;16:2177–2185.

21 Fang JL, Vaca CE: Detection of DNA adducts of acetaldehyde in peripheral white blood cells of alcohol abusers. Carcinogenesis 1997;18:627–632.

22 Matsuda T, Terashima I, Matsumoto Y, Yabushita H, Matsui S, Shibutani S: Effective utilization of N2-ethyl-2′-deoxyguanosine triphosphate during DNA synthesis catalyzed by mammalian replicative DNA polymerases. Biochemistry 1999;38:929–935.

23 Terashima I, Matsuda T, Fang TW, Suzuki N, Kobayashi J, Kohda K, Shibutani S: Miscoding potential of the N2-ethyl-2′-deoxyguanosine DNA adduct by the exonuclease-free Klenow fragment of Escherichia coli DNA polymerase I. Biochemistry 2001;40:4106–4114.

24 Perrino FW, Blans P, Harvey S, Gelhaus SL, McGrath C, Akman SA, Jenkins GS, LaCourse WR, Fishbein JC: The N2-ethylguanine and the O6-ethyl- and O6-methylguanine lesions in DNA: contrasting responses from the 'bypass' DNA polymerase eta and the replicative DNA polymerase alpha. Chem Res Toxicol 2003;16:1616–1623.

25 Hubscher U, Maga G, Spadari S: Eukaryotic DNA polymerases. Annu Rev Biochem 2002;71: 133–163.

26 Masutani C, Kusumoto R, Yamada A, Dohmae N, Yokoi M, Yuasa M, Araki M, Iwai S, Takio K, Hanaoka F: The XPV (xeroderma pigmentosum variant) gene encodes human DNA polymerase eta. Nature 1999;399:700–704.

27 Woodgate R: A plethora of lesion-replicating DNA polymerases. Genes Dev 1999;13:2191–2195.

28 Friedberg EC, Wagner R, Radman M: Specialized DNA polymerases, cellular survival, and the genesis of mutations. Science 2002;296:1627–1630.

29 Prakash S, Johnson RE, Prakash L: Eukaryotic translesion synthesis DNA polymerases: specificity of structure and function. Annu Rev Biochem 2005;74:317–353.

30 Wang M, McIntee EJ, Cheng G, Shi Y, Villalta PW, Hecht SS: Identification of DNA adducts of acetaldehyde. Chem Res Toxicol 2000;13:1149–1157.

31 Nath RG, Ocando JE, Chung FL: Detection of 1, N2-propanodeoxyguanosine adducts as potential endogenous DNA lesions in rodent and human tissues. Cancer Res 1996;56:452–456.

32 Chung FL, Zhang L, Ocando JE, Nath RG: Role of 1,N2-propanodeoxyguanosine adducts as endogenous DNA lesions in rodents and humans. IARC Sci Publ 1999;150:45–54.

33 Mao H, Schnetz-Boutaud NC, Weisenseel JP, Marnett LJ, Stone MP: Duplex DNA catalyzes the chemical rearrangement of a malondialdehyde deoxyguanosine adduct. Proc Natl Acad Sci USA 1999;96:6615–6620.

34 de los Santos C, Zaliznyak T, Johnson F: NMR characterization of a DNA duplex containing the major acrolein-derived deoxyguanosine adduct gamma –OH-1,–N2-propano-2′-deoxyguanosine. J Biol Chem 2001;276:9077–9082.

35 Eder E, Budiawan: Cancer risk assessment for the environmental mutagen and carcinogen croton-aldehyde on the basis of TD(50) and comparison with 1,N(2)-propanodeoxyguanosine adduct levels. Cancer Epidemiol Biomarkers Prev 2001;10:883–888.

36 Sako M, Yaekura I, Deyashiki Y: Chemo- and regio-selective modifications of nucleic acids by acetaldehyde and crotonaldehyde. Nucleic Acids Res Suppl 2002:21–22.

37 Sako M, Inagaki S, Esaka Y, Deyashiki Y: Histones accelerate the cyclic 1,N2-propanoguanine adduct-formation of DNA by the primary metabolite of alcohol and carcinogenic crotonaldehyde. Bioorg Med Chem Lett 2003;13:3497–3498.

38 Warters RL, Newton GL, Olive PL, Fahey RC: Radioprotection of human cell nuclear DNA by polyamines: radiosensitivity of chromatin is influenced by tightly bound spermine. Radiat Res 1999;151:354–362.

39 Chiu S, Oleinick NL: Radioprotection against the formation of DNA double-strand breaks in cellular DNA but not native cellular chromatin by the polyamine spermine. Radiat Res 1997;148: 188–192.

40 Chattopadhyay MK, Tabor CW, Tabor H: Polyamines protect *Escherichia coli* cells from the toxic effect of oxygen. Proc Natl Acad Sci USA 2003;100:2261–2265.

41 Ha HC, Sirisoma NS, Kuppusamy P, Zweier JL, Woster PM, Casero RA Jr: The natural polyamine spermine functions directly as a free radical scavenger. Proc Natl Acad Sci USA 1998;95: 11140–11145.

42 Newton GL, Aguilera JA, Ward JF, Fahey RC: Effect of polyamine-induced compaction and aggregation of DNA on the formation of radiation-induced strand breaks: quantitative models for cellular radiation damage. Radiat Res 1997;148:272–284.

43 Tabor CW, Tabor H: Polyamines. Annu Rev Biochem 1984;53:749–790.

44 Wallace HM: Polyamines and their role in human disease – an introduction. Biochem Soc Trans 2003;31:354–355.

45 Gerner EW, Meyskens FL Jr: Polyamines and cancer: old molecules, new understanding. Nat Rev Cancer 2004;4:781–792.

46 Bachrach U: Polyamines and cancer: minireview article. Amino Acids 2004;26:307–309.

47 Theruvathu JA, Jaruga P, Nath RG, Dizdaroglu M, Brooks PJ: Polyamines stimulate the formation of mutagenic 1,N2-propanodeoxyguanosine adducts from acetaldehyde. Nucleic Acids Res 2005;33:3513–3520.

48 Inagaki S, Esaka Y, Goto M, Deyashiki Y, Sako M: LC-MS study on the formation of cyclic 1,N2-propano guanine adduct in the reactions of DNA with acetaldehyde in the presence of histone. Biol Pharm Bull 2004;27:273–276.

49 Niedernhofer LJ, Riley M, Schnetz-Boutaud N, Sanduwaran G, Chaudhary AK, Reddy GR, Marnett LJ: Temperature-dependent formation of a conjugate between tris(hydroxymethyl) aminomethane buffer and the malondialdehyde-DNA adduct pyrimidopurinone. Chem Res Toxicol 1997;10:556–561.

50 Matsuda T, Kawanishi M, Yagi T, Matsui S, Takebe H: Specific tandem GG to TT base substitutions induced by acetaldehyde are due to intra-strand crosslinks between adjacent guanine bases. Nucleic Acids Res 1998;26:1769–1774.

51 Kozekov ID, Nechev LV, Moseley MS, Harris CM, Rizzo CJ, Stone MP, Harris TM: DNA interchain cross-links formed by acrolein and crotonaldehyde. J Am Chem Soc 2003;125: 50–61.

52 Lao Y, Hecht SS: Synthesis and properties of an acetaldehyde-derived oligonucleotide interstrand cross-link. Chem Res Toxicol 2005;18:711–721.

53 Czerny C, Eder E, Runger TM: Genotoxicity and mutagenicity of the alpha, beta-unsaturated carbonyl compound crotonaldehyde (butenal) on a plasmid shuttle vector. Mutat Res 1998;407: 125–134.

54 Kawanishi M, Matsuda T, Sasaki G, Yagi T, Matsui S, Takebe H: A spectrum of mutations induced by crotonaldehyde in shuttle vector plasmids propagated in human cells. Carcinogenesis 1998; 19:69–72.

55 Fernandes PH, Kanuri M, Nechev LV, Harris TM, Lloyd RS: Mammalian cell mutagenesis of the DNA adducts of vinyl chloride and crotonaldehyde. Environ Mol Mutagen 2005;45:455–459.

56 Kurtz AJ, Lloyd RS: 1,N2-deoxyguanosine adducts of acrolein, crotonaldehyde, and trans-4-hydroxynonenal cross-link to peptides via Schiff base linkage. J Biol Chem 2003;278:5970–5976.

57 Kuykendall JR, Bogdanffy MS: Reaction kinetics of DNA-histone crosslinking by vinyl acetate and acetaldehyde. Carcinogenesis 1992;13:2095–2100.

58 Hengstler JG, Bogdanffy MS, Bolt HM, Oesch F: Challenging dogma: thresholds for genotoxic carcinogens? The case of vinyl acetate. Annu Rev Pharmacol Toxicol 2003;43:485–520.

59 Hagmar L, Stromberg U, Tinnerberg H, Mikoczy Z: The usefulness of cytogenetic biomarkers as intermediate endpoints in carcinogenesis. Int J Hyg Environ Health 2001;204:43–47.

60 Obe G, Natarajan AT, Meyers M, Hertog AD: Induction of chromosomal aberrations in peripheral lymphocytes of human blood in vitro, and of SCEs in bone-marrow cells of mice in vivo by ethanol and its metabolite acetaldehyde. Mutat Res 1979;68:291–294.

61 Esmer C, Sanchez S, Ramos S, Molina B, Frias S, Carnevale A: DEB test for Fanconi anemia detection in patients with atypical phenotypes. Am J Med Genet A 2004;124:35–39.

62 Yang IY, Miller H, Wang Z, Frank EG, Ohmori H, Hanaoka F, Moriya M: Mammalian translesion DNA synthesis across an acrolein-derived deoxyguanosine adduct. Participation of DNA polymerase eta in error-prone synthesis in human cells. J Biol Chem 2003;278:13989–13994.

63 Washington MT, Minko IG, Johnson RE, Haracska L, Harris TM, Lloyd RS, Prakash S, Prakash L: Efficient and error-free replication past a minor-groove N2-guanine adduct by the sequential action of yeast Rev1 and DNA polymerase zeta. Mol Cell Biol 2004;24:6900–6906.

64 Washington MT, Minko IG, Johnson RE, Wolfle WT, Harris TM, Lloyd RS, Prakash S, Prakash L: Efficient and error-free replication past a minor-groove DNA adduct by the sequential action of human DNA polymerases iota and kappa. Mol Cell Biol 2004;24:5687–5693.

65 Yang IY, Hossain M, Miller H, Khullar S, Johnson F, Grollman A, Moriya M: Responses to the major acrolein-derived deoxyguanosine adduct in *Escherichia coli*. J Biol Chem 2001;276: 9071–9076.

66 Yang IY, Chan G, Miller H, Huang Y, Torres MC, Johnson F, Moriya M: Mutagenesis by acrolein-derived propanodeoxyguanosine adducts in human cells. Biochemistry 2002;41:13826–13832.

67 Minko IG, Zou Y, Lloyd RS: Incision of DNA-protein crosslinks by UvrABC nuclease suggests a potential repair pathway involving nucleotide excision repair. Proc Natl Acad Sci USA 2002;99: 1905–1909.

68 Adair GM, Rolig RL, Moore-Faver D, Zabelshansky M, Wilson JH, Nairn RS: Role of ERCC1 in removal of long non-homologous tails during targeted homologous recombination. Embo J 2000;19:5552–5561.

69 Niedernhofer LJ, Odijk H, Budzowska M, van Drunen E, Maas A, Theil AF, de Wit J, Jaspers NG, Beverloo HB, Hoeijmakers JH, Kanaar R: The structure-specific endonuclease Ercc1-Xpf is required to resolve DNA interstrand cross-link-induced double-strand breaks. Mol Cell Biol 2004;24:5776–5787.

70 Alter BP, Greene MH, Velazquez I, Rosenberg PS: Cancer in Fanconi anemia. Blood 2003;101: 2072.

71 Thompson LH: Unraveling the Fanconi anemia-DNA repair connection. Nat Genet 2005;37: 921–922.

72 Wang X, D'Andrea AD: The interplay of Fanconi anemia proteins in the DNA damage response. DNA Repair (Amst) 2004;3:1063–1069.

73 Matsushita N, Kitao H, Ishiai M, Nagashima N, Hirano S, Okawa K, Ohta T, Yu DS, McHugh PJ, Hickson ID, Venkitaraman AR, Kurumizaka H, Takata M: A FancD2-monoubiquitin fusion reveals hidden functions of Fanconi anemia core complex in DNA repair. Mol Cell 2005;19: 841–847.

74 Howlett NG, Taniguchi T, Olson S, Cox B, Waisfisz Q, De Die-Smulders C, Persky N, Grompe M, Joenje H, Pals G, Ikeda H, Fox EA, D'Andrea AD: Biallelic inactivation of BRCA2 in Fanconi anemia. Science 2002;297:606–609.

Acetaldehyde-DNA Adducts

75 Dumitrescu RG, Shields PG: The etiology of alcohol-induced breast cancer. Alcohol 2005;35: 213–225.
76 Wood RD, Mitchell M, Lindahl T: Human DNA repair genes, 2005. Mutat Res 2005;577: 275–283.
77 Clarkson SG, Wood RD: Polymorphisms in the human XPD (ERCC2) gene, DNA repair capacity and cancer susceptibility: an appraisal. DNA Repair (Amst) 2005;4:1068–1074.

P.J. Brooks
Section on Molecular Neurobiology
Laboratory of Neurogenetics
National Institute on Alcohol Abuse and Alcoholism
National Institutes of Health
5625 Fishers Lane, Room 3S32, MSC 9412
Bethesda, MD 20892–9412 (USA)
Tel. +1 301 496 7920, Fax +1 301 480 2839
E-Mail pjbrooks@mail.nih.gov

Cho CH, Purohit V (eds): Alcohol, Tobacco and Cancer.
Basel, Karger, 2006, pp 95–108

·······················
Ethanol and Liver Cancer

Iain H. McKillop, Laura W. Schrum

Department of Biology, University of North Carolina at Charlotte,
Charlotte, N.C., USA

Abstract

The incidence of hepatocellular carcinoma continues to increase on a global scale and
currently accounts for approximately 500,000 deaths/year. Unlike many other common malignancies hepatocellular carcinoma occurs predominantly within the context of known risk factors, of which hepatitis infection, exposure to aflatoxin and chronic ethanol abuse are the most
common. Despite the diversity in nature of the initial insult, resulting cirrhosis of the liver represents the most common precursor to hepatocellular carcinoma development. After ethanol
consumption the liver represents the major site of metabolism. Hepatic ethanol metabolism
via alcohol dehydrogenase leads to the production of acetaldehyde and free radicals that are
capable of binding to numerous cellular targets including DNA and components of cell signaling pathways. In addition to intracellular damage caused by ethanol metabolism, increased
acetaldehyde levels leads to the depletion of intracellular glutathione, an endogenous antioxidant involved in detoxification. Chronic excessive ethanol intake leads to induction of hepatic
microsomal, CYP2E1, an enzyme that metabolizes ethanol to acetaldehyde and, in doing so,
leads to further free radical production and aberrant cell function. In addition to ethanol
metabolism, CYP2E1 induction/activity has also been demonstrated to play an important role
in the activation of pro-carcinogens, changes in cell cycle, nutritional deficiencies and altered
immune system responses associated with chronic excessive ethanol intake. The central role
of ethanol metabolism in mediating many of the detrimental effects of ethanol consumption
has led to considerable interest in the study and use of dietary antioxidants to blunt these
effects. Included in this group are *S*-adenosyl-L-methionine and plant-derived flavanoids.

Hepatocellular Carcinoma

Hepatocellular carcinoma (HCC) currently represents the fifth most common
malignancy in men and the eighth in women accounting for a predicted 564,000
new cases diagnosed in the year 2000 and more than half a million deaths annually
[1]. The insidious nature of HCC is such that the major complicating factor in the

successful long-term treatment of this disease remains late detection and/or intra/extra-hepatic tumor metastasis [2, 3]. At present, surgical resection of the tumor mass remains the most common treatment but is restricted to 40–50% of patients diagnosed with long-term (5–7 years) success rates typically being 10–25% [2, 3]. Other therapies including intra-arterial embolism, percutaneous ethanol injection and cryoablation have been employed but with limited success, whereas the use of hepatic transplant is restricted by the availability of organs and the absence of metastasis prior to transplant [4–8].

As with other major organs, tumors of the liver can be either primary in nature or secondary following metastasis to the liver (or in less common cases, the direct spread of neoplasia from adjacent organs) [9, 10]. While primary hepatic tumors can arise from a variety of liver cell types, more than 80% of all primary tumors diagnosed are HCCs resulting from the transformation of hepatic parenchymal cells (hepatocytes) [9, 10]. Unlike several other common malignancies HCC rarely demonstrates common hereditary patterns of development. Despite a lack of common genetic or familial markers HCC is unusual in that it predominantly occurs within the realms of known risk factors of which underlying hepatic cirrhosis represents the most common precursor for future HCC development [9, 10]. In Asian and (sub-Saharan) African countries where viral hepatitis and/or exposure to aflatoxin is common, the incidence of HCC is reported to be as high as 20–150 cases/100,000, underlying cirrhosis occurring in approximately 50–60% of those diagnosed with HCC [10–12]. In contrast, in Western Europe and North America the incidence of HCC is considerably lower (1.5–3 cases/100,0000) yet underlying cirrhosis is associated with approximately 80–90% of patients who go on to develop HCC. While the increasing incidence of HCC diagnosis in Western Europe and North America correlates to an increase in viral hepatitis C, chronic excess ethanol consumption remains a major risk factor in the development of hepatic cirrhosis and progression to HCC [10, 11].

Ethanol and Liver Cancer

The correlation between chronic excessive ethanol consumption and hepatic disease has been well documented [13, 14]. Depending on the level and period of ethanol consumption hepatic complications can range from moderate steatosis (fatty liver) to acute alcoholic hepatitis, and if long term excessive ethanol consumption continues, the development of liver fibrosis and cirrhosis [15]. Despite the strong correlation between underlying cirrhosis (whether viral or chemically induced) and the development of HCC, delineating direct relationships between ethanol consumption and HCC development remain tenuous for several reasons.

Epidemiological studies demonstrate a strong dose–response relationship between ethanol consumption and the risk of cancer in men and women [16]. Yet, despite these correlations the risk of HCC development is higher in former drinkers for up to 10 years following cessation than those who continue to consume ethanol in excess [16, 17]. This paradoxical observation may be explained in part by fact that patients stop drinking following the early signs of liver disease, yet the underlying hepatic disease remains, and with it, the risk of cancer development. In contrast, patients who continue to consume ethanol in excess will often die from complications of alcohol use prior to the manifestation of HCC [17]. Similarly, continued excessive ethanol consumption causes continual hepatic insult that may in turn impede hepatic regeneration/repopulation. In contrast, the hepatic regeneration/repopulation following cessation of drinking may lead to the emergence of genetic alterations in the repopulating cells that have resulted from prior ethanol consumption. Of interest, epidemiological data also demonstrate that long-term ingestion of small amounts of alcohol do not increase the risk of HCC development yet larger 'toxic' amounts ingested over shorter periods of time do increase risk [16, 17]. In addition to the levels of ethanol consumed and the time periods of consumption, several other factors complicate our understanding of the direct role of ethanol as a risk factor for HCC development. Epidemiological studies report that ethanol dependent patients with viral hepatitis and patients with ethanol dependency prior to aflatoxin exposure have a poorer prognosis than patients who abstain from ethanol [17]. However, given the role of both viral hepatitis and aflatoxin as risk factors for HCC development it remains difficult to assess the relative contribution of these factors versus that of ethanol in the processes of hepatocyte transformation and/or progression of HCC following transformation. Analogous to the presence of superimposed factors known to increase the risk of HCC development, considerable debate has emerged regarding the role of additional factors not directly associated with ethanol consumption per se. Of these factors, the high incidence of cigarette smoking in ethanol dependent patients [18, 19] along with differences in dietary intake [20–22], the socioeconomic status and relative balance of diet [21, 23] and regional differences in the type of beverage consumed [24, 25] are identified as potential factors in determining HCC development.

Ethanol and Hepatic Damage

The development of HCC centers upon the initial transformation of the hepatocyte due to DNA damage induced via a variety of chemical and/or viral factors. However, the development of ethanol-related liver disease is a multistage process that, while centered upon the effects of ethanol on parenchymal cells, also involve other non-parenchymal cell populations and systemic, non-hepatic effects of ethanol. Following ingestion the hepatocytes of the liver represent the major site of ethanol metabolism. Ethanol metabolism occurs via three main pathways, alcohol dehydrogenase (ADH), microsomal ethanol oxidizing system (comprising predominantly of inducible cytochrome P450 2E1 (CYP2E1) and catalase [26–28]). ADH-dependent ethanol metabolism results in the production of acetaldehyde and free radicals [27]. These highly reactive compounds can bind rapidly to and damage various cell constituents including DNA [29, 30]. In addition, elevated intracellular acetaldehyde decreases DNA repair mechanisms and the methylation of cytosine [30, 31].

Following chronic excessive ethanol intake microsomal CYP2E1 induction occurs [28, 32, 33]. Induction of CYP2E1 leads to the metabolism of ethanol to acetaldehyde producing similar effects to those observed with ADH-dependent metabolism. In addition, CYP2E1 induction/ethanol metabolism leads to pronounced increases in free radical production and cellular damage through peroxidation [28, 32, 33]. While ethanol metabolism following CYP2E1 induction can have pronounced detrimental effects on the cell, the induction of CYP2E1 can also lead to significant indirect effects in the hepatocyte, many of which are associated with cellular transformation. CYP2E1 has been demonstrated to enhance the activation of procarcinogens, many of which are present in alcoholic beverages, tobacco smoke and the diet [23, 33, 34] and the distribution of carcinogens both intra and extrahepatically. The induction of CYP2E1 is also associated with changes in cell cycle duration and nutritional aberrations such as vitamin A, folate, zinc and selenium deficiency as well as altered immune system responses leading to increased susceptibility to viral infection (e.g. HBV & HCV) [29, 35].

In addition to the direct detrimental effects of ethanol/ethanol metabolism in the liver chronic ethanol consumption also acts to deplete the body's natural antioxidant supply, glutathione (GSH) [36–38]. After chronic ethanol intake depleted mitochondrial GSH occurs due, at least in part, to decreased GSH transport into the mitochondria from the cytosolic compartment, leading to increased hepatocyte susceptibility to further damage from reactive oxygen species (ROS) generation [36].

In addition to the direct effects of ethanol metabolism on hepatocyte integrity, prolonged excessive ethanol intake can lead to other systemic non-hepatic and hepatic, non-hepatocyte effects that are integral in the development of hepatic disease and the progression to HCC. Chronic excessive ethanol consumption alters the normal balance of GI bacterial flora and increases net GI permeability [39, 40]. The hepatic vascular architecture is such that this leads to increased lipopolysaccharide delivery to the liver via the hepatic portal vein/hepatic sinusoid. In this instance elevated intrahepatic lipopolysaccharide stimulates the activation of the livers resident macrophage population (Kupffer cells) and increases reactive oxygen species generation. Elevated ROS production in turn leads to the activation of nuclear factor-κB (NFκB) in the Kupffer cell population resulting in the synthesis and release of pro-inflammatory cytokines [40–42]. The elevated synthesis and release of pro-inflammatory cytokines, in conjunction with other factors may have pronounced subsequent effects on both Kupffer cell response and the response of other cell types within the liver [40]. Under normal circumstances stellate cells are largely quiescent and function as the main hepatic storage site of vitamin A [43]. However, these cells can be rapidly activated in response to hepatic insult including, pro-inflammatory cytokines, prolonged changes in hepatic circulation and/or elevated levels of toxins and/or oxidative stress [44]. After activation stellate cells undergo transformation to form dividing myofibroblastic-like cells involved in the deposition of collagen [43, 44]. If unchecked this continued deposition of collagen leads to hepatic scarring, decreased blood flow/liver function, cell death and eventually cirrhosis. The central role of Kupffer cells during this process is further highlighted experimentally by depletion of Kupffer cells using gadolinium chloride [45] or treatment with antibiotics to reduce GI bacteria [46] which leads to significantly decreased ethanol-related hepatic injury.

Ethanol and Cell Signaling Pathways in the Liver

The direct effects of alcohol metabolism on hepatocyte function and the indirect effects of chronic ethanol consumption play a significant role in the process of initial cell transformation and the development of underlying hepatic cirrhosis. However, it is becoming increasingly apparent that ethanol and/or the metabolism of ethanol plays a significant role in moderating cell signaling pathways that are central to normal and abnormal hepatocyte function, proliferation and apoptosis. A considerable body of evidence now exists demonstrating that ethanol is capable of altering the activity of a wide range of signaling cascades involved in both pro- and antimitogenic responses and the regulation of apoptosis. Given that transformation and subsequent progression of HCC is characterized by the inability of cells to respond to normal signaling mechanisms, direct and/or indirect effects of ethanol on the integrity and/or functionality of these pathways play a significant role in initial transformation and progression. Predictably, ethanol has been demonstrated to affect a wide range of signaling cascades including NFκB [47], the mitogen activated protein kinases (MAPKs) [48], phospholipase C and D [49, 50], adenylate cyclase [51], protein kinase C [52, 53], c-jun N-terminal kinase [54] and signal-transducing activators of transcription [55]. As is often the case in the absence of ethanol, the effects of ethanol on signaling cascade integrity in the hepatocyte appears to be both ligand specific and varies significantly depending on length of exposure to ethanol. Acute ethanol treatment of hepatocytes increases *p42/p44* MAPK and *c-jun* N-terminal kinase activity [56, 57], potentiates *p42/p44* MAPK signaling [58] and prolongs *p42/p44* MAPK activity in response to insulin EGF and HGF [55]. In contrast, acute ethanol treatment inhibits IL-6 and IFN-activated signal-transducing activators of transcription [59] while failing to alter vasopressin stimulated *p42/p44* MAPK activity [55, 58, 60]. Interestingly, the use of hepatic cells incapable of ethanol metabolism, inhibitors of ethanol metabolism or direct addition of acetaldeyde demonstrate that ethanol metabolism is not a significant factor in mediating the response to acute ethanol treatment. Chronic ethanol feeding also causes significant changes in specific hepatocyte receptor expression and function. Studies by Shaw et al. [61] demonstrate that chronic treatment with ethanol down regulates glucagon receptor expression and function after stimulation, effects that are attributed to increased oxidative stress. Similarly, chronic ethanol feeding inhibits EGF and insulin stimulated *p42/p44* MAPK activity [55], suppresses the potentiation of angiotensin II-dependent p*42/p44* MAPK activity [58] and increases hepatocyte responsiveness to endotoxin [62]. However, it is important to remember than in models of chronic ethanol exposure it is likely that other, non-hepatocyte cells also play a significant role in the response of hepatocytes to ligands [17, 40, 63].

Effect of Ethanol on Hepatocellular Carcinoma Progression

Significant advances are being made in understanding the underlying mechanisms by which chronic excessive ethanol intake leads to the onset and progression of hepatic diseases including HCC. Despite these findings several important questions remain to be addressed if we are to fully understand the etiology of HCC. Little doubt exists that the generation of acetaldehyde and ROS following ethanol metabolism is central during the process of hepatocyte DNA damage and cellular transformation. As previously described, the liver demonstrates a unique capacity to regenerate following a reduction in hepatic mass and/or function [64, 65]. While this is most dramatically observed following surgical resection, the repopulation of hepatic cells following deletion (via apoptosis or necrosis) is also of critical importance in maintaining functional hepatic mass. During the repopulation process coordination of cell signaling pathways is essential in regulating both mitogenesis and differentiation. Given that chronic ethanol ingestion is widely reported to affect hepatic signal transduction integrity it is highly likely that many of these pathways also play a central role in aberrations in regeneration, and if compromised affect the mitogenic capacity of cells in the liver.

During hepatic regeneration induced by surgical ablation altered expression of adenylyl cyclase linked G-protein signaling occurs in a time dependent manner [64, 66, 67]. In an animal model of chronic ethanol feeding and hepatectomy, G-protein signaling is perturbed leading to inhibited cAMP-dependent signaling [68]. Previous reports from our laboratory and others have demonstrated that both human and animal (rat) HCC is associated with increased expression and function of inhibitory G-proteins (Gi-proteins) [69, 70] and components of a *p42/p44* MAPK cascade [71–73]. Stimulation of Gi-proteins activates this *p42/p44* MAPK cascade leading to enhanced cell mitogenesis, an effect not seen in normal hepatocytes [74]. Exposure to moderate doses of ethanol (25 m*M* for 24 h) in turn causes further increases in Gi-protein expression and enhanced *p42/p44* MAPK signaling and growth [75]. Given that the effects of ethanol on signaling pathways in normal hepatocytes can be either direct or a result of ethanol metabolism we assessed the HCC cells ability to metabolize ethanol, their responsiveness to ethanol in the presence of an inhibitor of ethanol metabolism (4-methyl pyrazole, 4MP) and after the direct addition of acetaldehyde. Twenty four hours exposure of HCC cells to ethanol led to significant ethanol metabolism that was effectively blocked using 4MP. Inhibition of ethanol metabolism using 4MP also abrogated the effects of ethanol on G-protein-linked *p42/p44* MAPK signaling whereas the direct addition of acetaldehyde mimicked the effects of ethanol [75]. These data indicate that ethanol metabolism is an essential step in altering G-protein expression and function in HCC cells. Further evidence of this was observed when the human HepG2 cell line (which does not to express either ADH or CYP2E1) was treated with ethanol (1–100 m*M*) and failed to elicit changes in Gi-protein expression or their ability to activate a *p42/p44* MAPK cascade (unpublished observations, I.H. McKillop and C.M Schmidt, 1999). More recently, preliminary data from our laboratory suggest that in a murine model of chronic ethanol feeding (6 weeks) hepatic Gi-protein expression and *p42/p44* MAPK activity becomes significantly elevated after 4 weeks, a time that coincides with the induction of CYP2E1 and increased capacity to metabolize ingested ethanol (unpublished data). Collectively these data suggest a significant role for ethanol metabolism in normal hepatic regeneration, the hepatic response to chronic ethanol ingestion and the progression of HCC.

In addition to data addressing the effects of ethanol on specific signal transduction pathways that regulate proliferation and/or apoptosis increasing experimental data, in conjunction with detailed epidemiological data, indicate that drinking patterns and/or pre-existing conditions also play a significant role in the development and progression of HCC. Epidemiological analysis demonstrates that ethanol dependent patients with viral hepatitis and patients with ethanol dependency prior to aflatoxin exposure have a poorer prognosis than patients who abstain from ethanol [17]. However, what is not apparent is what effect moderate ethanol consumption has in patients who demonstrate these risk factors or the relative contribution ethanol makes at the mechanistic level in the presence of either viral hepatitis and/or aflatoxin exposure. Indeed, the failure to understand these events may become ever more important given the increasing levels of ethanol consumption in countries that are already associated with high levels of viral hepatitis and/or aflatoxin exposure. Similarly, patterns of ethanol consumption may play a significant role in determining the outcome of ethanol consumption on initial transformation and/or progression of HCC. This may be of particular significance given the alarming increase in 'binge drinking' that is emerging. In this instance, individuals with ostensibly healthy lifestyles (balanced diets, do not smoke and exercise regularly) consume large volumes of ethanol over relatively short (2–3 day) periods of time and then refrain from further consumption over the following days–weeks period [76]. In this instance a pattern of abstinence-excess develops in which the liver is subject to repeated cycles of hepato-toxic insult followed by recovery. This raises the possibility of increased cycles of ethanol-induced cellular damage, deletion and repopulation and may increase the risk genetic damage and transformation.

The use of Antioxidants in Reducing the Incidence of HCC

One of the major advances in understanding why prolonged excessive ethanol consumption affects the liver so profoundly is the identification of the role of ROS and oxidative stress in hepatic cell damage. Despite the range of direct and indirect effects of ethanol in the body, ROS generation is a unifying factor in the development and progression of hepatic damage both in the hepatocyte and other hepatic cell populations. This has led to intense interest in the potential use of antioxidants as a means to attenuate ethanol related hepatic injury. In recent years, S-adenosyl-L-methionine (SAMe) has generated considerable interest as an agent for use in protecting against the effects of ethanol on the liver. The liver is the major site of endogenous SAMe synthesis and degradation and is central in maintaining SAMe homeostasis [77]. Under normal conditions SAMe is important in the processes of transmethylation, trans-sulfuration and polyamine synthesis, processes that are involved in critical intracellular functions including nucleic acid/protein synthesis and DNA methylation [37, 78]. Furthermore, the role of SAMe in trans-sulfuration is central as a precursor in GSH synthesis [37] and thus makes a logical target for therapeutic intervention given the association between ethanol induced hepatic diseases and oxidative stress [40, 79]. Using a diverse range of experimental animal models of liver disease SAMe has been shown to be protective by restoring GSH levels, decreasing collagen synthesis and increasing hepatic function [80–86].

Additionally, investigators report MAT1A (which catalyzes the formation of SAMe from methionine) knockout mice are predisposed to liver injury [87]. In contrast, exogenous SAMe administration in clinical trials has generated considerable controversy. Long term SAMe treatment has been demonstrated to increase the survival of patients with alcoholic cirrhosis and/or delayed the need for liver transplantation [77]. In contrast, Rambaldi and Gluud [88] performed collective analysis of SAMe clinical trials and report these data could not demonstrate significant effects of SAMe on mortality, liver related mortality, mortality or liver transplantation and liver complications of patients with alcoholic liver disease, concluding that SAMe should not be used for alcoholic liver disease outside randomized clinical trials. While the analysis of Rambaldi and Gluud [88] seemingly discourages the use of SAMe as an hepatic antioxidant, a better understanding of the mechanisms of action of SAMe in animal models versus human beings is still required. In addition to the role of the SAMe–GSH axis in the liver, SAMe also acts as a methyl donor. Methylation is involved in numerous cellular processes including gene expression and cell growth [78, 89, 90]. SAMe has now been shown to be chemopreventative possibly through increased DNA methylation and inhibition of growth related genes including H-*ras*, K-*ras* and c-*myc* [89]. In liver fibrosis, increased DNA methylation of the α1(I) collagen gene caused decreased transcriptional activity [91]. Additionally, α2(I) collagen in cancer cells and collagen VI in transformed fibroblasts also showed repressed gene expression after methylation [92].

In addition to the interest in the use of SAMe in treating ethanol related liver disease renewed interest is now being shown in the potential use of natural plant-derived antioxidants. While many such antioxidants are under investigation, particular attention has been focused on the use of flavanoids derived from the plant *Silybum marianum* (milk thistle) [93] and polyphenols derived from tea *Camellia sinensis* [94–96]. Unlike SAMe, it is believed that these agents are not involved in the synthesis and repletion of GSH after ethanol ingestion but act through their antioxidant capacity to reduce hepatic damage in a variety of animal models of hepatic injury including those induced by ethanol [97–102]. While the antioxidant capacity of these agents is believed to underlie their biological effects other data also suggests potential antitumorigenic effects of these agents in hepatic and GI tumor progression [94–96, 101, 103].

References

1   Bosch FX, Ribes J, Cleries R, Diaz M: Epidemiology of hepatocellular carcinoma. Clin Liver Dis 2005;9:191–211.
2   Okuda K: Hepatocellular carcinoma–history, current status and perspectives. Dig Liver Dis 2002;34:613–616.

3   Sitzmann JV, Abrams R: Improved survival for hepatocellular cancer with combination surgery and multimodality treatment. Ann Surg 1993;217:149–154.
4   Fiorentini G, Poddie DB, Cantore M, Giovanis P, Guadagni S, De Giorgi U, Cariello A, et al: Locoregional therapy for liver metastases from colorectal cancer: the possibilities of intraarterial chemotherapy, and new hepatic-directed modalities. Hepatogastroenterology 2001;48:305–312.
5   Ido K, Isoda N, Sugano K: Microwave coagulation therapy for liver cancer: laparoscopic microwave coagulation. J Gastroenterol 2001;36:145–152.
6   Moroz P, Jones SK, Gray BN: Status of hyperthermia in the treatment of advanced liver cancer. J Surg Oncol 2001;77:259–269.
7   Okuda K: Hepatocellular carcinoma. J Hepatol 2000;32:225–237.
8   Yang ZW, Xu GL: Isolated hepatic perfusion: a regional therapy for liver cancer. Hepatobiliary Pancreat Dis Int 2004;3:12–16.
9   Aguayo A, Patt YZ: Liver cancer. Clin Liver Dis 2001;5:479–507.
10   Macdonald GA: Pathogenesis of hepatocellular carcinoma. Clin Liver Dis 2001;5:69–85.
11   Di Bisceglie AM, Rustgi VK, Hoofnagle JH, Dusheiko GM, Lotze MT: NIH conference. Hepatocellular carcinoma. Ann Intern Med 1988;108:390–401.
12   Parkin DM, Stjernsward J, Muir CS: Estimates of the worldwide frequency of twelve major cancers. Bull World Health Organ 1984;62:163–182.
13   El-Serag HB: Epidemiology of hepatocellular carcinoma. Clin Liver Dis 2001;5:87–107.
14   Longnecker MP: Alcohol consumption and risk of cancer in humans: an overview. Alcohol 1995;12:87–96.
15   Hoek JB, Pastorino JG: Cellular signaling mechanisms in alcohol-induced liver damage. Semin Liver Dis 2004;24:257–272.
16   Donato F, Tagger A, Gelatti U, Parrinello G, Boffetta P, Albertini A, Decarli A, et al: Alcohol and hepatocellular carcinoma: the effect of lifetime intake and hepatitis virus infections in men and women. Am J Epidemiol 2002;155:323–331.
17   Voigt MD: Alcohol in hepatocellular cancer. Clin Liver Dis 2005;9:151–169.
18   Yu H, Harris RE, Kabat GC, Wynder EL: Cigarette smoking, alcohol consumption and primary liver cancer: a case-control study in the USA. Int J Cancer 1988;42:325–328.
19   Shibata A, Hirohata T, Toshima H, Tashiro H: The role of drinking and cigarette smoking in the excess deaths from liver cancer. Jpn J Cancer Res 1986;77:287–295.
20   Holmes-McNary M: Impact factors on development of cirrhosis and subsequent hepatocellular carcinoma. Compend Contin Educ Dent 2001;22:19–33.
21   Kuper H, Tzonou A, Lagiou P, Mucci LA, Trichopoulos D, Stuver SO, Trichopoulou A: Diet and hepatocellular carcinoma: a case-control study in Greece. Nutr Cancer 2000;38:6–12.
22   Romeo R, Colombo M: The natural history of hepatocellular carcinoma. Toxicology 2002;181–182:39–42.
23   Lieber CS: Alcohol: its metabolism and interaction with nutrients. Annu Rev Nutr 2000;20:395–430.
24   Nanji AA, French SW: Hepatocellular carcinoma. Relationship to wine and pork consumption. Cancer 1985;56:2711–2712.
25   Yamada Y, Weller RO, Kleihues P, Ludeke BI: Effects of ethanol and various alcoholic beverages on the formation of O6-methyldeoxyguanosine from concurrently administered N-nitrosomethylbenzylamine in rats: a dose-response study. Carcinogenesis 1992;13:1171–1175.
26   Crabb DW, Bosron WF, Li TK: Ethanol metabolism. Pharmacol Ther 1987;34:59–73.
27   Ekstrom G, Ingelman-Sundberg M: Rat liver microsomal NADPH-supported oxidase activity and lipid peroxidation dependent on ethanol-inducible cytochrome P-450 (P-450IIE1). Biochem Pharmacol 1989;38:1313–1319.
28   Lieber CS, Leo MA: Metabolism of ethanol and some associated adverse effects on the liver and the stomach. Recent Dev Alcohol 1998;14:7–40.
29   Albano E, French SW, Ingelman-Sundberg M: Hydroxyethyl radicals in ethanol hepatotoxicity. Front Biosci 1999;4:D533–D540.
30   Dreosti IE, Ballard FJ, Belling GB, Record IR, Manuel SJ, Hetzel BS: The effect of ethanol and acetaldehyde on DNA synthesis in growing cells and on fetal development in the rat. Alcohol Clin Exp Res 1981;5:357–362.

31   Lieber CS: Biochemical mechanisms of alcohol-induced hepatic injury. Alcohol Alcohol Suppl 1991;1:283–290.

32   Morimoto M, Hagbjork AL, Nanji AA, Ingelman-Sundberg M, Lindros KO, Fu PC, Albano E, et al: Role of cytochrome P4502E1 in alcoholic liver disease pathogenesis. Alcohol 1993;10:459–464.

33   Dupont I, Lucas D, Clot P, Menez C, Albano E: Cytochrome P4502E1 inducibility and hydroxyethyl radical formation among alcoholics. J Hepatol 1998;28:564–571.

34   Guengerich FP, Shimada T, Yun CH, Yamazaki H, Raney KD, Thier R, Coles B, et al: Interactions of ingested food, beverage, and tobacco components involving human cytochrome P4501A2, 2A6, 2E1, and 3A4 enzymes. Environ Health Perspect 1994;102(suppl 9):49–53.

35   Djordjevic D, Nikolic J, Stefanovic V: Ethanol interactions with other cytochrome P450 substrates including drugs, xenobiotics, and carcinogens. Pathol Biol (Paris) 1998;46:760–770.

36   Fernandez-Checa JC, Kaplowitz N, Garcia-Ruiz C, Colell A: Mitochondrial glutathione: importance and transport. Semin Liver Dis 1998;18:389–401.

37   Lu SC: Regulation of glutathione synthesis. Curr Top Cell Regul 2000;36:95–116.

38   Lieber CS: S-adenosyl-L-methionine: its role in the treatment of liver disorders. Am J Clin Nutr 2002;76:1183S–1187S.

39   Bode C, Bode JC: Effect of alcohol consumption on the gut. Best Pract Res Clin Gastroenterol 2003;17:575–592.

40   Hoek JB, Pastorino JG: Ethanol, oxidative stress, and cytokine-induced liver cell injury. Alcohol 2002;27:63–68.

41   McClain CJ, Song Z, Barve SS, Hill DB, Deaciuc I: Recent advances in alcoholic liver disease. IV. Dysregulated cytokine metabolism in alcoholic liver disease. Am J Physiol Gastrointest Liver Physiol 2004;287:G497–G502.

42   Thurman RG, Bradford BU, Iimuro Y, Frankenberg MV, Knecht KT, Connor HD, Adachi Y, et al: Mechanisms of alcohol-induced hepatotoxicity: studies in rats. Front Biosci 1999;4:e42–e46.

43   Senoo H: Structure and function of hepatic stellate cells. Med Electron Microsc 2004;37:3–15.

44   Sato M, Suzuki S, Senoo H: Hepatic stellate cells: unique characteristics in cell biology and phenotype. Cell Struct Funct 2003;28:105–112.

45   Vollmar B, Ruttinger D, Wanner GA, Leiderer R, Menger MD: Modulation of kupffer cell activity by gadolinium chloride in endotoxemic rats. Shock 1996;6:434–441.

46   Thurman RG: II. Alcoholic liver injury involves activation of Kupffer cells by endotoxin. Am J Physiol 1998;275:G605–G611.

47   Yang SQ, Lin HZ, Yin M, Albrecht JH, Diehl AM: Effects of chronic ethanol consumption on cytokine regulation of liver regeneration. Am J Physiol 1998;275:G696–G704.

48   Reddy MA, Shukla SD: Potentiation of mitogen-activated protein kinase by ethanol in embryonic liver cells. Biochem Pharmacol 1996;51:661–668.

49   Hoek JB, Thomas AP, Rooney TA, Higashi K, Rubin E: Ethanol and signal transduction in the liver. FASEB J 1992;6:2386–2396.

50   Bocckino SB, Wilson PB, Exton JH: Ca2+-mobilizing hormones elicit phosphatidylethanol accumulation via phospholipase D activation. FEBS Lett 1987;225:201–204.

51   Hoffman PL, Tabakoff B: Ethanol and guanine nucleotide binding proteins: a selective interaction. FASEB J 1990;4:2612–2622.

52   Kruger H, Wilce PA, Shanley BC: Ethanol and protein kinase C in rat brain. Neurochem Int 1993;22:575–581.

53   Roivainen R, Hundle B, Messing RO: Protein kinase C and adaptation to ethanol. Exs 1994;71: 29–38.

54   Pandey SC: Acute and chronic ethanol consumption effects on the immunolabeling of Gq/11 alpha subunit protein and phospholipase C isozymes in the rat brain. J Neurochem 1996;67: 2355–2361.

55   Chen J, Ishac EJ, Dent P, Kunos G, Gao B: Effects of ethanol on mitogen-activated protein kinase and stress-activated protein kinase cascades in normal and regenerating liver. Biochem J 1998;334(Pt 3):669–676.

56   Lee YJ, Aroor AR, Shukla SD: Temporal activation of p42/44 mitogen-activated protein kinase and c-Jun N-terminal kinase by acetaldehyde in rat hepatocytes and its loss after chronic ethanol exposure. J Pharmacol Exp Ther 2002;301:908–914.

57   Tombes RM, Auer KL, Mikkelsen R, Valerie K, Wymann MP, Marshall CJ, McMahon M, et al: The mitogen-activated protein (MAP) kinase cascade can either stimulate or inhibit DNA synthesis in primary cultures of rat hepatocytes depending upon whether its activation is acute/phasic or chronic. Biochem J 1998;330(Pt 3):1451–1460.

58   Weng Y, Shukla SD: Ethanol alters angiotensin II stimulated mitogen activated protein kinase in hepatocytes: agonist selectivity and ethanol metabolic independence. Eur J Pharmacol 2000;398:323–331.

59   Chen J, Clemens DL, Cederbaum AI, Gao B: Ethanol inhibits the JAK-STAT signaling pathway in freshly isolated rat hepatocytes but not in cultured hepatocytes or HepG2 cells: evidence for a lack of involvement of ethanol metabolism. Clin Biochem 2001;34:203–209.

60   Aroor AR, Shukla SD: MAP kinase signaling in diverse effects of ethanol. Life Sci 2004;74:2339–2364.

61   Shaw S, Eng J, Jayatilleke E: Ethanol-induced free radical injury to the hepatocyte glucagon receptor. Alcohol 1995;12:273–277.

62   Koteish A, Yang S, Lin H, Huang X, Diehl AM: Chronic ethanol exposure potentiates lipopolysaccharide liver injury despite inhibiting Jun N-terminal kinase and caspase 3 activation. J Biol Chem 2002;277:13037–13044.

63   Casey CA, Nanji A, Cederbaum AI, Adachi M, Takahashi T: Alcoholic liver disease and apoptosis. Alcohol Clin Exp Res 2001;25:49S–53S.

64   Diehl AM, Rai RM: Liver regeneration 3: regulation of signal transduction during liver regeneration. FASEB J 1996;10:215–227.

65   Koniaris LG, McKillop IH, Schwartz SI, Zimmers TA: Liver regeneration. J Am Coll Surg 2003;197:634–659.

66   Diehl AM, Yang SQ, Wolfgang D, Wand G: Differential expression of guanine nucleotide-binding proteins enhances cAMP synthesis in regenerating rat liver. J Clin Invest 1992;89:1706–1712.

67   Yagami T, Kirita S, Matsushita A, Kawasaki K, Mizushima Y: Alterations in the stimulatory G protein of the rat liver after partial hepatectomy. Biochim Biophys Acta 1994;1222:81–87.

68   Diehl AM, Yang SQ, Cote P, Wand GS: Chronic ethanol consumption disturbs G-protein expression and inhibits cyclic AMP-dependent signaling in regenerating rat liver. Hepatology 1992;16:1212–1219.

69   McKillop IH, Wu Y, Cahill PA, Sitzmann JV: Altered expression of inhibitory guanine nucleotide regulatory proteins (Gi-proteins) in experimental hepatocellular carcinoma. J Cell Physiol 1998;175:295–304.

70   Schmidt CM, McKillop IH, Cahill PA, Sitzmann JV: Alterations in guanine nucleotide regulatory protein expression and activity in human hepatocellular carcinoma. Hepatology 1997;26: 1189–1194.

71   Schmidt CM, McKillop IH, Cahill PA, Sitzmann JV: Increased MAPK expression and activity in primary human hepatocellular carcinoma. Biochem Biophys Res Commun 1997;236:54–58.

72   McKillop IH, Schmidt CM, Cahill PA, Sitzmann JV: Altered expression of mitogen-activated protein kinases in a rat model of experimental hepatocellular carcinoma. Hepatology 1997;26: 1484–1491.

73   Huynh H, Nguyen TT, Chow KH, Tan PH, Soo KC, Tran E: Over-expression of the mitogen-activated protein kinase (MAPK) kinase (MEK)-MAPK in hepatocellular carcinoma: its role in tumor progression and apoptosis. BMC Gastroenterol 2003;3:19.

74   McKillop IH, Schmidt CM, Cahill PA, Sitzmann JV: Inhibitory guanine nucleotide regulatory protein activation of mitogen-activated protein kinase in experimental hepatocellular carcinoma in vitro. Eur J Gastroenterol Hepatol 1999;11:761–768.

75   McKillop IH, Vyas N, Schmidt CM, Cahill PA, Sitzmann JV: Enhanced Gi-protein-mediated mitogenesis following chronic ethanol exposure in a rat model of experimental hepatocellular carcinoma. Hepatology 1999;29:412–420.

76   Gill JS: Reported levels of alcohol consumption and binge drinking within the UK undergraduate student population over the last 25 years. Alcohol Alcohol 2002;37:109–120.

77   Mato JM, Corrales FJ, Lu SC, Avila MA: S-Adenosylmethionine: a control switch that regulates liver function. FASEB J 2002;16:15–26.

78   Lu SC: S-Adenosylmethionine. Int J Biochem Cell Biol 2000;32:391–395.

79 Fataccioli V, Andraud E, Gentil M, French SW, Rouach H: Effects of chronic ethanol administration on rat liver proteasome activities: relationship with oxidative stress. Hepatology 1999;29:14–20.

80 Rojkind M, Rojkind MH, Cordero-Hernandez J: In vivo collagen synthesis and deposition in fibrotic and regenerating rat livers. Coll Relat Res 1983;3:335–347.

81 Maher JJ, McGuire RF: Extracellular matrix gene expression increases preferentially in rat lipocytes and sinusoidal endothelial cells during hepatic fibrosis in vivo. J Clin Invest 1990;86: 1641–1648.

82 Muller A, Machnik F, Zimmermann T, Schubert H: Thioacetamide-induced cirrhosis-like liver lesions in rats – usefulness and reliability of this animal model. Exp Pathol 1988;34:229–236.

83 Zern MA, Leo MA, Giambrone MA, Lieber CS: Increased type I procollagen mRNA levels and in vitro protein synthesis in the baboon model of chronic alcoholic liver disease. Gastroenterology 1985;89:1123–1131.

84 Tsukamoto H, Towner SJ, Ciofalo LM, French SW: Ethanol-induced liver fibrosis in rats fed high fat diet. Hepatology 1986;6:814–822.

85 Garcia-Trevijano ER, Iraburu MJ, Fontana L, Dominguez-Rosales JA, Auster A, Covarrubias-Pinedo A, Rojkind M: Transforming growth factor beta1 induces the expression of alpha1(I) pro-collagen mRNA by a hydrogen peroxide-C/EBP beta-dependent mechanism in rat hepatic stellate cells. Hepatology 1999;29:960–970.

86 Muriel P, Castro V: Effects of S-adenosyl-L-methionine and interferon-alpha2b on liver damage induced by bile duct ligation in rats. J Appl Toxicol 1998;18:143–147.

87 Lu SC, Alvarez L, Huang ZZ, Chen L, An W, Corrales FJ, Avila MA, et al: Methionine adenosyl-transferase 1A knockout mice are predisposed to liver injury and exhibit increased expression of genes involved in proliferation. Proc Natl Acad Sci USA 2001;98:5560–5565.

88 Rambaldi A, Gluud C: S-adenosyl-L-methionine for alcoholic liver diseases. Cochrane Database Syst Rev 2001:CD002235.

89 Pascale RM, Simile MM, Seddaiu MA, Daino L, Vinci MA, Pinna G, Bennati S, et al: Chemoprevention of rat liver carcinogenesis by S-adenosyl-L-methionine: is DNA methylation involved? Basic Life Sci 1993;61:219–237.

90 Mato JM, Alvarez L, Ortiz P, Pajares MA: S-adenosylmethionine synthesis: molecular mechanisms and clinical implications. Pharmacol Ther 1997;73:265–280.

91 Thompson JP, Simkevich CP, Holness MA, Kang AH, Raghow R: In vitro methylation of the promoter and enhancer of Pro alpha 1(I) collagen gene leads to its transcriptional inactivation. J Biol Chem 1991;266:2549–2556.

92 Kopp MU, Winterhalter KH, Trueb B: DNA methylation accounts for the inhibition of collagen VI expression in transformed fibroblasts. Eur J Biochem 1997;249:489–496.

93 Mereish KA, Bunner DL, Ragland DR, Creasia DA: Protection against microcystin-LR-induced hepatotoxicity by Silymarin: biochemistry, histopathology, and lethality. Pharm Res 1991;8:273–277.

94 Chandra S, De Mejia Gonzalez E: Polyphenolic compounds, antioxidant capacity, and quinone reductase activity of an aqueous extract of Ardisia compressa in comparison to mate (Ilex paraguariensis) and green (Camellia sinensis) teas. J Agric Food Chem 2004;52:3583–3589.

95 Jimenez-Lopez JM, Cederbaum AI: Green tea polyphenol epigallocatechin-3-gallate protects HepG2 cells against CYP2E1-dependent toxicity. Free Radic Biol Med 2004;36:359–370.

96 Qin G, Gopalan-Kriczky P, Su J, Ning Y, Lotlikar PD: Inhibition of aflatoxin B1-induced initiation of hepatocarcinogenesis in the rat by green tea. Cancer Lett 1997;112:149–154.

97 Campos R, Garrido A, Guerra R, Valenzuela A: Silybin dihemisuccinate protects against glutathione depletion and lipid peroxidation induced by acetaminophen on rat liver. Planta Med 1989;55:417–419.

98 Szilard S, Szentgyorgyi D, Demeter I: Protective effect of Legalon in workers exposed to organic solvents. Acta Med Hung 1988;45:249–256.

99 Feher J, Lang I, Nekam K, Gergely P, Muzes G: In vivo effect of free radical scavenger hepatoprotective agents on superoxide dismutase (SOD) activity in patients. Tokai J Exp Clin Med 1990;15:129–134.

100 Halim AB, El-Ahmady O, Hassab-Allah S, Abdel-Galil F, Hafez Y, Darwish A: Biochemical effect of antioxidants on lipids and liver function in experimentally-induced liver damage. Ann Clin Biochem 1997;34(Pt 6):656–663.

Ethanol and Liver Cancer

101 Hakova H, Misurova E: Therapeutical effect of silymarin on nucleic acids in the various organs of rats after radiation injury. Radiats Biol Radioecol 1996;36:365–370.

102 de Groot H, Dehmlow C, Raven U: Tissue injury by free radicals and the protective effects of flavonoids. Methods Find Exp Clin Pharmacol 1996;18(suppl B):23–25.

103 Magliulo E, Carosi PG, Minoli L, Gorini S: Studies on the regenerative capacity of the liver in rats subjected to partial hepatectomy and treated with silymarin. Arzneimittelforschung 1973;23: (suppl):161–167.

Iain H. McKillop
Department of Biology
University of North Carolina at Charlotte
Charlotte, NC 28223 (USA)
Tel. +1 704 687 4050, Fax +1 704 687 3128
E-Mail imckillo@email.uncc.edu

Cho CH, Purohit V (eds): Alcohol, Tobacco and Cancer.
Basel, Karger, 2006, pp 109–118

....

Alcohol, Reactive Oxygen Species, Pancreatitis and Pancreatic Cancer

Stephen J. Pandol[a–c], *Mouad Edderkaoui*[a–c], *Jong Kyun Lee*[d],
Richard Hu[a–c], *Aurelia Lugea*[a–c], *James Sul*[a–c], *Minoti Apte*[e],
Eva Vaquero[f], *Izumi Ohno*[a–c], *Ilya Gukovsky*[a–c], *Vay Liang W. Go*[a,c],
Anna S. Gukovskaya[a–c]

[a]Pancreatic Research Group, [b]VA Greater Los Angeles Health Care System and
[c]University of California, Los Angeles, Calif., USA; [d]Sungkyunkwan University
School of Medicine, Samsung Medical Center, Seoul, Korea; [e]The University of
New South Wales, Sydney, Australia; [f]Hospital General Vall d'Hebron,
Barcelona, Spain

Abstract

Although the pathogenesis of pancreatic cancer is poorly understood, evidence from epidemiologic studies indicates that chronic pancreatitis from any etiology, including alcohol abuse, significantly increases the risk of development of pancreatic adenocarcinoma. Such results suggest that the tissue microenvironment of chronic pancreatitis provides factors that facilitate the initiation and promotion of cancer. In the present review, we compile evidence to support the concept that reactive oxygen species and NADPH oxidase (a reactive oxygen species-generating system) play a key role in the development of pancreatitis as well as in the initiation and promotion of pancreatic cancer.

Pancreatic cancer is the fourth leading cause of cancer mortality in both men and women in the United States with an estimated incidence of 30,000 cases in 2003 [1, 2]. Although pancreatic malignant neoplasm can arise in either endocrine or exocrine pancreas, more than 95% of all pancreatic cancers are ductal adenocarcinoma. Pancreatic ductal adenocarcinoma is among the most fatal of cancers, with a 5-year survival rate of less than 4%. There is substantial geographic variation in the incidence of pancreatic cancer, with the highest rates in economically developed countries, such as the United States and Europe, and the lowest rates in Africa and some Asian countries [2, 3]. The risk factors for pancreatic ductal adenocarcinoma include age (1 per 100,000 at age 40 years, rising to greater than 100 per 100,000 at age 80 years), race (highest among African-Americans), obesity and inactivity, diabetes mellitus, and chronic alcoholic or hereditary pancreatitis. Dietary factors, such as high energy intake and low fruit and vegetable intake, are contributing risk factors [3–7].

The results of numerous prospective cohort and case-control studies for alcohol consumption and pancreatic cancer risk have been inconsistent, with many confounding variables present in various study investigations. However, heavy alcohol consumption is the major cause of acute and chronic pancreatitis [2, 8–10]. Chronic pancreatitis has been clearly demonstrated to increase pancreatic cancer risk [2, 11, 12]. Chronic heavy alcohol consumption can interact with other risk factors (e.g. smoking), which can ultimately affect the multi-step process of carcinogenesis and lead to the development of pancreatic cancer.

Although the pathogenesis of pancreatic cancer is poorly understood, there is substantial evidence that chronic pancreatitis from any cause is a major risk factor for pancreatic cancer [2, 3, 11, 13]. Chronic pancreatitis is a progressive disorder of chronic inflammation, fibrosis and loss of parenchymal cells in the organ. Approximately 70–90% of the cases of chronic pancreatitis result from alcohol abuse while the remainder of the cases are associated with genetic disorders (i.e. hereditary pancreatitis and cystic fibrosis) and unknown causes (i.e. idiopathic) [13, 14]. The fact that there is an increase in incidence of pancreatic cancer in chronic pancreatitis (regardless of the predisposing cause) suggests that the chronic inflammation and fibrosis of chronic pancreatitis provide a favorable environment for the oncogenic process. This effect of chronic inflammatory diseases providing an environment for the initiation and promotion of cancer is referred to as a 'landscaping' disorder [15].

Many factors have been proposed to participate in the abnormal microenvironment leading to increased risk of oncogenic transformation including various cytokines, reactive oxygen species (ROS) and inflammatory signals such as NF-κB and cyclooxygenase-2 [13, 16, 17]. In addition, ROS have been proposed to mediate genetic changes associated with malignant transformation [16]. Recent evidence demonstrates a key role for ROS in mediating prosurvival and anti-apoptotic responses in both pancreatic cancer cells and pancreatic stellate cells (PSCs). The latter have now been shown to play central roles in the mechanism of chronic pancreatitis and pancreatic cancer, contributing to both the fibrosis and inflammatory responses in these disorders. These findings suggest that ROS play a central role in both providing an environment conducive to the initiation of the oncogenic process as well as facilitating the growth and metastasis of pancreatic adenocarcinoma once initiated. Evidence to support this hypothesis is discussed in further detail below.

Reactive Oxygen Species Generation and
Anti-Oxidant Systems in Pancreatic Cancer

ROS are highly reactive O_2 metabolites that include superoxide radical (O_2^-), hydrogen peroxide (H_2O_2), and hydroxyl radical $(OH^\bullet)$ [18]. Cellular ROS levels are determined by the balance between ROS generation and their neutralization by anti-oxidant systems. ROS levels in cancer cells are greater than in normal cells [19]. One reason for this is that the activities of key anti-oxidant enzymes, such as Mn- and CuZn-superoxide dismutases (SOD) and catalase, are nearly always low in cancer cells due to gene mutations or modifications in protein structure [20]. In particular, the anti-oxidant activity of pancreatic cancer is lower than in normal pancreas, probably due to the low expression of Mn-SOD [21].

ROS were long thought to promote cell death. Indeed, exogenous H_2O_2 stimulates death in many normal and cancer cells, probably by causing plasma membrane damage (leading to necrosis) or mild mitochondrial damage (resulting in apoptosis) [22, 23]. However, recent data indicate that ROS may also have a pro-survival role, depending on the magnitude and pattern of ROS response [24]. While severe oxidative stress causes cell death, low endogenous doses of ROS act as mitogens and pro-survival factors [24, 25]. In fact, we have found [26, 27] that in pancreatic cancer cells, ROS mediate the anti-apoptotic and pro-survival effects of both growth hormones and extracellular matrix proteins, two key factors that are involved in the growth of this cancer.

Our results [26, 27] showed that growth factors (in particular, insulin-like growth factor-I) as well as extracellular matrix proteins (such as fibronectin and laminin) stimulate the formation of ROS in pancreatic cancer cells through activation of a membrane NADPH oxidase. We further found that the NADPH oxidase plays an essential role in suppression of apoptosis promoted by growth factors and extracellular matrix proteins in the pancreatic cancer cells. That is, inhibiting NADPH oxidase by various approaches facilitates cell death by apoptosis.

The intracellular targets that mediate the anti-apoptotic and pro-survival effects in pancreatic cancer cells may include key signaling molecules such as transcription factor NF-κB, MAP kinases, and tyrosine phosphatases [25, 28–30].

Although less well studied, preliminary results suggest a key role for ROS generated from an NADPH oxidase system in the activation and proliferation of PSCs [31]. As discussed below, PSCs play a key role in the mechanism of chronic pancreatitis, the major risk factor for pancreatic cancer. Further, recent studies show the presence and importance of stellate cells in promoting the desmoplastic reaction and the growth of the cancer cells in pancreatic adenocarcinoma [32–34].

Reactive Oxygen Species, Stellate Cells and
Chronic Alcoholic Pancreatitis

As noted earlier, the most common etiology of chronic pancreatitis is alcohol abuse [14]. Chronic pancreatitis is characterized by chronic inflammation, fibrosis, and loss of parenchymal cells, eventually leading to exocrine and endocrine insufficiency. Patients with this disease suffer from debilitating chronic pain, maldigestion and diabetes. Alcohol abuse may cause episodes of recurrent acute pancreatitis with increasing amounts of fibrosis, chronic inflammation and parenchymal cell loss with each successive episode. This sequence of events has been recently reproduced in animal models of repeated episodes of acute pancreatitis [35–37]. Especially interesting to the hypothesis presented here, Matsumura et al. [38] demonstrated that repeated administration of a SOD inhibitor, diethyldithiocarbamate, caused lipid peroxidation and fibrosis in rat pancreas.

On the other hand, these processes may proceed silently over years with progressive fibrosis and loss of function, and without symptoms of acute pancreatitis. Of particular note, progression of chronic pancreatitis can continue even after complete cessation of alcohol intake indicating that the processes are self-sustaining once a certain stage of the disease is reached [14]. This occurrence is of particular relevance to the present discussion, given that patients with chronic pancreatitis are known to be at significant increased risk for pancreatic cancer [12].

Although the elucidation of the mechanisms underlying chronic pancreatitis are incomplete, considerable progress has been made in our understanding of the fibrosis process resulting from the identification and characterization of PSCs starting in 1998 [33, 39, 40]. Studies with these cells suggest that they play a key role in chronic pancreatitis in a manner analogous to hepatic stellate cells in hepatic fibrosis [41, 42].

In normal pancreas, quiescent PCSs are identified using antibodies to desmin, a cytoskeletal protein and stellate cell selective marker [39]. They are present in the periacinar space with long cytoplasmic processes encircling the base of the acinus. Similar to hepatic stellate cells, in their quiescent state, PSCs store significant amounts of vitamin A as lipid droplets in their cytoplasm.

There is general acceptance that during pancreatic injury, PSCs are activated in a manner similar to hepatic stellate cells [43]. Activation consists of processes of transformation to a myofibroblastic phenotype with loss of the vitamin A stores, expression of the cytoskeletal protein α-smooth muscle actin, cell proliferation and production and secretion of large amounts of extracellular matrix proteins including collagen, fibronectin and laminin [44]. Activation can be mediated by cytokines such as transforming growth factor β, platelet-derived growth factor [33], TNFα and interleukins acting via (i) a paracrine pathway (secondary to the production and secretion of these agents by pancreatic parenchymal cells and inflammatory cells) or (ii) an autocrine pathway due to endogenous production of cytokines by PSCs themselves [45, 46]. Autocrine activation of PSCs may account for the continued progression of chronic pancreatitis processes in individuals even after cessation of alcohol abuse.

Support for a role for PSCs in the pathogenesis of pancreatic fibrosis in chronic pancreatitis comes from investigations of pancreatic tissue in patients with chronic pancreatitis and from animal experimental models [47, 48]. In both situations, PSCs have been shown to be present (as identified by α-smooth muscle actin expression) in fibrotic areas of the pancreas. Furthermore, in situ hybridization studies for collagen mRNA has demonstrated that PSCs are the predominant source of collagen in fibrotic areas. Such findings provide strong evidence that PSCs play a major role in the development of pancreatic fibrosis.

In addition to the fact that stellate cell activation can occur with cytokines as indicated above, there is evidence that alcohol and alcohol metabolism leads to stellate cell activation [39, 45, 49]. These studies demonstrate that PSCs exhibit alcohol dehydrogenase (ADH, a major ethanol oxidizing enzyme) activity suggesting a capacity to metabolize ethanol; that exposure to both ethanol and acetaldehyde leads to PSC activation associated with generation of ROS (as measured by intracellular lipid peroxidation); and that the anti-oxidant, vitamin E prevents these effects of alcohol and acetaldehyde. Signaling pathways mediating the ethanol and acetaldehyde-induced activation of PSCs have now been identified. Recent studies demonstrate that the MAP kinase pathway (extracellular signal-regulated kinase 1/2, c-Jun N-terminal kinase and p38 MAP kinase) and the transcription factor, activator protein-1 mediate PSC activation by ethanol and its metabolites [34, 39]. Importantly, the induction of these signaling pathways by ethanol and acetaldehyde and the consequent activation of PSCs is blocked by the anti-oxidants N-acetylcysteine and vitamin E [39, 50], providing further support for the role of ROS in PSC activation.

The above in vitro findings are corroborated by in vivo studies using tissues from patients with chronic pancreatitis and animal models of chronic pancreatitis [39, 48, 51]. Casini et al. [48] have demonstrated a marked increase in lipid peroxidation in the pancreas of patients with chronic pancreatitis as measured by immunocytochemistry using antibodies to 4-hydroxynonenal – protein adducts. The adducts were highly localized to acinar cells adjacent to areas of PSCs and fibrosis. Evidence of increased oxidant stress has also been demonstrated in animal models of pancreatitis. Indeed, a recent report by Gomez et al. [51] has shown that the anti-oxidant vitamin E attenuates the production of ROS as well as stellate cell activation and pancreatic fibrosis in the pancreas of an animal model of chronic pancreatitis. Like adenocarcinoma, in chronic pancreatitis tissue there is decreased anti-oxidant enzyme activity as well [52].

Thus, in vitro and in vivo evidence to date supports a role for ROS in the pathogenesis of chronic pancreatitis.

The discussion above focuses on ROS in chronic pancreatitis. Further, Gukovskaya et al. [53] and others [54–63] have shown that ROS play a key pathologic role in acute pancreatitis as well. Gukovskaya et al. [53] showed that the neutrophil NADPH oxidase-generated ROS mediates intrapancreatic trypsin activation. Thus, ROS play key roles in the pathogenesis of both acute and chronic pancreatitis.

Mechanisms of NADPH oxidase activation. Lessons from Phagocytic Cells

As best characterized in leukocytes, NADPH oxidase is a multi-component and membrane-bound enzyme system that uses electrons derived from intracellular NADPH to generate superoxide anion which dismutates to H_2O_2 and other ROS that are used for host defense against bacterial and fungal pathogens [64]. In this system there is a membrane-associated component (flavocytochrome b_{558}, a.k.a. cyt b) consisting of two subunits (gp91[phox] and p22[phox]). When the phagocyte is stimulated, the small GTPase Rac2 as well as cytosolic proteins, p47[phox] and p67[phox], are induced to assemble with the membrane-bound components resulting in enzyme activation. The roles of the components in the system have been demonstrated using genetic deletion experiments. For example, genetic deletions in Rac2 or p47[phox] result in significantly reduced or absent ROS production in neutrophils to various stimuli [53, 65–67]. p47[phox] is the major regulatory subunit which exists in an equilibrium between a free state and a state complexed with p67[phox] [64]. With neutrophil stimulation, p47[phox] is phosphorylated on multiple sites through the action of several kinases resulting in the translocation of the phosphorylated p47[phox]–p67[phox] (and possibly p40[phox]) complex to the membrane where it associates with cyt b-Rac2 to form an active enzyme complex [68]. In the mechanism of activation p67[phox] contains an activation domain for cyt b which facilitates the critical electron transfer step in ROS production while Rac2 is thought to act as an adapter to provide the proper orientation for cyt b and p67[phox] [69]. Observations that there are stimulus-dependent ROS at modest levels in non-phagocytic cells and that p22[phox] is present in almost all tissues examined have prompted a search for gp91[phox] or homologs in other tissues. Over the past few years, several homologs of gp91[phox] have been identified in various tissues. These have been called NOXes (non-phagocytic oxidases). Nox2 is gp91[phox]. The others are called Nox1, Nox3, Nox4 and Nox5 [70]. Specific tissue distribution for these enzymes has been determined and their functions are under active investigation [64, 70].

Regarding pancreatic cancer, we have recently determined that Nox4 and Nox5 along with p22[phox] are expressed in pancreatic cancer cells and mediate growth factor-mediated resistance to apoptosis of the cancer cells [26]. Although the members of NADPH oxidase components in the pancreatic stellate cell have not been established, hepatic stellate cells express gp91^{phox}, p22^{phox}, p47^{phox} and p67^{phox} [71]. Thus, the NADPH oxidase systems and their roles in both pancreatitis and pancreatic cancer warrant further study.

Conclusion

Information in this review suggests key roles for ROS in the mechanism of pancreatitis and pancreatitis cancer. ROS are necessary signals for cell survival and proliferation of both pancreatic cancer and pancreatic stellate cells. Further, there is a ROS-rich environment in the tissue of chronic pancreatitis due to the presence of both activated stellate cells and chronic inflammatory cells as well as a decreased anti-oxidant system. Given that increased ROS levels may induce pro-carcinogenic genetic changes and also promote pro-survival and anti-apoptotic responses in pancreatic cancer cells, it is highly likely that ROS mediate the promotion of cancer in chronic pancreatitis tissue. Chronic pancreatitis provides an ideal environment for the initiation and promotion of cancer. Finally, as argued in this article, the NADPH oxidase systems may represent important mechanisms for generating ROS in both pancreatitis and pancreatic cancer.

References

1   Jemal A, Murray T, Samuels A, Ghafoor A, Ward E, Thun MJ: Cancer statistics, 2003. CA Cancer J Clin 2003;53:5–26.
2   Go VL, Gukovskaya A, Pandol SJ: Alcohol and pancreatic cancer. Alcohol 2005;35:205–211.
3   World Cancer Research Fund & American Institute for Cancer Research. Food, Nutrition and the Prevention of Cancer: A Global Perspective. Washington, DC, American Institute for Cancer Research, 1997.
4   Gapstur SM, Gann P: Is pancreatic cancer a preventable disease? JAMA 2001;286:967–968.
5   International Agency for Research on Cancer. IARC Handbooks of Cacner Prevention Vol. 6: Weight Control and Physical Activity. Lyon, France, IARC Press, 2002.
6   International Agency for Research on Cancer. IARC Handbooks of Cacner Prevention: Fruit and Vegetables. Lyon, France, IARC Press, 2003, vol 8.
7   Ries LAG, Eisner MP, Kosary CL, Hankey BF, Miller BA, Clegg L, Mariotto A, Feuer EJ, Edwards BK (eds): SEER Cancer Statistics Review, 1975–2001. National Cancer Institute, 2004.
8   Dufour MC, Adamson MD: The epidemiology of alcohol-induced pancreatitis. Pancreas 2003;27: 286–290.
9   Durbec JP, Sarles H: Multicenter survey of the etiology of pancreatic diseases. Relationship between the relative risk of developing chronic pancreaitis and alcohol, protein and lipid consumption. Digestion 1978;18:337–350.
10 Swaroop VS, Chari ST, Clain JE: Severe acute pancreatitis. JAMA 2004;291:2865–2868.
11 Gordis L, Gold EB: Epidemiology and etiology of pancreas cancer in the Pancreas: Biology, Pathobiology, and Disease. New York, Raven Press, 1993, pp 837–855.
12 Lowenfels AB, Maisonneuve P, Cavallini G, Ammann RW, Lankisch PG, Andersen JR, Dimagno EP, Andren-Sandberg A, Domellof L: International Pancreatitis Study Group: Pancreatitis and the risk of pancreatic cancer. N Engl J Med 1993;328:1433–1437.

13   Whitcomb DC: Inflammation and cancer V. Chronic pancreatitis and pancreatic cancer. Am J Physiol Gastrointest Liver Physiol 2004;287:G315–G319.

14   Forsmark CE: Chronic pancreatitis; in Feldman M, Friedman LS, Sleisenger MH (eds): Sleisenger & Fordtran's Gastrointestinal & Liver Disease Pathophysiology/Diagnosis/Management. London, Saunders, 2002, pp 943–969.

15   Kinzler KW, Vogelstein B: Landscaping the cancer terrain. Science 1998;280:1036–1037.

16   Farrow B, Evers BM: Inflammation and the development of pancreatic cancer. Surg Oncol 2002;10:153–169.

17   Albazaz R, Verbeke CS, Rahman SH, McMahon MJ: Cyclooxygenase-2 expression associated with severity of PanIN lesions: a possible link between chronic pancreatitis and pancreatic cancer. Pancreatology 2005;5:361–369.

18   Thannickal VJ, Fanburg BL: Reactive oxygen species in cell signaling. Am J Physiol Lung Cell Mol Physiol 2000;279:L1005–L1028.

19   Szatrowski TP, Nathan CF: Production of large amounts of hydrogen peroxide by human tumor cells. Cancer Res 1991;51:794–798.

20   Huang Y, He T, Domann FE: Decreased expression of manganese superoxide dismutase in transformed cells is associated with increased cytosine methylation of the SOD2 gene. DNA Cell Biol 1999;18:643–652.

21   Cullen JJ, Weydert C, Hinkhouse MM, Ritchie J, Domann FE, Spitz D, Oberley LW: The role of manganese superoxide dismutase in the growth of pancreatic adenocarcinoma. Cancer Res 2003;63:1297–1303.

22   Cai J, Jones DP: Mitochondrial redox signaling during apoptosis. J Bioenerg Biomembr 1999;31:327–334.

23   Simon HU, Haj-Yehia A, Levi-Schaffer F: Role of reactive oxygen species (ROS) in apoptosis induction. Apoptosis 2000;5:415–418.

24   Brar SS, Corbin Z, Kennedy TP, Hemendinger R, Thornton L, Bommarius B, Arnold RS, Whorton AR, Sturrock AB, Huecksteadt TP, Quinn MT, Krenitsky K, Ardie KG, Lambeth JD, Hoidal JR: NOX5 NAD(P)H oxidase regulates growth and apoptosis in DU 145 prostate cancer cells. Am J Physiol Cell Physiol 2003;285:C353–C369.

25   Chiarugi P, Cirri P: Redox regulation of protein tyrosine phosphatases during receptor tyrosine kinase signal transduction. Trends Biochem Sci 2003;28:509–514.

26   Vaquero EC, Edderkaoui M, Pandol SJ, Gukovsky I, Gukovskaya AS: Reactive oxygen species produced by NAD(P)H oxidase inhibit apoptosis in pancreatic cancer cells. J Biol Chem 2004; 279:34643–34654.

27   Edderkaoui M, Hong P, Vaquero EC, Lee JK, Fischer L, Friess H, Buchler MW, Lerch MM, Pandol SJ, Gukovskaya AS: Extracellular matrix stimulates reactive oxygen species production and increases pancreatic cancer cell survival through 5-lipoxygenase and NADPH oxidase. Am J Physiol Gastrointest Liver Physiol 2005;289:G1137–G1147.

28   Green DR: Death and NF-kappaB in T cell activation: life at the edge. Mol Cell 2003;11: 551–552.

29   Torres M, Forman HJ: Redox signaling and the MAP kinase pathways. Biofactors 2003;17: 287–296.

30   Zhang Y, Chen F: Reactive oxygen species (ROS), troublemakers between nuclear factor-kappaB (NF-kappaB) and c-Jun NH(2)-terminal kinase (JNK). Cancer Res 2004;64:1902–1905.

31   Hu R, Wang YL, Hong P, Edderkaoui M, Shahsahebi M, Lugea A, Apte MV, Gukovskaya AS, Pandol SJ: Pancreatic stellate cell NADPH oxidase system and its regulation by ethanol and growth factors. Gastroenterology 2005;128:A-631.

32   Apte MV, Park S, Phillips PA, Santucci N, Goldstein D, Kumar RK, Ramm GA, Buchler M, Friess H, McCarroll JA, Keogh G, Merrett N, Pirola R, Wilson JS: Desmoplastic reaction in pancreatic cancer: role of pancreatic stellate cells. Pancreas 2004;29:179–187.

33   Bachem MG, Schneider E, Gross H, Weidenbach H, Schmid RM, Menke A, Siech M, Beger H, Grunert A, Adler G: Identification, culture, and characterization of pancreatic stellate cells in rats and humans. Gastroenterology 1998;115:421–432.

34   Bachem MG, Schunemann M, Ramadani M, Siech M, Beger H, Buck A, Zhou S, Schmid-Kotsas A, Adler G: Pancreatic carcinoma cells induce fibrosis by stimulating proliferation and matrix synthesis of stellate cells. Gastroenterology 2005;128:907–921.

35 Neuschwander-Tetri BA, Burton FR, Presti ME, Britton RS, Janney CG, Garvin PR, Brunt EM, Galvin NJ, Poulos JE: Repetitive self-limited acute pancreatitis induces pancreatic fibrogenesis in the mouse. Dig Dis Sci 2000;45:665–674.

36 Deng X, Wang L, Elm MS, Gabazadeh D, Diorio GJ, Eagon PK, Whitcomb DC: Chronic alcohol consumption accelerates fibrosis in response to cerulein-induced pancreatitis in rats. Am J Pathol 2005;166:93–106.

37 Perides G, Tao X, West N, Sharma A, Steer ML: A mouse model of ethanol dependent pancreatic fibrosis. Gut 2005;54:1461–1467.

38 Matsumura N, Ochi K, Ichimura M, Mizushima T, Harada H, Harada M: Study on free radicals and pancreatic fibrosis–pancreatic fibrosis induced by repeated injections of superoxide dismutase inhibitor. Pancreas 2001;22:53–57.

39 Apte MV, Haber PS, Applegate TL, Norton ID, McCaughan GW, Korsten MA, Pirola RC, Wilson JS: Periacinar stellate shaped cells in rat pancreas: identification, isolation, and culture. Gut 1998;43:128–133.

40 Saotome T, Inoue H, Fujimiya M, Fujiyama Y, Bamba T: Morphological and immunocytochemical identification of periacinar fibroblast-like cells derived from human pancreatic acini. Pancreas 1997;14:373–382.

41 Li D, Friedman SL: Liver fibrogenesis and the role of hepatic stellate cells: new insights and prospects for therapy. J Gastroenterol Hepatol 1999;14:618–633.

42 Friedman SL: Seminars in medicine of the Beth Israel Hospital, Boston. The cellular basis of hepatic fibrosis. Mechanisms and treatment strategies. N Engl J Med 1993;328:1828–1835.

43 Pandol SJ: Are we studying the correct state of the stellate cell to elucidate mechanisms of chronic pancreatitis? Gut 2005;54:744–745.

44 Mews P, Phillips P, Fahmy R, Korsten M, Pirola R, Wilson J, Apte M: Pancreatic stellate cells respond to inflammatory cytokines: potential role in chronic pancreatitis. Gut 2002;50:535–541.

45 Masamune A, Kikuta K, Satoh M, Satoh A, Shimosegawa T: Alcohol activates activator protein-1 and mitogen-activated protein kinases in rat pancreatic stellate cells. J Pharmacol Exp Ther 2002;302:36–42.

46 Masamune A, Kikuta K, Suzuki N, Satoh M, Satoh K, Shimosegawa T: A c-Jun NH2-terminal kinase inhibitor SP600125 (anthra[1,9-cd]pyrazole-6 (2H)-one) blocks activation of pancreatic stellate cells. J Pharmacol Exp Ther 2004;310:520–527.

47 Haber PS, Keogh GW, Apte MV, Moran CS, Stewart NL, Crawford DH, Pirola RC, McCaughan GW, Ramm GA, Wilson JS: Activation of pancreatic stellate cells in human and experimental pancreatic fibrosis. Am J Pathol 1999;155:1087–1095.

48 Casini A, Galli A, Pignalosa P, Frulloni L, Grappone C, Milani S, Pederzoli P, Cavallini G, Surrenti C: Collagen type I synthesized by pancreatic periacinar stellate cells (PSC) co-localizes with lipid peroxidation-derived aldehydes in chronic alcoholic pancreatitis. J Pathol 2000;192: 81–89.

49 Apte MV, Phillips PA, Fahmy RG, Darby SJ, Rodgers SC, McCaughan GW, Korsten MA, Pirola RC, Naidoo D, Wilson JS: Does alcohol directly stimulate pancreatic fibrogenesis? Studies with rat pancreatic stellate cells. Gastroenterology 2000;118:780–794.

50 McCarroll JA, Phillips P, Pirola R, Wilson J, Apte M: Oxidant stress induces the MAP kinase pathway in pancreatic stellate cells. Gastroenterology 2003,24:A-616.

51 Gomez JA, Molero X, Vaquero E, Alonso A, Salas A, Malagelada JR: Vitamin E attenuates biochemical and morphological features associated with the development of chronic pancreatitis. Am J Physiol Gastrointest Liver Physiol 2004;287:162–169.

52 Cullen JJ, Mitros FA, Oberley LW: Expression of antioxidant enzymes in diseases of the human pancreas: another link between chronic pancreatitis and pancreatic cancer. Pancreas 2003;26: 23–27.

53 Gukovskaya AS, Vaquero E, Zaninovic V, Gorelick FS, Lusis AJ, Brennan ML, Holland S, Pandol SJ: Neutrophils and NADPH oxidase mediate intrapancreatic trypsin activation in murine experimental acute pancreatitis. Gastroenterology 2002;122:974–984.

54 Telek G, Scoazec JY, Chariot J, Ducroc R, Feldmann G, Roz C: Cerium-based histochemical demonstration of oxidative stress in taurocholate-induced acute pancreatitis in rats. A confocal laser scanning microscopic study. J Histochem Cytochem 1999;47:1201–1212.

55 Ito T, Nakao A, Kishimoto W, Nakano M, Takagi H: The involvement and sources of active oxygen in experimentally induced acute pancreatitis. Pancreas 1996;12:173–177.

56 Peralta J, Reides C, Garcia S, Llesuy S, Pargament G, Carreras MC, Catz S, Poderoso JJ: Oxidative stress in rodent closed duodenal loop pancreatitis. Int J Pancreatol 1996;19:61–69.

57 Dabrowski A, Gabryelewicz A: Oxidative stress. An early phenomenon characteristic of acute experimental pancreatitis. Int J Pancreatol 1992;12:193–199.

58 Schoenberg MH, Buchler M, Baczako K, Bultmann B, Younes M, Gasper M, Kirchmayr R, Beger HG: The involvement of oxygen radicals in acute pancreatitis. Klin Wochenschr 1991;69: 1025–1031.

59 Dabrowski A, Gabryelewicz A, Chyczewski L: The effect of platelet activating factor antagonist (BN 52021) on cerulein-induced acute pancreatitis with reference to oxygen radicals. Int J Pancreatol 1991;8:1–11.

60 Nonaka A, Manabe T, Tamura K, Asano N, Imanishi K, Tobe T: Changes of xanthine oxidase, lipid peroxide and superoxide dismutase in mouse acute pancreatitis. Digestion 1989;43:41–46.

61 Wisner J, Green D, Ferrell L, Renner I: Evidence for a role of oxygen derived free radicals in the pathogenesis of caerulein induced acute pancreatitis in rats. Gut 1988;29:1516–1523.

62 Blind PJ, Marklund SL, Stenling R, Dahlgren ST: Parenteral superoxide dismutase plus catalase diminishes pancreatic edema in sodium taurocholate-induced pancreatitis in the rat. Pancreas 1988;3:563–567.

63 Guice KS, Miller DE, Oldham KT, Townsend CM Jr, Thompson JC: Superoxide dismutase and catalase: a possible role in established pancreatitis. Am J Surg 1986;151:163–169.

64 Bokoch GM, Knaus UG: NADPH oxidases: not just for leukocytes anymore! Trends Biochem Sci 2003;28:502–508.

65 Dorseuil O, Vazquez A, Lang P, Bertoglio J, Gacon G, Leca G: Inhibition of superoxide production in B lymphocytes by rac antisense oligonucleotides. J Biol Chem 1992;267:20540–20542.

66 Roberts AW, Kim C, Zhen L, Lowe JB, Kapur R, Petryniak B, Spaetti A, Pollock JD, Borneo JB, Bradford GB, Atkinson SJ, Dinauer MC, Williams DA: Deficiency of the hematopoietic cell-specific Rho family GTPase Rac2 is characterized by abnormalities in neutrophil function and host defense. Immunity 1999;10:183–196.

67 Kim C, Dinauer MC: Rac2 is an essential regulator of neutrophil nicotinamide adenine dinucleotide phosphate oxidase activation in response to specific signaling pathways. J Immunol 2001;166:1223–1232.

68 Lapouge K, Smith SJ, Groemping Y, Rittinger K: Architecture of the p40–p47–p67phox complex in the resting state of the NADPH oxidase. A central role for p67phox. J Biol Chem 2002;277: 10121–10128.

69 Bokoch GM, Diebold BA: Current molecular models for NADPH oxidase regulation by Rac GTPase. Blood 2002;100:2692–2696.

70 Cheng G, Cao Z, Xu X, van Meir EG, Lambeth JD: Homologs of gp91phox: cloning and tissue expression of Nox3, Nox4, and Nox5. Gene 2001;269:131–140.

71 Bataller R, Schwabe RF, Choi YH, Yang L, Paik YH, Lindquist J, Qian T, Schoonhoven R, Hagedorn CH, Lemasters JJ, Brenner DA: NADPH oxidase signal transduces angiotensin II in hepatic stellate cells and is critical in hepatic fibrosis. J Clin Invest 2003;112:1383–1394.

Stephen J. Pandol, MD
VAGLAHS, Los Angeles, 11301 Wilshire Boulevard
Building 258, Room 340
Los Angeles, CA 90073 (USA)
Tel. +1 310 478 3711 ext. 49417, Fax +1 310 268 4578
E-Mail stephen.pandol@med.va.gov

Cho CH, Purohit V (eds): Alcohol, Tobacco and Cancer.
Basel, Karger, 2006, pp 119–139

·······················

Alcohol and Breast Cancer Risk

Ramona G. Dumitrescu, Peter G. Shields

Lombardi Comprehensive Cancer Center, Georgetown University Medical Center,
Washington, D.C., USA

Abstract

Breast cancer is the most common cancer in women in the United States, other than
non-melanotic skin cancer and is the second leading cause of death in women. Breast cancer
is a multifactorial disease, resulting from gene–environmental interactions. Among the best-
known breast cancer risk factors, most of them cannot be controlled or modified. However,
alcohol consumption is a modifiable risk factor that represents the leading non-hormonally
related breast cancer risk factor. Alcohol intake is associated with increased breast cancer
risk by the majority of epidemiological studies and experimental animal studies. Conflicting
data has been generated about the relationship between the breast cancer risk and alcohol
type, time of use and dose. Ethanol may cause breast cancer through different mechanisms,
such as the induction of mutations by the acetaldehyde, changes in estrogen levels and estrogen receptor response, through oxidative damage by generating reactive oxygen species
and/or the effect on one-carbon metabolism pathways on DNA methylation and DNA synthesis. Genetic variants for all these potential pathways involved in alcohol-induced breast
cancers might increase a woman's breast cancer risk. Definite studies on breast cancer etiology would lead to improved public health recommendations, allowing women to make individual choices about their lifestyle and might guide to better prevention strategies.

According to American Cancer Society, breast cancer represents the second leading cause of death in women and is the most frequently diagnosed cancer in women in United States, other than non-melanotic skin cancer [1].
Well-established risk factors, such as older age, family history of breast cancer
and risk factors related to a woman's reproductive life explain only a proportion
of the breast cancers [2]. Most of these risk factors cannot be controlled or
modified and the etiology of the disease is still not completely understood.
Alcohol drinking, however, is a modifiable risk factor. Alcohol consumption
is consistently associated with increased risk for breast cancer in both

premenopausal [3, 4] and postmenopausal women [3] in the majority of epidemiological studies during the past two decades. Within the United States, approximately 14,000 women are diagnosed each year with breast cancer attributable to alcohol consumption, accounting for a population attributable risk of 2.1% [5]. The attributable risk is higher in Italy e.g. where there is greater alcohol intake; the attributable proportion is as high as 10% among women [6]. A recent combined analysis of 53 epidemiological studies has shown that the relative risk of breast cancer was 1.32 (95% CI = 1.19–1.45) for women with an intake of 35–44 g alcohol per day, compared with the abstainers and 1.46 (95% CI = 1.33–1.61) for consumers of more than 45 g alcohol per day [7]. Breast cancer risk increases by 9% for each additional 10–12 g of alcohol intake on daily basis (equivalent to 1 drink/day) [8, 9] and by 41% for 2–5 drinks/day [8]. Remarkably, the increased risk for breast cancer in association with even moderate alcohol consumption is consistent in different age groups, geographic regions or different levels of consumption [10]. Experimental animal studies (mice and rats), with few exceptions have shown that ethanol administrated systematically is tumorigenic [11–13].

Conflicting data has been generated about the relationship between the breast cancer risk and alcohol type, time of use and dose. Associations are found for a particular alcohol type that is consistent with whatever the type of beverage is most frequently consumed in different populations. A recent prospective population-based cohort study has shown that the type of alcohol intake (beer, wine, or spirits) influences breast cancer risk in relationship to menopausal status [4]. Premenopausal women who had an intake of more than 27 drinks per week had a high relative risk for breast cancer. Interestingly, there was the greatest risk for consumption spirits; more than six spirits per week increased breast cancer risk among postmenopausal women [4]. Other studies observed that breast cancer risk does not vary with the type of alcoholic beverage consumed [10]. The data on the age at which women began drinking and the influence on breast cancer risk also have differed. Some studies found no association between the age at first alcohol consumption and breast cancer risk [10, 14] while others found that an early start to drinking increases a woman's risk of developing breast cancer even beyond menopause [15, 16]. Another study has shown that breast cancer risk increases with the number of years of drinking [17]. Recently, a prospective cohort study has observed that among postmenopausal women recent alcohol consumption is a predictor of the breast cancer risk, independent of the exposure to alcohol at younger age [18].

Alcohol may cause breast cancer through different mechanisms, such as the induction of mutations by the acetaldehyde, changes in the estrogens metabolism and response, through oxidative damage and/or the effect on one-carbon metabolism pathways. Differences in diet and genetic polymorphisms for all

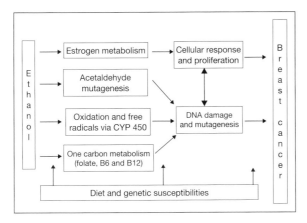

Fig. 1. Hypothesized mechanistic relationships between ethanol and breast cancer.

these potential pathways involved in alcohol-induced breast cancers might influence a woman's risk for developing breast cancer (fig. 1).

Ethanol Metabolism and Mutagenesis

Ethanol Metabolism

It has been shown that the ethanol is metabolized in different organs, including the breast [19, 20]. Alcohol dehydrogenase (ADH) and to a lesser extent cytochrome P450 2E1 (CYP2E1), oxidizes the ethanol to acetaldehyde. Although alcohol beverage consumption is considered a known human carcinogen by the 11th Report on Carcinogens [21], increasing evidence suggests that it might be acetaldehyde rather than ethanol, directly involved in carcinogenesis [22]. Experiments in cell culture systems and in animal models have shown that acetaldehyde causes point mutations, chromosomal damage including sister-chromatid exchanges and chromosomal aberrations [11], DNA cross-linking [23, 24] and aneuploidy [25]. Acetaldehyde also induces G to A transitions and mutations at A:T base pairs in the *p53* gene of esophageal tumors [26]. Moreover, acetaldehyde can interfere with DNA repair machinery, by a direct inhibition of the O6 methyl-guanyltransferase, involved in the repair of the DNA adducts caused by alkylating agents [27] and by impairing the repair of chromosomal breaks induced in response to DNA damage [28] in human lymphocytes. Acetaldehyde is able to bind rapidly to cellular proteins and DNA, leading to protein and DNA adducts, which affect normal cellular function and morphology. The acetaldehyde adducts have been detected in both animal models and humans [29, 30]. It has been hypothesized that the formation of the DNA adducts (e.g. N^2-ethyl-2'-deoxyguanosine) is the mechanism by which acetaldehyde causes mutations and/or replication errors and may be related to the development of human cancers induced by alcohol/acetaldehyde [31, 32]. Recently, it has been proposed that acetaldehyde is converted to crotonaldehyde (CrA) in the presence of polyamines and forms α-methyl-γ-hydro-1, N^2-propano-2'-deoxyguanosine (Cr-PdG) adducts that may be responsible for the carcinogenic effect of alcohol consumption [33].

Acetaldehyde is converted to acetate by aldehyde dehydrogenase (ALDH), xanthine oxidoreductase (XOR) and by aldehyde oxidase (AOX). The reaction catalyzed by XOR and AOX enzymes can generate reactive oxygen species (ROS), such as the superoxide anion (O_2^-), hydroxyl radical (•OH) and hydrogen peroxide (H_2O_2), but not the reaction catalyzed by ALDH. The major oxidative DNA damage product is 8-hydroxy-2'-guanosine, an oxidized form of guanine, which causes mutations A:T to C:C to G:C to T:A transversion [34]. Furthermore, in the female breast, it has been observed that the •OH and not O_2^- or H_2O_2 can induce •OH DNA adducts, base deletions, single and double strand breaks [35, 36]. The level of •OH adducts, such as 8-hydoxy-deoxyguanosine (8-OH-dG) has been shown to be approximately 10-fold higher in cancerous breast tissue than in normal breast tissue [37]. Similar observations have been made for malondialdehyde-DNA adducts when breast cancer tissue has been compared with normal surrounding tissue and breast tissue from reduction mammoplasty subjects [38]. These data suggest that the formation of the DNA adducts can be used as premalignant markers of breast cancer [38, 39] indicating the role of ROS in breast carcinogenesis. Chronic alcohol consumption induces oxidative stress and generates lipid peroxidation products such as malondialdehyde and 4-hydroxy-2-nonenal [40], reactive aldehydes which interact with cellular proteins and generate protein adducts detected in livers of ethanol-fed animals and humans consuming alcohol [41].

Genetic Variants of Genes Involved in Ethanol Metabolism

Breast cancer risk may be modulated by genetic polymorphisms and/or mutations of genes involved in the ethanol metabolism. Several polymorphisms of alcohol dehydrogenase 1B and 1C (*ADH1B* and *ADH1C*) may modulate the ethanol elimination rates and acetaldehyde levels among various populations [42]. For *ADH1B* polymorphism, the *ADH1B*2* allele encodes for a high activity isoenzyme (~40 times more active than the enzyme encoded by *1 allele) which has a high frequency in Asians but low in Caucasians. The presence of this allele has a protective effect against alcoholism among Caucasians because of the high amount of acetaldehyde generated and its toxic effects (e.g. alcohol-induced headaches after 1 or 2 drinks, severe hangovers) [43, 44]. For a *ADH1C* (*ADH3*, old nomenclature) gene polymorphism, the *ADH1C*1* allele encodes an enzyme with higher activity (~2.5 times more active) compared with the enzyme resulted from the transcription of *ADH1C*2*. Premenopausal women homozygous for *ADH1C*1* allele (*ADH1C*1,1* genotype) have been found to be at 1.8 times higher risk for developing breast cancer than women with another genotype [45]. Studies conducted in our laboratory found that fast oxidizers for this polymorphism have an increased risk of premenopausal breast cancer and oral cavity cancer [46, 47]. However, another breast cancer study did not replicate this positive association, but it has been observed that among fast oxidizers, the rapid elimination of alcohol may result in a decrease of androgen and sex hormone-binding globulin levels in association with alcohol consumption [48]. Recently, it has been shown that *ADH1C*1,1* genotype is an independent risk factor for the development of alcohol-associated esophageal, hepatocellular, head and neck tumors among heavy drinkers [49]. In the *ADH2* gene (*ADH4*, old nomenclature), a polymorphism in position −75 of the promoter and a substitution Ile to Val at amino acid 308, may influence the metabolism of ethanol [50, 51]. Mitochondrial class 2 ALDH (*ALDH2*) is mainly involved in acetaldehyde oxidation processes and in humans is polymorphic, with *ALDH2*1* and *ALDH2*2* alleles. Blood acetaldehyde levels are 20 times higher in the homozygous *ALDH2*2* than in individuals with an *ALDH2*1* allele [22]. A 96 bp insertion polymorphism in the regulatory region of *CYP2E1* has been shown to increase metabolic activity in drinkers and obese African-American women [52]. Studying the *CYP2E1* Rsal polymorphism in a Korean population it has been suggested that the breast cancer risk for alcohol-consuming women may be modified by this polymorphism [53]. Moreover, it has also been observed that homozygous *CYP2E1*1/*1* and carriers of at least one *ALDH2*2* allele have a higher degree of DNA damage, because of a low metabolism rate, which may be lead to the accumulation of ethanol and/or acetaldehyde. This damage increases the micronuclei frequency in human lymphocytes and may play a role in alcohol-induced carcinogenesis [54].

Genetic Variants of DNA Repair Genes and Ethanol

The acetaldehyde-induced DNA damage can be repaired through nucleotide excision repair pathway [55].

The effect of alcohol on breast cancer risk may be modified by several genetic polymorphisms in DNA repair genes. For example, *XRCC1 Arg399Gln* polymorphism is associated with alcohol-induced esophageal cancers [56], lung cancer risk [57] and with *p53* mutations [58]. Moreover, alcohol drinkers with allelic variants of the *XRCC1* gene at codons 194, 280 and 399 have an increased risk for colorectal cancer [59]. In a breast cancer study, *XRCC1* 399 *Arg/Arg* genotype is associated with breast cancer risk, especially among white women, occupationally exposed to high-dose of radiation to the chest [60]. There are polymorphisms in *XPD* (also known as *ERCC2*) gene which can modulate cancer risk in different populations. An *XPD Lys751Gln* polymorphism in exon 23 has shown to influence the number of chromatid breaks induced after DNA damage, e.g. persons who were homozygous for the *751Lys* allele had a higher number of chromatid aberrations, compared with those carriers of a *751Gln* allele [61]. Recently, it has been shown that the presence of at least one variant *Gln* allele for the *XPD Lys751Gln* polymorphism was associated with a 20% increased risk of developing breast tumors. The increase in breast cancer risk has been observed only among current smokers, homozygotes of the variant allele (*Gln/Gln*) [62]. A study conducted in a German population has found that the haplotype encoding *XPD/ERCC2 Asp312/Gln751* is linked to increased breast cancer risk in these women [63]. In contrast, in three case-control studies, common polymorphisms in *ERCC2* have been not associated with breast cancer risk [64]. Studying other DNA-repair genetic polymorphisms and breast cancer risk it has been observed that increasing number of combined variant alleles of *XRCC1 Arg194Trp*, *XRCC3 Thr241Met*, and *ERCC4/XPF Arg415Gln* contributes to breast cancer susceptibility [65].

Ethanol-Induced Oxidative Stress and Mitochondria

The major intracellular source of ROS is the mitochondrial respiratory chain, which generates one of the ROS generated by the ethanol metabolism, O_2^- [66, 67]. The DNA damage induced by ethanol may impair mitochondrial function, increasing the oxidative stress in the cell, which may lead to accumulation of cell damage. The uncontrolled formation of ROS by mitochondria disrupts the normal activation of the mitochondrial permeability transition, which leads to an increased sensitivity of the cells to pro-apoptotic or death signals [67]. On the other hand, mitochondria are protected against oxidative stress by antioxidant enzymes such as manganese superoxide dismutase (Mn-SOD) and glutathione peroxidase (GPX). Thus, mitochondria can both create and clear free radicals. For example, O_2^- can be catalyzed to H_2O_2 and oxygen (O_2) by the Mn-SOD inside the mitochondria [68]. Moreover, the formation of free radicals during the oxidation of acetaldehyde by xanthine oxidase is inhibited by the presence of Mn-SOD [69]. It has also been shown that the amount of DNA damage (8-OH-dG) induced by high concentrations of hydroxy estradiols is significantly reduced by the presence of endogenous antioxidants, including Mn-SOD [70, 71]. Moreover, the estradiol-induced mitochondrial ROS can modulate G1 to S transition of the cell cycle [72]. Furthermore, an amino acid change from a valine to alanine, in the -9 position of *Mn-SOD* gene alters the structure of the protein, affecting the mitochondrial transport of human Mn-SOD [73]. We found in our breast cancer case-control study, that premenopausal women homozygous for the variant allele Ala for the *Mn-SOD Val/Ala* polymorphism have an increased risk of breast cancer. The risk is more pronounced among women with low intake of fruits and vegetables and of dietary ascorbic acid and alpha-tocopherol [74]. Similar findings have been reported by other studies conducted in Finish Caucasian and Chinese women [75, 76]. It has also been suggested that the presence of the variant allele Ala may modify breast cancer risk especially among current smokers [77], and that it may be implicated in breast carcinogenesis in young women [78].

Glutathione peroxidase 1 (GPX1) prevents the oxidative damage to DNA, proteins and lipids by detoxifying hydrogen and lipid peroxides as part of the enzymatic antioxidant defense system. A genetic polymorphism of *GPX1* gene in codon 198 (*Pro198Leu*) encodes an isoenzyme of *GPX1* expressed in erythrocytes, breast and other tissues. The presence of the variant allele for *GPX1 Pro198Leu* polymorphism has been associated with increased lung cancer risk among Caucasians male [79]. Moreover, the leucine allele is associated with breast cancer and the loss of heterozygosity at *GPX1* locus is observed in approximately 36% of the breast tumors [80]. Similar findings were reported recently in a nested case-control study, but interestingly, the alcohol intake was associated with increased GPX activity only in the carriers of the wild type allele and not the variant allele [81].

Long-term ethanol exposure results in a decrease of the mitochondrial reduced glutathione (GSH) observed in both animals and humans [82], with almost no effect on cytosolic GSH levels, leading to mitochondrial damage [83, 84].

Ethanol exposure of different cells lines has been shown to influence different genes expression and protein synthesis in numerous tissues [85]. For example, the human mammary epithelial cell line MCF-10F treated with ethanol and acetaldehyde before exposure to the environmental carcinogen benzo[*a*]pyrene have shown an increased benzo[*a*]pyrene (B[*a*]P)–DNA adduct formation. This effect was probably due to a decrease in protein expression of the phase II detoxification enzyme glutathione-*S*-transferase P1(GSTP1) [85]. Additionally, the genetic polymorphisms in enzymes involved in ROS metabolism may influence individual susceptibility to breast cancer, related to oxidative stress. For example, patients with genotype of high *GSTP1* activity have lower levels of 8-OHdG in DNA of breast cancer tissues than others [86].

Furthermore, breast cancer risk in Korean women is increased by the *GSTP1* Val allele and the effect is modified by the alcohol intake [87].

Mitochondrial mutations are present in approximately 40% of breast tumors [88–90]. Ethanol-induced oxidative stress targets the mitochondrial DNA (mt DNA) [67, 91], and therefore may cause mt DNA mutations. Animal studies have shown that chronic ethanol exposure leads to increased oxidative damage to mt DNA, especially in older animals, and to an increased number of strand breaks [92]. Moreover, a mtDNA *G10398A* polymorphism that affects the structure of Complex I in the mitochondrial electron transport chain, important in the free radicals formation has been associated with increased risk of invasive breast cancer in African-American women [93].

Alcohol Drinking and Estrogens

Estrogens are critical for the cellular proliferation of both normal and neoplastic breast epithelial cells. The mechanisms that may be responsible for estrogens-induced breast carcinogenesis, include: (1) receptor-mediated hormonal activity promoting cellular proliferation, leading to accumulation of the genetic damage; (2) the activation of the cytochrome P450 metabolic pathways, which may increase the mutation rates and leads to genotoxicity; and (3) the induction of aneuploidy [94]. Furthermore, higher levels of sex hormones (total estradiol, free estradiol, non-sex hormone-binding globulin-bound estradiol, estrone, estrone sulfate, androstenedione, dehydroepiandrosterone, dehydroepiandrosterone sulfate, and testosterone) have been shown to be consistently associated with increased breast cancer risk in postmenopausal women [95].

Alcohol drinking interferes with estrogen metabolism in multiple ways. Alcohol consumption has been shown to influence the menstrual cycle length, so that lean and physically active women who do not consume alcohol are more likely to have variable or long menstrual cycles, reflecting the anovulation or low levels of estrogens [96]. There are data suggesting that acute and chronic alcohol consumption increases serum and urinary estrogen levels in premenopausal women who do not use any hormones [97, 98]. However, in postmenopausal women, alcohol drinking has a more pronounced risk effect in hormonal replacement therapy (HRT) users [99]. It has also been shown that follicle-stimulating hormone decreases with increasing alcohol intake in premenopausal women [100]. Moreover, women using HRT have an increased risk for breast cancer and this risk increases with the duration of use [101] and that continuous combined HRT is associated with increased breast cancer risk among current users [102]. Alcohol drinking also has been shown to increase endogenous levels of estrogens and dehydroepiandrosterone sulfate in post-menopausal women, suggesting that the alcohol consumption may increase breast cancer risk at least partially through an effect on sex steroid levels [103]. Interestingly, another study found that the increased risk for breast cancer in hormone users did not vary with the alcohol intake [104]. Alcohol use is associated with decreased bone loss in postmenopausal women by increasing the circulating levels of estrogen and by reducing bone remodeling [105]. It has also been shown that alcohol decreases serum IGF-I in premenopausal women and that IGF-I level increases over the course of the menstrual cycle independent of the alcohol [106]. Moreover, the ethanol stimulates cellular proliferation, estrogen receptor α (ER-α) and aromatase expression in human breast cancer cells [107] and down-regulates the expression of the tumor suppressor gene BRCA1, a potent inhibitor of ER-α activity, suggesting the involvement of these pathways in the etiology of alcohol-related breast cancer [108]. Among postmenopausal women, alcohol intake has been associated with risk of lobular carcinomas and ER+/PR+ (ER/PR positive) tumors [109, 110] or both ER+PR+ and ER+PR− (ER positive/PR negative) tumors [111]. Interestingly, women who use postmenopausal hormones had an increased risk for the development of ER+ tumors in association with alcohol consumption [112]. In contrast, alcohol use appears to be associated only with ER−/PR− (ER/PR negative) breast cancers in the Iowa Women's Health Study cohort [113].

Experimental animal studies indicate that ethanol is involved in mammary tumorigenesis [12] and enhance dimethylbenzanthracene-induced mammary tumors in rat [114]. The ethanol effects on the rodent mammary gland might be due to increasing the susceptibility to carcinogen-induced damage through changes of the mammary gland structural development and by stimulating the cell proliferation in the terminal end buds of the mammary gland [3].

Genetic Variants for Genes Involved in Estrogen Metabolism

Genetic variants of several genes involved in estrogen synthesis and metabolism, such as CYP17, CYP19, CYP1A1, CYP1B1 and the catechol-o-methyltransferase (COMT) are associated with breast cancer risk. The presence of the variant allele (A2 allele) for a polymorphism in the 5'-untranslated region upstream of the transcription start region of CYP17 gene is associated with higher serum and plasma levels of steroid hormones and with increased risk of advanced breast cancer [115–117], whereas A1/A1 genotype is associated with a later age at menarche among the controls and with reduced risk for developing breast cancer [116]. However, other studies found a weak [118] or no association between the variant A2 allele and breast cancer risk [119]. A tetranucleotide $(TTTA)_{10}$ simple tandem repeat polymorphism of CYP19 gene has been associated with breast cancer risk, although other studies did not support this finding or found an association with longer repeats CYP19 [117, 119]. In a separate study, the $(TTTA)_7$ variant was associated with ER-positive, but not ER-negative breast tumors [120]. The variant allele m2 for the *CYP1A1* polymorphism has been significantly associated with breast cancer risk [121], and especially in combination with *GSTM1* homozygous null deletion genotype [122]. Some studies showed that the *CYP1B1* gene polymorphism in codon 432 increases breast cancer risk, especially after different environmental exposures [123, 124] whereas others did not find any association [125, 126]. The low-activity genotype for the *COMT Val158Met* polymorphism has been associated with increased breast cancer risk in premenopausal women [127] especially in combination with *GSTM1* null genotype among HRT users [128] but other studies could not find any correlation [129, 130]. How these genetic polymorphisms might interact with alcohol consumption remains to be studied.

Breast Density and Alcohol Consumption

A well-known breast cancer risk factor is increased mammographic breast density [131] and it has associated with a 4–6-fold increased breast cancer risk [132, 133]. Among factors that increase breast tissue density by mammography, such as nulliparity, late age at first birth, younger age, low body mass [134], HRT use [135, 136] and family history of breast cancer [137, 138], alcohol is also associated with increased breast density. It has been observed that breast mammographic density increases with alcohol consumption in both pre-menopausal and postmenopausal women [3, 134].

Several genetic polymorphisms have been associated with mammographic breast density. For example, in premenopausal women a 7/7 *UGT1A1* genotype (compared to the 6/6) is associated with lower breast density, whereas in post-menopausal women this genotype together with short *AIB1* alleles are associated with greater breast density [139]. Moreover, low-activity *COMT* and *CYP1A2* variant alleles are associated with lower mammographic density [140]. Women carriers of two A alleles for the *IGFBP3* gene have a 5-fold increased risk of having an increased breast density, compared with women with no A allele [141].

Folate Metabolism, One-Carbon Metabolism Genes, Methylation and Breast Cancer Risk

An inverse association of breast cancer risk with folate intake or circulating levels of folate has been observed, especially in interaction with higher alcohol intake [142–144]. The association between folate intake and breast cancer risk can be modified by the *MTHFR C677T* polymorphism (TT genotype) [145].

Moreover, the risk of postmenopausal breast cancer in women with family history may be reduced by higher levels of folate, in the absence of alcohol consumption [146]. Therefore, adequate folate intake may reduce the risk of breast cancer, especially ER− tumors [147], in women with high alcohol consumption [148]. The alcohol affects folate status in tissues that rapidly proliferate [149]. Poor folate intake, in combination with alcohol consumption impairs erythropoiesis by several potential mechanisms such as blockage of the folate release from hepatocytes, intestinal malabsorption of folate, inhibition of DNA methyltransferase or methionine synthetase (*MS*) and by cellular depletion of the overall folate [149]. Furthermore, even moderate alcohol intake may reduce vitamin B12 serum levels among healthy, postmenopausal women [150]. Folate and vitamin B12 are critical for DNA methylation and nucleotide synthesis. Therefore, folate deficiency may increase the risk of malignancy by causing DNA hypomethylation and/or inducing massive incorporation of uracil during DNA synthesis [151–153]. In animal models, it has been shown that folate supplementation has a protective effect against mt DNA deletions, induced after cancer chemotherapy [154] and reduces the DNA strand breaks in the *p53* gene [155]. In rats, chronic alcohol exposure leads to genomic DNA hypomethylation [156]. Moreover, ethanol exposure together with folate deficiency may act together to decrease the liver SAM/*S*-adenosylhomocysteine ratio and to increase liver *S*-adenosyshomocysteine, DNA strand breaks, urinary 8-oxo-2′-deoxyguanosine, plasma homocysteine, and aspartate transaminase [157].

Genetic Variants of One-Carbon Metabolism Genes

Several genetic polymorphisms in one-carbon metabolism genes, such as methylenetetrahydrofolate reductase (*MTHFR*) and *MS* have a modifier effect on breast cancer risk. For example, homozygous for the variant allele T for the *MTHFR C677T* polymorphism with low levels of folate (dietary or total folate intake) have an increased breast cancer risk [158, 159] and show genomic DNA hypomethylation [160]. Studies conducted in our laboratory found that postmenopausal women with the *MTHFR 677TT* genotype, have increased breast cancer risk, especially when their alcohol intake is high and the dietary folate, vitamin B_6 and B_{12} is low (unpublished data). Interestingly, homozygotes for the wild-type allele C, HRT users, who consume more than 10 g alcohol/day have an increased breast cancer risk [161]. DNA methylation status is influenced by the interaction between *MTHFR C677T* polymorphism and folate levels [162].

MS is critical in producing methionine, required for the production of *S*-adenosylmethionine, the universal methyl group donor. In animal studies, it has been shown that ethanol and more recently the acetaldehyde inhibits MS activity [163]. Interestingly, the ethanol potently interferes with IGF-1 activation of *MS* and blocked its effect on DNA methylation [164]. The presence of the variant allele G for *MS A2756G* polymorphism is associated with modest homocysteine reduction [165], DNA hypomethylation [166] and reduced risk of colorectal cancer in men who consumed less than a drink/day [167]. Furthermore, the homozygous for the variant allele *MTHFR 677T* and *MS 2756G* are associated with constitutive global genomic hypomethylation in normal tissues and low levels of CpG hypermethylation of tumor suppressor genes [168]. In a breast cancer case-control study conducted in our laboratory it has been observed that *MS A2756G* polymorphism is associated with an overall reduced breast cancer risk in premenopausal women and a reduced risk of *p16* promoter hypermethylation (unpublished data).

p53 Mutational Spectra

In breast cancer, *p53* gene, which is critical for cell cycle control, DNA repair, cellular differentiation and apoptosis [169, 170] is mutated in 15–50% of the tumors [171]. *p53* mutations rate is influenced by mutagens exposure and by genetic susceptibility [172, 173]. Furthermore, *p53* mutations are associated with worse prognosis and resistance of the tumors to chemotherapy [174]. Studying gene-expression signature, derived from differences between *p53* mutant (mt) and wild-type (wt) breast tumors, it has been observed that *p53* expression signature predicts breast cancer outcome better than *p53* mutation status alone [174]. Interestingly, more than 33% of the *p53* mutations are found at CpG dinucleotides sites (e.g. C→T transitions). Therefore, the most important mutational targets in *p53* are methylated CpG dinucleotides (e.g. hotspots such as codons 175, 213, 245, 248, 273, and 282 contain methylated CpGs) [175]. *p53* gene mutational spectrum varies by race and geographic location, in different US populations [172, 176, 177], in Russian breast cancer patients [178], in Hispanic women [179], in African-Brazilian women (A:T–G:C transversion and G:C–C:G transition), [180] and in breast cancer patients in Taiwan (G:C–C:G transversions) [181], suggesting a complex gene–environment interaction.

Studying the interaction between the alcohol and *p53* mutations in a breast cancer case-control study, we observed that premenopausal women with higher alcohol intake (16 or more drinks/month) have an increased chance of having tumors with *p53* mutations whereas in postmenopausal women, tumors with *p53* mutations were associated with higher folate intake [182]. However, other studies could not find a similar relationship [183–185]. Mechanistically, it has been shown that methylation of cytosines determines the hot spots in the human *p53* gene and that there is hydrolytic deamination of 5-methylcytosine in double-stranded DNA [186, 187], explaining the occurrence of mutational hot spots at the CpG dinucleotides. Moreover, it has been shown that acetaldehyde is able to inhibit fetal DNA methyltransferase activity, leading to DNA hypomethylation in mice [188] and to induce G/A transitions and A:T mutations in the *p53* gene [26].

Conclusions

In conclusion, the mechanisms by which ethanol may induce breast cancer are related to acetaldehyde mutagenesis, estrogen metabolic pathways, folate metabolism and *p53* gene regulation. Genetic polymorphisms for these pathways might increase a woman's breast cancer risk. Definitive studies on breast carcinogenesis would lead to improved health recommendations, allow women to make individual choices about lifestyle and risk, place alcohol drinking into a broader context of interactions with other choices such diet, HRT etc., and lead to more rationale prevention strategies.

Acknowledgement

This work was supported by the grant BC022346:DAMD17–03–1-0446, Breast Cancer of Excellence from US Department of Defense.

References

1   American Cancer Society. Breast Cancer Facts and Figures 2005–2006 (http://www.cancer.org/downloads/STT/CAFF2005BrF.pdf). 2005.
2   Madigan MP, Ziegler RG, Benichou J, Byrne C, Hoover RN: Proportion of breast cancer cases in the United States explained by well-established risk factors. J Natl Cancer Inst 1995;87: 1681–1685.
3   Singletary KW, Gapstur SM: Alcohol and breast cancer – review of epidemiologic and experimental evidence and potential mechanisms. JAMA 2001;286:2143–2151.
4   Petri AL, Tjonneland A, Gamborg M, et al: Alcohol intake, type of beverage, and risk of breast cancer in pre- and postmenopausal women. Alcohol Clin Exp Res 2004;28:1084–1090.
5   Tseng M, Weinberg CR, Umbach DM, Longnecker MP: Calculation of population attributable risk for alcohol and breast cancer (United States). Cancer Causes Control 1999;10:119–123.
6   Mezzetti M, La VC, Decarli A, Boyle P, Talamini R, Franceschi S: Population attributable risk for breast cancer: diet, nutrition, and physical exercise. J Natl Cancer Inst 1998;90:389–394.
7   Hamajima N, Hirose K, Tajima K, et al: Alcohol, tobacco and breast cancer – collaborative reanalysis of individual data from 53 epidemiological studies, including 58,515 women with breast cancer and 95,067 women without the disease. Br J Cancer 2002;87:1234–1245.
8   Smith-Warner SA, Spiegelman D, Yaun SS, et al: Alcohol and breast cancer in women: a pooled analysis of cohort studies. JAMA 1998;279:535–540.
9   Ellison RC, Zhang YQ, McLennan CE, Rothman KJ: Exploring the relation of alcohol consumption to risk of breast cancer. American J Epidemiol 2001;154:740–747.
10 Terry MB, Zhang FF, Kabat G, et al: Lifetime alcohol intake and breast cancer risk. Ann Epidemiol 2005.

11   Poschl G, Seitz HK: Alcohol and cancer. Alcohol 2004;39:155–165.
12   Watabiki T, Okii Y, Tokiyasu T, et al: Long-term ethanol consumption in ICR mice causes mammary tumor in females and liver fibrosis in males. Alcohol Clin Exp Res 2000;24: 117S–122S.
13   Singletary KW, McNary MQ: Influence of ethanol intake on mammary gland morphology and cell proliferation in normal and carcinogen-treated rats. Alcohol Clin Exp Res 1994;18: 1261–1266.
14   Longnecker MP, Newcomb PA, Mittendorf R, et al: Risk of breast cancer in relation to lifetime alcohol consumption. J Natl Cancer Inst 1995;87:923–929.
15   Harvey EB, Schairer C, Brinton LA, Hoover RN, Fraumeni JF Jr: Alcohol consumption and breast cancer. J Natl Cancer Inst 1987;78:657–661.
16   van't VP, Kok FJ, Hermus RJ, Sturmans F: Alcohol dose, frequency and age at first exposure in relation to the risk of breast cancer. Int J Epidemiol 1989;18:511–517.
17   Bowlin SJ, Leske MC, Varma A, Nasca P, Weinstein A, Caplan L: Breast cancer risk and alcohol consumption: results from a large case-control study. Int J Epidemiol 1997;26:915–923.
18   Tjonneland A, Christensen J, Thomsen BL, et al: Lifetime alcohol consumption and post-menopausal breast cancer rate in Denmark: a prospective cohort study. J Nutr 2004;134:173–178.
19   Saleem MM, Al Tamer YY, Skursky L, Al Habbal Z: Alcohol dehydrogenase activity in the human tissues. Biochem Med 1984;31:1–9.
20   Wright RM, McManaman JL, Repine JE: Alcohol-induced breast cancer: a proposed mechanism. Free Radic Biol Med 1999;26:348–354.
21   U.S.: Department of Health and Human Services PHSNTP: Report on Carcinogens, ed 11. (http://ntp.niehs.nih.gov/ntp/roc/eleventh/profiles/s007alco.pdf). 2005.
22   Seitz HK, Matsuzaki S, Yokoyama A, Homann N, Vakevainen S, Wang XD: Alcohol and cancer. Alcohol Clin Exp Res 2001;25(5 suppl ISBRA):137S–143S.
23   Blasiak J, Trzeciak A, Malecka-Panas E, Drzewoski J, Wojewodzka M: In vitro genotoxicity of ethanol and acetaldehyde in human lymphocytes and the gastrointestinal tract mucosa cells. Toxicol In Vitro 2000;14:287–295.
24   Kuykendall JR, Bogdanffy MS: Formation and stability of acetaldehyde-induced crosslinks between poly-lysine and poly-deoxyguanosine. Mutat Res 1994;311:49–56.
25   Migliore L, Cocchi L, Scarpato R: Detection of the centromere in micronuclei by fluorescence in situ hybridization: its application to the human lymphocyte micronucleus assay after treatment with four suspected aneugens. Mutagenesis 1996;11:285–290.
26   Noori P, Hou SM: Mutational spectrum induced by acetaldehyde in the HPRT gene of human T lymphocytes resembles that in the p53 gene of esophageal cancers. Carcinogenesis 2001;22: 1825–1830.
27   Espina N, Lima V, Lieber CS, Garro AJ: In vitro and in vivo inhibitory effect of ethanol and acetaldehyde on O6-methylguanine transferase. Carcinogenesis 1988;9:761–766.
28   Blasiak J: Ethanol and acetaldehyde impair the repair of bleomycin-damaged DNA in human lymphocytes. Cytobios 2001;106:141–149.
29   Jeong KS, Soh Y, Jeng J, Felder MR, Hardwick JP, Song BJ: Cytochrome P450 2E1 (CYP2E1)-dependent production of a 37-kDa acetaldehyde-protein adduct in the rat liver. Arch Biochem Biophys 2000;384:81–87.
30   Nakamura K, Iwahashi K, Furukawa A, et al: Acetaldehyde adducts in the brain of alcoholics. Arch Toxicol 2003;77:591–593.
31   Fang JL, Vaca CE: Detection of DNA adducts of acetaldehyde in peripheral white blood cells of alcohol abusers. Carcinogenesis 1997;18:627–632.
32   Matsuda T, Terashima I, Matsumoto Y, Yabushita H, Matsui S, Shibutani S: Effective utilization of N2-ethyl-2′-deoxyguanosine triphosphate during DNA synthesis catalyzed by mammalian replicative DNA polymerases. Biochemistry 1999;38:929–935.
33   Theruvathu JA, Jaruga P, Nath RG, Dizdaroglu M, Brooks PJ: Polyamines stimulate the formation of mutagenic 1,N2-propanodeoxyguanosine adducts from acetaldehyde. Nucleic Acids Res 2005;33:3513–3520.
34   Kohen R, Nyska A: Oxidation of biological systems: oxidative stress phenomena, antioxidants, redox reactions, and methods for their quantification. Toxicol Pathol 2002;30:620–650.

35 Malins DC, Holmes EH, Polissar NL, Gunselman SJ: The etiology of breast cancer. Characteristic alteration in hydroxyl radical-induced DNA base lesions during oncogenesis with potential for evaluating incidence risk. Cancer 1993;71:3036–3043.

36 Lubec G: The hydroxyl radical: from chemistry to human disease. J Investig Med 1996;44:324–346.

37 Musarrat J, Arezina-Wilson J, Wani AA: Prognostic and aetiological relevance of 8-hydroxyguanosine in human breast carcinogenesis. Eur J Cancer 1996;32A:1209–1214.

38 Li D, Zhang W, Sahin AA, Hittelman WN: DNA adducts in normal tissue adjacent to breast cancer: a review. Cancer Detect Prev 1999;23:454–462.

39 Malins DC, Polissar NL, Nishikida K, Holmes EH, Gardner HS, Gunselman SJ: The etiology and prediction of breast cancer. Fourier transform-infrared spectroscopy reveals progressive alterations in breast DNA leading to a cancer-like phenotype in a high proportion of normal women. Cancer 1995;75:503–517.

40 Chen J, Petersen DR, Schenker S, Henderson GI: Formation of malondialdehyde adducts in livers of rats exposed to ethanol: role in ethanol-mediated inhibition of cytochrome c oxidase. Alcohol Clin Exp Res 2000;24:544–552.

41 Tuma DJ, Casey CA: Dangerous byproducts of alcohol breakdown – focus on adducts. Alcohol Res Health 2003;27:285–290.

42 Pastino GM, Flynn EJ, Sultatos LG: Genetic polymorphisms in ethanol metabolism: issues and goals for physiologically based pharmacokinetic modeling. Drug Chem Toxicol 2000;23: 179–201.

43 Wall TL, Shea SH, Luczak SE, Cook TA, Carr LG: Genetic associations of alcohol dehydrogenase with alcohol use disorders and endophenotypes in white college students. J Abnorm Psychol 2005;114:456–465.

44 Borras E, Coutelle C, Rosell A, et al: Genetic polymorphism of alcohol dehydrogenase in europeans: the ADH2*2 allele decreases the risk for alcoholism and is associated with ADH3*1. Hepatology 2000;31:984–989.

45 Coutelle C, Hohn B, Benesova M, et al: Risk factors in alcohol associated breast cancer: alcohol dehydrogenase polymorphism and estrogens. Int J Oncol 2004;25:1127–1132.

46 Freudenheim JL, Ambrosone CB, Moysich KB, et al: Alcohol dehydrogenase 3 genotype modification of the association of alcohol consumption with breast cancer risk. Cancer Causes Control 1999;10:369–377.

47 Harty LC, Caporaso NE, Hayes RB, et al: Alcohol dehydrogenase 3 genotype and risk of oral cavity and pharyngeal cancers [see comments]. J Natl Cancer Inst 1997;89:1698–1705.

48 Hines LM, Hankinson SE, Smith-Warner SA, et al: A prospective study of the effect of alcohol consumption and ADH3 genotype on plasma steroid hormone levels and breast cancer risk. Cancer Epidemiol Biomarkers Prev 2000;9:1099–1105.

49 Homann N, Stickel F, Konig IR, et al: Alcohol dehydrogenase 1C*1 allele is a genetic marker for alcohol-associated cancer in heavy drinkers. Int J Cancer 2005.

50 Edenberg HJ, Jerome RE, Li M: Polymorphism of the human alcohol dehydrogenase 4 (ADH4) promoter affects gene expression. Pharmacogen 1999;9:25–30.

51 Stromberg P, Svensson S, Hedberg JJ, Nordling E, Hoog JO: Identification and characterisation of two allelic forms of human alcohol dehydrogenase 2. Cell Mol Life Sci 2002;59:552–559.

52 McCarver DG, Byun R, Hines RN, Hichme M, Wegenek W: A genetic polymorphism in the regulatory sequences of human CYP2E1: association with increased chlorzoxazone hydroxylation in the presence of obesity and ethanol intake. Toxicol Appl Pharmacol 1998;152:276–281.

53 Choi JY, Abel J, Neuhaus T, et al: Role of alcohol and genetic polymorphisms of CYP2E1 and ALDH2 in breast cancer development. Pharmacogen 2003;13:67–72.

54 Ishikawa H, Miyatsu Y, Kurihara K, Yokoyama K: Gene–environmental interactions between alcohol-drinking behavior and ALDH2 and CYP2E1 polymorphisms and their impact on micronuclei frequency in human lymphocytes. Mutat Res 2006;584:1–9.

55 Matsuda T, Kawanishi M, Yagi T, Matsui S, Takebe H: Specific tandem GG to TT base substitutions induced by acetaldehyde are due to intra-strand crosslinks between adjacent guanine bases. Nucleic Acids Res 1998;26:1769–1774.

56 Lee JM, Lee YC, Yang SY, et al: Genetic polymorphisms of XRCC1 and risk of the esophageal cancer. Int J Cancer 2001;95:240–246.

57   Divine KK, Gilliland FD, Crowell RE, et al: The XRCC1 399 glutamine allele is a risk factor for adenocarcinoma of the lung. Mutat Res 2001;461:273–278.

58   Hsieh LL, Chien HT, Chen IH, et al: The XRCC1 399Gln polymorphism and the frequency of p53 mutations in Taiwanese oral squamous cell carcinomas. Cancer Epidemiol Biomarkers Prev 2003;12:439–443.

59   Hong YC, Lee KH, Kim WC, et al: Polymorphisms of XRCC1 gene, alcohol consumption and colorectal cancer. Int J Cancer 2005;116:428–432.

60   Duell EJ, Millikan RC, Pittman GS, et al: Polymorphisms in the DNA repair gene XRCC1 and breast cancer. Cancer Epidemiol Biomarkers Prev 2001;10:217–222.

61   Lunn RM, Helzlsouer KJ, Parshad R, et al: XPD polymorphisms: effects on DNA repair proficiency. Carcinogenesis 2000;21:551–555.

62   Terry MB, Gammon MD, Zhang FF, et al: Polymorphism in the DNA repair gene XPD, polycyclic aromatic hydrocarbon-DNA adducts, cigarette smoking, and breast cancer risk. Cancer Epidemiol Biomarkers Prev 2004;13:2053–2058.

63   Justenhoven C, Hamann U, Pesch B, et al: ERCC2 genotypes and a corresponding haplotype are linked with breast cancer risk in a German population. Cancer Epidemiol Biomarkers Prev 2004;13:2059–2064.

64   Kuschel B, Chenevix-Trench G, Spurdle AB, et al: Common polymorphisms in ERCC2 (Xeroderma pigmentosum D) are not associated with breast cancer risk. Cancer Epidemiol Biomarkers Prev 2005;14:1828–1831.

65   Smith TR, Levine EA, Perrier ND, et al: DNA-repair genetic polymorphisms and breast cancer risk. Cancer Epidemiol Biomarkers Prev 2003;12(11 Pt 1):1200–1204.

66   Ishii H, Kurose I, Kato S: Pathogenesis of alcoholic liver disease with particular emphasis on oxidative stress. J Gastroenterol Hepatol 1997;12:S272–S282.

67   Hoek JB, Cahill A, Pastorino JG: Alcohol and mitochondria: a dysfunctional relationship. Gastroenterology 2002;122:2049–2063.

68   Millan-Crow LA, Cruthirds DL: Invited review: manganese superoxide dismutase in disease. Free Radic Res 2001;34:325–336.

69   Albano E, Clot P, Comoglio A, Dianzani MU, Tomasi A: Free radical activation of acetaldehyde and its role in protein alkylation. FEBS Lett 1994;348:65–69.

70   Mobley JA, Bhat AS, Brueggemeier RW: Measurement of oxidative DNA damage by catechol estrogens and analogues in vitro. Chem Res Toxicol 1999;12:270–277.

71   Ambrosone CB: Oxidants and antioxidants in breast cancer. Antioxid Redox Signal 2000;2:903–917.

72   Felty Q, Singh KP, Roy D: Estrogen-induced G1/S transition of G0-arrested estrogen-dependent breast cancer cells is regulated by mitochondrial oxidant signaling. Oncogene 2005;24:4883–4893.

73   Shimoda-Matsubayashi S, Matsumine H, Kobayashi T, Nakagawa-Hattori Y, Shimizu Y, Mizuno Y: Structural dimorphism in the mitochondrial targeting sequence in the human manganese superoxide dismutase gene. A predictive evidence for conformational change to influence mitochondrial transport and a study of allelic association in Parkinson's disease. Biochem Biophys Res Commun 1996;226:561–565.

74   Ambrosone CB, Freudenheim JL, Thompson PA, et al: Manganese superoxide dismutase (MnSOD) genetic polymorphisms, dietary antioxidants, and risk of breast cancer. Cancer Res 1999;59:602–606.

75   Mitrunen K, Sillanpaa P, Kataja V, et al: Association between manganese superoxide dismutase (MnSOD) gene polymorphism and breast cancer risk. Carcinogenesis 2001;22:827–829.

76   Cai Q, Shu XO, Wen W, et al: Genetic polymorphism in the manganese superoxide dismutase gene, antioxidant intake, and breast cancer risk: results from the Shanghai breast cancer study. Breast Cancer Res 2004;6:R647–R655.

77   Tamimi RM, Hankinson SE, Spiegelman D, Colditz GA, Hunter DJ: Manganese superoxide dismutase polymorphism, plasma antioxidants, cigarette smoking, and risk of breast cancer. Cancer Epidemiol Biomarkers Prev 2004;13:989–996.

78   Bergman M, Ahnstrom M, Palmeback WP, Wingren S: Polymorphism in the manganese superoxide dismutase (MnSOD) gene and risk of breast cancer in young women. J Cancer Res Clin Oncol 2005;131:439–444.

79 Ratnasinghe D, Tangrea JA, Andersen MR, et al: Glutathione peroxidase codon 198 polymorphism variant increases lung cancer risk. Cancer Res 2000;60:6381–6383.

80 Hu YJ, Diamond AM: Role of glutathione peroxidase 1 in breast cancer: loss of heterozygosity and allelic differences in the response to selenium. Cancer Res 2003;63:3347–3351.

81 Ravn-Haren G, Olsen A, Tjonneland A, et al: Associations between GPX1 Pro198Leu polymorphism, erythrocyte GPX activity, alcohol consumption and breast cancer risk in a prospective cohort study. Carcinogenesis 2005.

82 Lieber CS: Role of oxidative stress and antioxidant therapy in alcoholic and nonalcoholic liver diseases. Adv Pharmacol 1997;38:601–628.

83 Cederbaum AI: Effects of alcohol on hepatic mitochondrial function and DNA. Gastroenterology 1999;117:265–269.

84 Bailey SM, Patel VB, Young TA, Asayama K, Cunningham CC: Chronic ethanol consumption alters the glutathione/glutathione peroxidase-1 system and protein oxidation status in rat liver. Alcohol Clin Exp Res 2001;25:726–733.

85 Barnes SL, Singletary KW, Frey R: Ethanol and acetaldehyde enhance benzo[a]pyrene-DNA adduct formation in human mammary epithelial cells. Carcinogenesis 2000;21:2123–2128.

86 Matsui A, Ikeda T, Enomoto K, et al: Increased formation of oxidative DNA damage, 8-hydroxy-2′-deoxyguanosine, in human breast cancer tissue and its relationship to GSTP1 and COMT genotypes. Cancer Lett 2000;151:87–95.

87 Kim SU, Lee KM, Park SK, et al: Genetic polymorphism of glutathione S-transferase P1 and breast cancer risk. J Biochem Mol Biol 2004;37:582–585.

88 Richard SM, Bailliet G, Paez GL, Bianchi MS, Peltomaki P, Bianchi NO: Nuclear and mitochondrial genome instability in human breast cancer. Cancer Res 2000;60:4231–4237.

89 Bianchi MS, Bianchi NO, Bailliet G: Mitochondrial DNA mutations in normal and tumor tissues from breast cancer patients. Cytogenet Cell Genet 1995;71:99–103.

90 Tan DJ, Bai RK, Wong LJ: Comprehensive scanning of somatic mitochondrial DNA mutations in breast cancer. Cancer Res 2002;62:972–976.

91 Wieland P, Lauterburg BH: Oxidation of mitochondrial proteins and DNA following administration of ethanol. Biochem Biophys Res Commun 1995;213:815–819.

92 Cahill A, Stabley GJ, Wang X, Hoek JB: Chronic ethanol consumption causes alterations in the structural integrity of mitochondrial DNA in aged rats. Hepatology 1999;30:881–888.

93 Canter JA, Kallianpur AR, Parl FF, Millikan RC: Mitochondrial DNA G10398A polymorphism and invasive breast cancer in African-American women. Cancer Res 2005;65:8028–8033.

94 Russo J, Russo IH: Genotoxicity of steroidal estrogens. Trends Endocrinol Metab 2004;15:211–214.

95 The Endogenous Hormones and Breast Cancer Collaborative Group: Endogenous sex hormones and breast cancer in postmenopausal women: reanalysis of nine prospective studies. Oxford, Cancer Research U.K. Epidemiology Unit, University of Oxford, 2002.

96 Cooper GS, Sandler DP, Whelan EA, Smith KR: Association of physical and behavioral characteristics with menstrual cycle patterns in women age 29–31 years. Epidemiology 1996;7:624–628.

97 Reichman ME, Judd JT, Longcope C, et al: Effects of alcohol consumption on plasma and urinary hormone concentrations in premenopausal women. J Natl Cancer Inst 1993;85:722–727.

98 Martin CA, Mainous AG III, Curry T, Martin D: Alcohol use in adolescent females: correlates with estradiol and testosterone. Am J Addict 1999;8:9–14.

99 Ginsburg ES: Estrogen, alcohol and breast cancer risk. J Steroid Biochem Mol Biol 1999;69:299–306.

100 Verkasalo PK, Thomas HV, Appleby PN, Davey GK, Key TJ: Circulating levels of sex hormones and their relation to risk factors for breast cancer: a cross-sectional study in 1092 pre- and postmenopausal women (United Kingdom). Cancer Causes Control 2001;12:47–59.

101 Collaborative Group on Hormonal Factors in Breast Cancer: Breast cancer and hormone replacement therapy: collaborative reanalysis of data from 51 epidemiological studies of 52,705 women with breast cancer and 108,411 women without breast cancer. Lancet 1997;350:1047–1059.

102 Weiss LK, Burkman RT, Cushing-Haugen KL, et al: Hormone replacement therapy regimens and breast cancer risk(1). Obstet Gynecol 2002;100:1148–1158.

103 Onland-Moret NC, Peeters PH, van der Schouw YT, Grobbee DE, van Gils CH: Alcohol and endogenous sex steroid levels in postmenopausal women: a cross-sectional study. J Clin Endocrinol Metab 2005;90:1414–1419.

104 Ursin G, Tseng CC, Paganini-Hill A, et al: Does menopausal hormone replacement therapy interact with known factors to increase risk of breast cancer? J Clin Oncol 2002;20:699–706.

105 Turner RT, Sibonga JD: Effects of alcohol use and estrogen on bone. Alcohol Res Health 2001;25:276–281.

106 Lavigne JA, Wimbrow HH, Clevidence BA, et al: Effects of alcohol and menstrual cycle on insulin-like growth factor-I and insulin-like growth factor binding protein-3. Cancer Epidemiol Biomarkers Prev 2004;13:2264–2267.

107 Etique N, Chardard D, Chesnel A, Merlin JL, Flament S, Grillier-Vuissoz I: Ethanol stimulates proliferation, ERalpha and aromatase expression in MCF-7 human breast cancer cells. Int J Mol Med 2004;13:149–155.

108 Fan S, Meng Q, Gao B, et al: Alcohol stimulates estrogen receptor signaling in human breast cancer cell lines. Cancer Res 2000;60:5635–5639.

109 Enger SM, Ross RK, Paganini-Hill A, Longnecker MP, Bernstein L: Alcohol consumption and breast cancer oestrogen and progesterone receptor status. Br J Cancer 1999;79:1308–1314.

110 Li CI, Malone KE, Porter PL, Weiss NS, Tang MT, Daling JR: The relationship between alcohol use and risk of breast cancer by histology and hormone receptor status among women 65–79 years of age. Cancer Epidemiol Biomarkers Prev 2003;12:1061–1066.

111 Rusiecki JA, Holford TR, Zahm SH, Zheng T: Breast cancer risk factors according to joint estrogen receptor and progesterone receptor status. Cancer Detect Prev 2005;29:419–426.

112 Suzuki R, Ye W, Rylander-Rudqvist T, Saji S, Colditz GA, Wolk A: Alcohol and postmenopausal breast cancer risk defined by estrogen and progesterone receptor status: a prospective cohort study. J Natl Cancer Inst 2005;97:1601–1608.

113 Gapstur SM, Potter JD, Drinkard C, Folsom AR: Synergistic effect between alcohol and estrogen replacement therapy on risk of breast cancer differs by estrogen/progesterone receptor status in the Iowa Women's Health Study. Cancer Epidemiol Biomarkers Prev 1995;4:313–318.

114 Singletary K, McNary M, Odoms A, Nelshoppen J, Wallig M: Ethanol consumption and DMBA-induced mammary carcinogenesis in rats. Nutr Cancer 1991;16:13–21.

115 Feigelson HS, Shames LS, Pike MC, Coetzee GA, Stanczyk FZ, Henderson BE: Cytochrome P450c17alpha gene (CYP17) polymorphism is associated with serum estrogen and progesterone concentrations. Cancer Res 1998;58:585–587.

116 Feigelson HS, Coetzee GA, Kolonel LN, Ross RK, Henderson BE: A polymorphism in the CYP17 gene increases the risk of breast cancer. Cancer Res 1997;57:1063–1065.

117 Dunn BK, Wickerham DL, Ford LG: Prevention of hormone-related cancers: breast cancer. J Clin Oncol 2005;23:357–367.

118 Ye Z, Parry JM: The CYP17 MspA1 polymorphism and breast cancer risk: a meta-analysis. Mutagenesis 2002;17:119–126.

119 Huber JC, Schneeberger C, Tempfer CB: Genetic modelling of the estrogen metabolism as a risk factor of hormone-dependent disorders. Maturitas 2002;42:1–12.

120 Miyoshi Y, Ando A, Hasegawa S, et al: Association of genetic polymorphisms in CYP19 and CYP1A1 with the oestrogen receptor-positive breast cancer risk. Eur J Cancer 2003;39:2531–2537.

121 Zhang Y, Wise JP, Holford TR, et al: Serum polychlorinated biphenyls, cytochrome P-450 1A1 polymorphisms, and risk of breast cancer in Connecticut women. Am J Epidemiol 2004;160:1177–1183.

122 Chacko P, Joseph T, Mathew BS, Rajan B, Pillai MR: Role of xenobiotic metabolizing gene polymorphisms in breast cancer susceptibility and treatment outcome. Mutat Res 2005;581:153–163.

123 Zheng W, Xie DW, Jin F, et al: Genetic polymorphism of cytochrome P450–1B1 and risk of breast cancer. Cancer Epidemiol Biomarkers Prev 2000;9:147–150.

124 Saintot M, Malaveille C, Hautefeuille A, Gerber M: Interaction between genetic polymorphism of cytochrome P450–1B1 and environmental pollutants in breast cancer risk. Eur J Cancer Prev 2004;13:83–86.

125 Thyagarajan B, Brott M, Mink P, et al: CYP1B1 and CYP19 gene polymorphisms and breast cancer incidence: no association in the ARIC study. Cancer Lett 2004;207:183–189.

126 Wen W, Cai Q, Shu XO, et al: Cytochrome P450 1B1 and catechol-O-methyltransferase genetic polymorphisms and breast cancer risk in Chinese women: results from the shanghai breast cancer study and a meta-analysis. Cancer Epidemiol Biomarkers Prev 2005;14:329–335.

127 Sazci A, Ergul E, Utkan NZ, Canturk NZ, Kaya G: Catechol-O-methyltransferase Val 108/158 Met polymorphism in premenopausal breast cancer patients. Toxicology 2004;204:197–202.

128 Mitrunen K, Kataja V, Eskelinen M, et al: Combined COMT and GST genotypes and hormone replacement therapy associated breast cancer risk. Pharmacogen 2002;12:67–72.

129 Le ML, Donlon T, Kolonel LN, Henderson BE, Wilkens LR: Estrogen metabolism-related genes and breast cancer risk: the multiethnic cohort study. Cancer Epidemiol Biomarkers Prev 2005;14: 1998–2003.

130 Modugno F, Zmuda JM, Potter D, et al: Estrogen metabolizing polymorphisms and breast cancer risk among older white women. Breast Cancer Res Treat 2005;93:261–270.

131 Wolfe JN: Breast patterns as an index of risk for developing breast cancer. Am J Roentgenol 1976;126:1130–1137.

132 Byrne C, Schairer C, Wolfe J, et al: Mammographic features and breast cancer risk: effects with time, age, and menopause status. J Natl Cancer Inst 1995;87:1622–1629.

133 Boyd NF, Byng JW, Jong RA, et al: Quantitative classification of mammographic densities and breast cancer risk: results from the Canadian national breast screening study. J Natl Cancer Inst 1995;87:670–675.

134 Vachon CM, Kuni CC, Anderson K, Anderson VE, Sellers TA: Association of mammographically defined percent breast density with epidemiologic risk factors for breast cancer (United States). Cancer Causes Control 2000;11:653–662.

135 Sala E, Warren R, McCann J, Duffy S, Luben R, Day N: High-risk mammographic parenchymal patterns, hormone replacement therapy and other risk factors: a case-control study. Int J Epidemiol 2000;29:629–636.

136 Freedman M, San Martin J, O'Gorman J, et al: Digitized mammography: a clinical trial of postmenopausal women randomly assigned to receive raloxifene, estrogen, or placebo. J Natl Cancer Inst 2001;93:51–56.

137 Pankow JS, Vachon CM, Kuni CC, et al: Genetic analysis of mammographic breast density in adult women: evidence of a gene effect. J Natl Cancer Inst 1997;89:549–556.

138 Vachon CM, King RA, Atwood LD, Kuni CC, Sellers TA: Preliminary sibpair linkage analysis of percent mammographic density. J Natl Cancer Inst 1999;91:1778–1779.

139 Haiman CA, Hankinson SE, De V, et al: Polymorphisms in steroid hormone pathway genes and mammographic density. Breast Cancer Res Treat 2003;77:27–36.

140 Maskarinec G, Lurie G, Williams AE, Le ML: An investigation of mammographic density and gene variants in healthy women. Int J Cancer 2004;112:683–688.

141 Lai JH, Vesprini D, Zhang W, Yaffe MJ, Pollak M, Narod SA: A polymorphic locus in the promoter region of the IGFBP3 gene is related to mammographic breast density. Cancer Epidemiol Biomarkers Prev 2004;13:573–582.

142 Negri E, La Vecchia C, Franceschi S: Re: dietary folate consumption and breast cancer risk. J Natl Cancer Inst 2000;92:1270–1271.

143 Sellers TA, Kushi LH, Cerhan JR, et al: Dietary folate intake, alcohol, and risk of breast cancer in a prospective study of postmenopausal women. Epidemiology 2001;12:420–428.

144 Zhang SM: Role of vitamins in the risk, prevention, and treatment of breast cancer. Curr Opin Obstet Gynecol 2004;16:19–25.

145 Shrubsole MJ, Gao YT, Cai Q, et al: MTHFR polymorphisms, dietary folate intake, and breast cancer risk: results from the Shanghai breast cancer study. Cancer Epidemiol Biomarkers Prev 2004;13:190–196.

146 Sellers TA, Grabrick DM, Vierkant RA, et al: Does folate intake decrease risk of postmenopausal breast cancer among women with a family history? Cancer Causes Control 2004;15: 113–120.

147 Zhang SM, Hankinson SE, Hunter DJ, Giovannucci EL, Colditz GA, Willett WC: Folate intake and risk of breast cancer characterized by hormone receptor status. Cancer Epidemiol Biomarkers Prev 2005;14:2004–2008.

148 Tjonneland A, Christensen J, Olsen A, et al: Folate intake, alcohol and risk of breast cancer among postmenopausal women in Denmark. Eur J Clin Nutr 2006;60:280–286.

149 Giovannucci E: Alcohol, one-carbon metabolism, and colorectal cancer: recent insights from molecular studies. J Nutr 2004;134:2475S–2481S.

150   Laufer EM, Hartman TJ, Baer DJ, et al: Effects of moderate alcohol consumption on folate and vit-
      amin B status in postmenopausal women. Eur J Clin Nutr 2004;58:1518–1524.
151   Blount BC, Mack MM, Wehr CM, et al: Folate deficiency causes uracil misincorporation
      into human DNA and chromosome breakage: implications for cancer and neuronal damage.
      PNAS 1997, pp 3290–3295.
152   Duthie SJ: Folic acid deficiency and cancer: mechanisms of DNA instability. Br Med Bull
      1999;55:578–592.
153   Friso S, Choi SW, Girelli D, et al: A common mutation in the 5,10-methylenetetrahydrofolate
      reductase gene affects genomic DNA methylation through an interaction with folate status. Proc
      Natl Acad Sci USA 2002;99:5606–5611.
154   Branda RF, Brooks EM, Chen Z, Naud SJ, Nicklas JA: Dietary modulation of mitochondrial DNA
      deletions and copy number after chemotherapy in rats. Mutat Res 2002;501:29–36.
155   Kim YI, Shirwadkar S, Choi SW, Puchyr M, Wang Y, Mason JB: Effects of dietary folate on DNA
      strand breaks within mutation-prone exons of the p53 gene in rat colon. Gastroenterology
      2000;119:151–161.
156   Choi SW, Stickel F, Baik HW, Kim YI, Seitz HK, Mason JB: Chronic alcohol consumption induces
      genomic but not p53-specific DNA hypomethylation in rat colon. J Nutr 1999;129:1945–1950.
157   Halsted CH, Villanueva JA, Devlin AM, et al: Folate deficiency disturbs hepatic methionine
      metabolism and promotes liver injury in the ethanol-fed micropig. Proc Natl Acad Sci USA
      2002;99:10072–10077.
158   Chen J, Gammon MD, Chan W, et al: One-carbon metabolism, MTHFR polymorphisms, and risk
      of breast cancer. Cancer Res 2005;65:1606–1614.
159   Bailey LB: Folate, methyl-related nutrients, alcohol, and the MTHFR 677C–>T polymorphism
      affect cancer risk: intake recommendations. J Nutr 2003;133(suppl 1):3748S–3753S.
160   Stern LL, Mason JB, Selhub J, Choi SW: Genomic DNA hypomethylation, a characteristic of most
      cancers, is present in peripheral leukocytes of individuals who are homozygous for the C677T
      polymorphism in the methylenetetrahydrofolate reductase gene. Cancer Epidemiol Biomarkers
      Prev 2000;9:849–853.
161   Le ML, Haiman CA, Wilkens LR, Kolonel LN, Henderson BE: MTHFR polymorphisms, diet,
      HRT, and breast cancer risk: the multiethnic cohort study. Cancer Epidemiol Biomarkers Prev
      2004;13:2071–2077.
162   Friso S, Choi SW: Gene-nutrient interactions and DNA methylation. J Nutr 2002;132
      (suppl):2382S–2387S.
163   Barak AJ, Beckenhauer HC, Tuma DJ: Methionine synthase: a possible prime site of the ethanolic
      lesion in liver. Alcohol 2002;26:65–67.
164   Waly M, Olteanu H, Banerjee R, et al: Activation of methionine synthase by insulin-like growth
      factor-1 and dopamine: a target for neurodevelopmental toxins and thimerosal. Mol Psychiatry
      2004;9:358–370.
165   Harmon DL, Shields DC, Woodside JV, et al: Methionine synthase D919G polymorphism is a sig-
      nificant but modest determinant of circulating homocysteine concentrations. Genet Epidemiol
      1999;298–309.
166   Matsuo K, Suzuki R, Hamajima N, et al: Association between polymorphisms of folate- and
      methionine-metabolizing enzymes and susceptibility to malignant lymphoma. Blood 2001;97:
      3205–3209.
167   Ma J, Stampfer MJ, Christensen B, et al: A polymorphism of the methionine synthase gene:
      association with plasma folate, vitamin B-12, homocyst(e)ine, and colorectal cancer risk. Cancer
      Epidemiol Biomarkers Prev 1999;8:825–829.
168   Paz MF, Avila S, Fraga MF, et al: Germ-line variants in methyl-group metabolism genes and sus-
      ceptibility to DNA methylation in normal tissues and human primary tumors. Cancer Res
      2002;62:4519–4524.
169   Leclerc D, Campeau E, Goyette P, et al: Human methionine synthase: cDNA cloning and identifi-
      cation of mutations in patients of the cblG complementation group of folate/cobalamin disorders.
      Hum Mol Genet 1996;5:1867–1874.
170   Vogelstein B, Kinzler KW: Achilles' heel of cancer? Nature 2001;412:865–866.

171 Olivier M, Hainaut P: TP53 mutation patterns in breast cancers: searching for clues of environmental carcinogenesis. Semin Cancer Biol 2001;11:353–360.

172 Hartmann A, Blaszyk H, Kovach JS, Sommer SS: The molecular epidemiology of p53 gene mutations in human breast cancer. Trends Genet 1997;13:27–33.

173 Shields PG, Harris CC: Cancer risk and low penetrance susceptibility genes in gene-environment interactions. J Clin Oncol 2000;18:2309–2315.

174 Miller LD, Smeds J, George J, et al: An expression signature for p53 status in human breast cancer predicts mutation status, transcriptional effects, and patient survival. Proc Natl Acad Sci USA 2005;102:13550–13555.

175 Pfeifer GP: p53 mutational spectra and the role of methylated CpG sequences. Mutat Res 2000;450:155–166.

176 Sommer SS, Cunningham J, McGovern RM, et al: Pattern of p53 gene mutations in breast cancers of women of the midwestern United States. J Natl Cancer Inst 1992;84:246–252.

177 Blaszyk H, Vaughn CB, Hartmann A, et al: Novel pattern of p53 gene mutations in an American black cohort with high mortality from breast cancer. Lancet 1994;343:1195–1197.

178 Lambrinakos A, Yakubovskaya M, Babon JJ, et al: Novel TP53 gene mutations in tumors of Russian patients with breast cancer detected using a new solid phase chemical cleavage of mismatch method and identified by sequencing. Hum Mutat 2004;23:186–192.

179 Lai H, Lai S, Ma F, Meng L, Trapido E: Prevalence and spectrum of p53 mutations in white Hispanic and non-Hispanic women with breast cancer. Breast Cancer Res Treat 2003;81:53–60.

180 Nagai MA, Schaer BH, Zago MA, et al: TP53 mutations in primary breast carcinomas from white and African-Brazilian patients. Int J Oncol 2003;23:189–196.

181 Chen FM, Hou MF, Wang JY, et al: High frequency of G/C transversion on p53 gene alterations in breast cancers from Taiwan. Cancer Lett 2004;207:59–67.

182 Freudenheim JL, Bonner M, Krishnan S, et al: Diet and alcohol consumption in relation to p53 mutations in breast tumors. Carcinogenesis 2004;25:931–939.

183 Gammon MD, Hibshoosh H, Terry MB, et al: Cigarette smoking and other risk factors in relation to p53 expression in breast cancer among young women. Cancer Epidemiol Biomarkers Prev 1999;8:255–263.

184 Simao TA, Ribeiro FS, Amorim LM, et al: TP53 mutations in breast cancer tumors of patients from Rio de Janeiro, Brazil: association with risk factors and tumor characteristics. Int J Cancer 2002;101:69–73.

185 van der KK, Rookus MA, Peterse HL, Van Leeuwen FE: p53 protein overexpression in relation to risk factors for breast cancer. Am J Epidemiol 1996;144:924–933.

186 Denissenko MF, Chen JX, Tang MS, Pfeifer GP: Cytosine methylation determines hot spots of DNA damage in the human P53 gene. Proc Natl Acad Sci USA 1997;94:3893–3898.

187 Shen JC, Rideout WM III, Jones PA: The rate of hydrolytic deamination of 5-methylcytosine in double-stranded DNA. Nucleic Acids Res 1994;22:972–976.

188 Garro AJ, McBeth DL, Lima V, Lieber CS: Ethanol consumption inhibits fetal DNA methylation in mice: implications for the fetal alcohol syndrome. Alcohol Clin Exp Res 1991;15:395–398.

Prof. Peter G. Shields, MD, Dir.
Lombardi Cancer Center, Georgetown University Medical Center
3800 Reservoir Rd. NW, LL (s) Level, Room 150, Box 571465
Washington, DC 20057–1465 (USA)
Tel. +1 202 687 0003, Fax +1 202 687 0004
E-Mail pgs2@georgetown.edu

Cho CH, Purohit V (eds): Alcohol, Tobacco and Cancer.
Basel, Karger, 2006, pp 140–159

....

Alcohol and Tobacco Smoke in Retinoid Metabolism and Signaling: Implications for Carcinogenesis

Heather Mernitz, Xiang-Dong Wang

Nutrition and Cancer Biology Laboratory, Jean Mayer United States Department of
Agriculture Human Nutrition Research Center on Aging at Tufts University,
Boston, Mass., USA

Abstract

Considerable evidence demonstrates that retinoids (retinol, retinoic acid, and retinyl
ester, which are also the important metabolites from provitamin A carotenoids), may be
effective in the prevention and treatment of a variety of human chronic diseases, including
cancer. Substantial work has been done investigating the mechanisms by which tobacco
smoke and excessive alcohol intake interfere with retinoid metabolism and signaling.
Exposure to cigarette smoke subjects tissues to increased reactive oxygen species, which can
induce cytochrome P450 enzymes and result in the degradation of retinoic acid, the bioactive
form of vitamin A. Further, oxidative stress can result in cleavage of β-carotene at positions
other than the central double bond, decreasing the production of retinoic acid from this vitamin A precursor. This leads to aberrant retinoid signaling through nuclear retinoid receptors,
while at the same time cigarette smoke also causes dysregulated signaling through the
mitogen-activated protein kinase signaling pathways. Alcohol acts as a competitive inhibitor
of vitamin A oxidation to retinoic acid involving alcohol dehydrogenases and acetaldehyde
dehydrogenases, induces cytochrome P450 enzymes (particularly CYP2E1) that degrade
retinol and retinoic acid, and alters retinoid homeostasis by increasing vitamin A mobilization from liver to extrahepatic tissues. Moreover, this alcohol-impaired retinoid homeostasis
interferes with retinoic acid signaling by decreasing target gene expression and interfering
with retinoic acid cross-talk with the mitogen-activated protein kinase pathways. The overall
effect of both cigarette smoke and chronic, excessive alcohol intake is dysregulated apoptosis and uncontrolled cellular proliferation, which can act to promote the process of carcinogenesis. Nutritional interventions that serve to restore normal retinoid signaling and
functioning may offer protection at the cellular level and represent a means to modify cancer
risk in high-risk human populations.

Chronic and excessive alcohol intake is associated with an increased risk of several cancers (e.g. oral cavity, larynx, esophagus, liver, lung, colorectal, and breast) [1–8] and tobacco smoke is clearly the leading risk factor for lung cancer and a contributing factor in several other cancers (e.g. oral cavity, larynx, esophagus, pancreas, stomach, colon, kidney, cervix, and bladder) [4, 9]. Carcinogenesis is a multistage process consisting of initiation, promotion, and progression. Initiation is rapid and occurs with high frequency (e.g. with exposure to a carcinogen, such as high levels of acetaldehyde from excessive alcohol intake and 4-(N-Methyl-N-nitrosamino)-1-(3-pyridyl)-1-butanone/polycyclic aromatic hydrocarbons from tobacco smoke). Promotion is a long-term process that requires chronic exposure to a tumor promoter (e.g. nutritional deficiencies induced by chronic alcohol consumption or smoke-related biochemical and molecular alterations). Thus, multiple mechanisms act together to play a role in carcinogenesis at both initiation and promotion stages. One mechanism by which alcohol and tobacco smoke may contribute to carcinogenesis is through alteration of vitamin A metabolism and signaling. Retinoids, naturally occurring and synthetic vitamin A metabolites and analogs, play an important role in regulating cellular growth and differentiation [10]. Vitamin A deficiency has been associated with increased incidence of cancer in epidemiological and animal studies [11, 12], and aberrant expression of nuclear retinoid receptors has been documented in various types of cancerous and pre-cancerous lesions (reviewed in [13]). In this chapter, we will first summarize normal retinoid metabolism and signaling and then go on to discuss alterations in retinoid metabolism and aberrant retinoid signaling in the presence of tobacco smoke exposure and chronic, excessive alcohol consumption.

Retinoid Metabolism and Mode of Action

Retinoid Metabolism

Vitamin A can be consumed directly from the diet, usually in the form of retinol or retinyl esters from meat and dairy foods, or made by the body through enzymatic cleavage of certain carotenoids, the darkly colored pigments found in fruits and vegetables [14]. Provitamin A carotenoids (β-carotene, α-carotene, and β-cryptoxanthin) are absorbed intact and can be cleaved within intestinal cells by a central cleavage enzyme, carotene-15,15′-monooxygenase, to form to retinal, which is then reduced to retinol by intestinal retinal reductase [15, 16]. Further, an excentric cleavage pathway has been shown in which additional carotene monooxygenases cleave carotenoids at positions outside the central double bond to form β-apocarotenals, which can be oxidized to β-apocarotenoic acids and then further oxidized through a β-oxidation-like process to form retinoic acid [17–19].

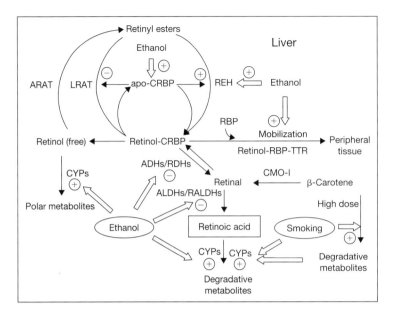

Fig. 1. Simplified schematic illustration of the possible effects of tobacco smoke and chronic, excessive alcohol intake on hepatic retinoid metabolism (see text for details). ADH = Alcohol dehydrogenase; ALDH = aldehyde dehydrogenase; ARAT = acyl-CoA:retinol acyltransferase; CMO-I = carotene-15,15′-monooxygenase; CRBP = cellular retinol-binding protein; CYPs = cytochrome P450 enzymes; LRAT = lecithin:retinol acyltransferase; RALDH = retinal dehydrogenase; RBP = plasma retinol-binding protein; RDH = retinol dehydrogenase; REH = retinyl ester hydrolase; TTR = transthyretin. Adapted with permission from reference [60].

Vitamin A is stored mainly in the liver in the form of retinyl esters, and can undergo hydrolysis to all-*trans*-retinol, the circulating form of vitamin A in the body (fig. 1) [14]. The majority of retinol, retinal, and retinoic acid in the body are found bound to retinoid binding proteins, which appear to facilitate optimal transport and metabolism of the retinoids [20]. Retinol can be oxidized to retinal by short-chain dehydrogenase/reductases and medium chain alcohol dehydrogenases (ADHs) [21]. Retinal is then irreversibly oxidized by to form retinoic acid by several retinal dehydrogenases (RALDHs), members of the aldehyde dehydrogenase (ALDH) family [22]. The all-*trans* and 9-*cis*-isomers of retinoic acid are the transcriptionally active forms of vitamin A that bind to nuclear retinoid receptors, ligand-inducible trans-activating transcription factors, to exert effects on cellular proliferation and differentiation. Retinol and retinoic acid can be further oxidized by an array of cytochrome P450 enzymes (CYPs) to degradative metabolites, including various oxo- and hydroxy-forms. Conjugation with glucuronic acid can give rise to retinol- and retinoic acid-glucuronides, water-soluble metabolites that help eliminate retinoids from the body [14]. Several groups have reported the existence of a retinoid-inducible cytochrome P450-related retinoic acid hydroxylase, CYP26, essential for retinoid catabolism [23–26].

Retinoid Function and Mode of Action

Retinoids are known to regulate a large number of biologic processes including reproduction and embryonic development, vision, bone formation, metabolism, hematopoiesis, differentiation, proliferation, and apoptosis [27–31]. The mechanism by which retinoids are able to elicit these effects lies in their ability to regulate gene expression at specific target sites within the body. Both retinoic acid receptors (RARα, RARβ, and RARγ) and retinoid X receptors (RXRα, RXRβ, and RXRγ) function as transcription factors, regulating gene expression by binding as dimeric complexes to the retinoic acid response element and the retinoid X response element located in the 5′ promoter region of susceptible genes [32]. Ligand binding causes a conformational change in the retinoid receptors, allowing for the dissociation of corepressors (e.g. NCoR, SMRT, histone deacetylase-containing complexes) and recruitment of coactivators (e.g. CBP/p300, ACTR, DRIP/TRAP), some of which have histone acetyltransferase activity for chromatin decondensation and others which establish contact with the basal transcriptional machinery [33, 34]. All-*trans*-retinoic acid can bind and transactivate only RAR, whereas 9-*cis*-retinoic acid can bind and transactivate both RAR and RXR [35].

The ability of retinoids to modulate cell growth and proliferation has led to investigation into their potential use as chemopreventive and chemotherapeutic agents against cancer [13, 32, 34]. Retinoids have been shown to suppress carcinogenesis in a variety of tissues including the skin, bladder, prostate, breast, and lung [28, 31, 36, 37]. Loss of RARβ expression is associated with tumor progression in a variety of cancers, and expression of RARβ is associated with the growth-inhibitory effects of retinoids [13, 34]. One possible mechanism behind this effect may involve cross-talk with the activator protein-1 (AP-1, the c-Jun and c-Fos heterodimer) signaling pathway (fig. 2). RARβ constitutively represses AP-1, and other retinoid receptors have been shown to cross-talk with AP-1 in an inhibitory manner as well [34]. AP-1 target genes are involved in inflammation and cellular proliferation, so inhibition of AP-1 activity may be one explanation behind the antiproliferative effects of retinoids. Additional mechanisms may include retinoid-induced ubiquitination and proteolysis of G1 cyclins and/or regulation of cyclin-dependent kinase inhibitors, both leading to cell cycle arrest [38–42].

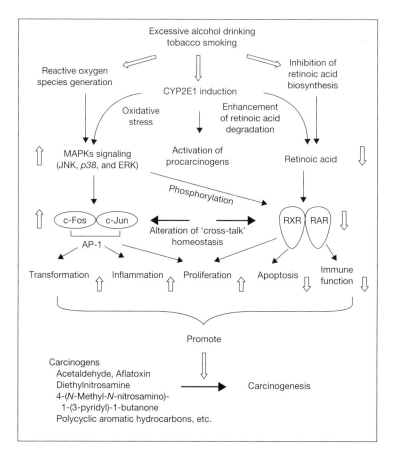

Fig. 2. Simplified Schematic illustration of the possible effects of tobacco smoke and chronic, excessive alcohol intake on retinoid signal transduction (see text for details). AP-1 = Activator-protein-1; CYP2E1 = cytochrome P450 2E1; ERK = extracellular-signal-regulated protein kinase; JNK = jun N-terminal kinase; MAPKs = mitogen activated protein kinases; RAR = retinoic acid receptor; RXR = retinoid X receptor. Adapted with permission from reference [60].

Alteration of Retinoid Metabolism by Tobacco Smoke and Alcohol Consumption

Cigarette Smoking Increases Oxidative Degradation of
β-Carotene and Retinoic Acid

Cigarette smoking is associated with substantially decreased plasma levels of β-carotene, despite only slightly lower intakes of carotenoids in smokers compared to nonsmokers [43–47]. This suggests that plasma levels of β-carotene among smokers may be reduced below levels due to dietary differences alone [47]. Several hypotheses have been proposed to explain this increased metabolism of carotenoids in the tissues of smokers, including increased induction of metabolic enzymes, excentric cleavage of β-carotene into harmful oxidative products, and oxidative degradation of cellular antioxidants (e.g. ascorbic acid, α-tocopherol) that normally serve to stabilize the reduced form of β-carotene (fig. 1).

We have shown that the free radical-rich environment in smoke-exposed lungs alters β-carotene metabolism and produces undesirable oxidative metabolites [48–50], which can facilitate the binding of metabolites of benz[a]pyrene to DNA [51], down-regulate RAR, [48], up-regulate AP-1 activity [50], induce carcinogen-activating enzymes [52], enhance the induction of BALB/c 3T3 cell transformation by benz[a]pyrene [53], inhibit gap junction communication in A549 lung cancer cells [54], and impair mitochondrial functions [55]. Exposure to cigarette smoke enhances oxidative excentric cleavage of β-carotene in ferret lung post-nuclear fractions after incubation with $10 \mu M$ β-carotene [48], and studies in our lab and others have demonstrated that the oxidative excentric cleavage products formed are involved in the carcinogenic process [51, 53, 56]. The mechanism involves induction of cytochrome P450 enzymes, CYP1A1 and 1A2, by carotenoid oxidative metabolites [56, 57], and increased activation of carcinogens leading to increased formation of carcinogen-DNA adducts [51]. The covalent modification of DNA bases by chemicals can alter the structure and biological processing of the DNA by cellular proteins governing replication, transcription, and repair. If not repaired or repaired incorrectly, these modifications may eventually lead to mutations and ultimately cancer, especially if the adduct is located in an oncogene or tumor suppressor gene.

We have observed that the combination of ascorbic acid, α-tocopherol, and β-carotene can inhibit the smoke-enhanced production of excentric cleavage metabolites of β-carotene, increase the formation of retinal and retinoic acid, and decrease the smoke-induced catabolism of retinoic acid [56, 58], and these combined antioxidants may provide protection against lung cancer in vivo by maintaining normal levels of retinoic acid [59]. These studies and the known biochemical interactions of β-carotene, vitamin E, and vitamin C suggest that the combination of nutrients, rather than individual agents, could be an effective chemopreventive strategy against lung cancer in smokers.

Alcohols and Retinoids Compete for Metabolic
Dehydrogenase Enzymes

Interference with vitamin A nutritional status and metabolism is one of the major alterations caused by chronic alcohol intake [60], and daily consumption of alcohol is related to reduced β-carotene levels, independent of smoking status [43, 47]. Further, we have shown that high doses of ethanol significantly reduced both plasma and liver retinoic acid levels in rats, compared to rats pair-fed a diet matched for energy and vitamin A content [61].

Alcohol and retinol metabolism involve a similar two-step oxidation process. Both are first oxidized to their aldehyde forms, which are subsequently oxidized to their respective carboxylic acids. Although the oxidation of retinol is catalyzed by both cytosolic and microsomal retinol dehydrogenase, ADHs and ALDHs are also implicated in the conversion of vitamin A to retinoic acid [62]. The reaction velocity for retinol conversion to retinal, the rate-limiting step in the synthesis of retinoic acid, is dramatically reduced in the presence of alcohol [22]. Competitive inhibition by alcohol is further demonstrated by the finding that biosynthesis of retinoic acid after retinol administration was reduced by 82% in *ADH* null mutant mice (*ADH1$^{-/-}$*) and 87% in wild-type mice treated with ethanol [63]. In addition, it has been reported that ethanol inhibits the oxidation of retinol into retinoic acid in human gastric and esophageal mucosa and in rat liver and colon mucosa, those primary organs affected by cancer in alcoholic patients [63–67]. These findings clearly demonstrate that retinoic acid biosynthesis can be impaired by ethanol through competition for ADH and ALDH enzymes, which may contribute to the increased risk for developing alcohol-related cancers (fig. 1).

The metabolism of ethanol leads to the generation of acetaldehyde, which is carcinogenic and mutagenic and binds to DNA and proteins, resulting in secondary hyperproliferation [68, 69]. Heavy alcohol drinking and smoking have previously been shown to synergistically increase salivary acetaldehyde, the first metabolite of ethanol and a constituent of tobacco smoke [70, 71]. Acetaldehyde has also been shown to inhibit the generation of retinoic acid from vitamin A in human prenatal tissue [72]. Inactivating polymorphisms of *ALDH2* lead to significantly slowed acetaldehyde metabolism and are strongly associated with esophageal squamous cell carcinoma in Asian drinkers [73]. Patients with *ALDH2* polymorphisms also have a higher risk for multiple cancers, especially of the oropharynx and stomach [74]. When the *ALDH2* polymorphism occurs in combination with a less-active form of *ADH2*, this risk is further increased [75]. For Caucasians, *ADH1* polymorphisms (*ADH1C*1*) resulting in increased enzyme activity and increased acetaldehyde production lead to a higher risk of upper aerodigestive tract and liver cancers in individuals considered to be heavy alcohol drinkers [76]. However, whether the presence of inactivating *ADH* and *ALDH* polymorphisms leads to slower conversion of retinol to retinal and/or retinal to retinoic acid is currently unknown.

Alcohol Enhances Catabolism of Vitamin A by
Inducing Cytochrome P450 2E1

Chronic alcohol ingestion has been shown to upregulate cytochrome P450 2E1 (CYP2E1), which is associated with the production of reactive oxygen species and enhanced activation of procarcinogens present in alcoholic beverages, in tobacco smoke, and in the diet [68, 77]. Further, CYP enzyme induction leads to generation of acetaldehyde in peripheral tissues [77], and can lead to excessive catabolism of retinoic acid [56]. Chronic alcohol intake has been shown to increase catabolism of retinol and retinoic acid into more polar metabolites in the liver [78]. Work from our laboratory has shown that the enhanced catabolism of retinol and retinoic acid in alcohol-fed rats can be inhibited by chlormethiazole, an inhibitor of CYP2E1 both in vitro and in vivo [79, 80], indicating that CYP2E1 is the major enzyme responsible for the alcohol-enhanced catabolism of retinoids in hepatic tissue after exposure to alcohol (fig. 1). It is possible that CYP2E1 enzyme induction in chronic intermittent drinking could continue to be a factor mediating oxidative stress, destroying retinol and retinoic acid even after alcohol is cleared. The finding that CYP2E1 is also present and inducible by alcohol in the esophagus, forestomach, and surface epithelium of the proximal colon [81] may explain why chronic and excessive alcohol intake is a risk not only for hepatic, but also for extrahepatic cell proliferation and carcinogenesis. Treatment with CYP2E1 inhibitors, such as chlormethiazole, have been shown to protect against alcohol-induced liver injury [82–84]. Whether the restoration of hepatic vitamin A status by chlormethiazole contributes to this protective effect needs further study. In addition, high doses of vitamin A supplementation over a period of months are hepatotoxic and chronic alcohol consumption enhances this intrinsic hepatotoxicity [85]. Therefore, a better understanding of the alcohol–retinoid interaction and the molecular mechanism(s) involved is needed before pursuing retinoids in the prevention of alcohol-related carcinogenesis in humans, particularly with respect to the detrimental effects of polar metabolites of vitamin A, which have been reported to cause hepatocyte death via loss of mitochondrial membrane potential [86].

Alcohol Increases Vitamin A Mobilization from Liver
to Extrahepatic Tissues

Alcohol has been shown to increase vitamin A mobilization from the liver to other organs (fig. 1) [87, 88]. While the exact mechanisms remain unclear, several possible explanations have been proposed. First, it is well known that cellular retinol-binding protein (CRBP) is necessary to maintain normal retinyl ester synthesis and storage. Alcohol treatment has been reported to increase expression of CRBP mRNA in the liver and snout of rats [80, 89]. Elevated apo-CRBP (unbound to retinol) has a stimulatory effect on retinyl ester hydrolysis and an inhibitory effect on retinol esterification, particularly in vitamin A deficiency [20]. Therefore, it is possible that elevated apo-CRBP levels in the liver of alcohol-fed rats could stimulate hepatic retinyl ester hydrolase. This hypothesis is supported by findings that the reduction of hepatic retinyl esters in alcohol-fed rats is not due to impairment of retinol esterification [80], and the hydrolysis of retinyl ester can be stimulated by alcohol in rat liver in vitro [90]. Because the oxidation of retinol into retinal and retinoic acid is inhibited by alcohol and the distribution of retinol between retinol-binding protein (RBP) in plasma and CRBP in cytosol is in equilibrium, retinol produced from alcohol-accelerated hydrolysis of retinyl ester in liver could be transported into peripheral tissues. Another hypothesis is that low levels of retinoic acid in the liver and peripheral tissues after chronic alcohol exposure may function as a feedback signal to regulate hepatic vitamin A metabolism [91]. This concept is supported by the demonstration that the restoration of plasma retinoic acid levels to a normal level by chlormethiazole in alcohol-fed rats was correlated with a reduction in retinyl palmitate levels in the plasma and an increase in the retinyl palmitate in the liver [80]. Further, retinoic acid treatment increases retinol esterification in the liver of vitamin A-deficient mice and rats [92]. However, the restoration of retinoic acid levels by chlormethiazole treatment did not increase the expression or activity of hepatic lecithin:retinol acyltransferase in alcohol-fed rats [80]. This finding is consistent with a previous observation that retinoic acid treatment did not increase hepatic concentrations of retinyl ester in alcohol-fed rats [64]. It seems that, in the situation of excessive and chronic alcohol consumption, the mechanism of this regulation of retinoic acid is more complicated and involves multiple factors, such as regulation of enzymes, lipoprotein secretion, and CRBP and RBP function. More studies are needed to determine the exact mechanism of alcohol-induced mobilization of hepatic vitamin A.

Alteration of Retinoid Signaling Pathways by Alcohol and Smoking

Alcohol and Smoke Exposure Diminish Expression of Retinoid-Responsive Genes

Studies have shown that decreased expression of retinoid receptors is a frequent event in non-small cell lung cancer, with reductions in RARβ expression noted in a variety of cancer cells and tissues [93–97]. Down-regulation of retinoid receptors could interfere with retinoid signal transduction, resulting in enhanced cell proliferation and dysregulated apoptosis and, potentially, malignant transformation (fig. 2). Further, recent evidence suggests that the RARβ subtypes, RARβ2 and RARβ4, may have contrasting biological effects (tumor suppressor and tumor promoter, respectively) in human carcinogenesis [98]. The down-regulation of all retinoid subclasses suggests a fundamental dysregulation of the retinoid pathway in lung cancer [93]. Conversely, restoration of RARβ2 in a RARβ-negative lung cancer cell line has been reported to inhibit tumorigenicity in nude mice [99] and retinoic acid can reverse benzy[*a*]pyrene diol epoxide suppression of RARβ protein by increasing transcription of RARβ in immortalized esophageal epithelial cells [100] and lung cancer cells [101]. In a small human trial, daily treatment with 9-*cis*-retinoic acid for three months was able to restore RARβ expression in the bronchial epithelium of former smokers [102].

Provitamin A carotenoids, such as β-carotene and its excentric cleavage metabolites, can serve as direct precursors for all-*trans*- and 9-*cis*-retinoic acid [19, 103]. We have demonstrated that low dose β-carotene supplementation (equivalent to 6 mg/day in humans) partially prevented the decrease in lung retinoic acid levels and inhibited the smoke-induced lung precancerous lesions observed in smoke-exposed ferrets [50]. Previously, β-carotene supplementation was shown to inhibit lung carcinogenesis in A/J mice [104] and prevent skin carcinoma formation by upregulating RARβ [105]. We have observed that the down-regulation of RARβ by smoke-borne carcinogens was completely reversed by treatment with either β-carotene or its oxidative metabolite, apo-14′-carotenoic acid, in normal bronchial epithelium cells [106]. We further demonstrated that the transactivation of the RARβ2 promoter by β-apo-14'-carotenoic acid appears to occur, in large part, via metabolism to all-*trans*-retinoic acid [106]. The molecular mode of the action of provitamin A carotenoids can be mediated by retinoic acid transcriptionally activating a series of genes with distinct antiproliferative or proapoptotic activity or by induction of apoptosis, eliminating cells with unrepairable alterations in the genome or killing neoplastic cells.

Expression of RARβ mRNA is down-regulated by ethanol, even in the presence of retinol [107], and RARβ expression is low in tumorigenic hepatoma cell lines [108, 109]. In our studies, we found that hepatic expression of RARα, RARβ, and RARγ were not affected by chronic alcohol feeding, but retinoic acid concentrations were significantly lower in the alcohol-fed rats [61]. It is possible that expression of RARs is not affected unless severe retinoic acid deficiency occurs. However, lower retinoic acid concentrations may cause a functional decrease in retinoid receptor activity due to low levels of ligand, and this could potentially diminish retinoid signal transduction. This result is supported by the observation that supplemental all-*trans*-retinoic acid in alcohol-fed rats greatly increased the levels of retinoid-responsive gene, mitogen-activated kinase phosphatase-1 (MKP-1), in liver tissue [110]. This induction of MKP-1 attenuated the alcohol-induced phosphorylation of Jun N-terminal kinases (JNKs), which have been shown to be required for tumorigenesis, in a multistep carcinogenesis model in mice lacking the *JNK2* gene [111].

Smoke Exposure and Chronic Ethanol Consumption
Interfere with Retinoic Acid Cross-Talk with the
Mitogen-Activated Protein Kinase Signaling Pathway

Cigarette smoke exposure is a major risk factor for lung cancer since it promotes genomic instability and the development of neoplasia by modulating molecular pathways involved in cell differentiation, cell proliferation and apoptosis. Jun N-terminal kinase (JNK), extracellular-signal-regulated protein kinase (ERK), and *p38* mitogen-activated protein kinase belong to mitogen-activated protein kinase (MAPK) family that are activated by phosphorylation in response to extracellular stimuli and environmental stress, and may play an important role in carcinogenesis (fig. 2) [112, 113]. It has been reported that components of smoke or smoke exposure itself can increase the phosphorylation of JNK and ERK in cell models [114, 115]. JNK was shown to phosphorylate c-Jun on sites Ser-63 and Ser-73 and increase AP-1 transcription activity, mediating cell proliferation and apoptosis [112, 113]. AP-1, a transcription factor formed as a c-Jun homodimer or heterodimer with c-Fos, mediates signals from a variety of sources of proliferative stimuli including growth factors, cytokines, and oxidative stress. It has been shown that components of AP-1 are important in modulating carcinogenesis, and transactivation of AP-1-dependent genes is required for tumor promotion [116]. On the other hand, MAPK phosphatases (MKPs), a family of dual-specificity protein phosphatases, can dephosphorylate both phospho-threonine and phospho-tyrosine residues to inactivate JNK, ERK and *p38* both in vitro and in vivo [117, 118]. It has been shown that phosphorylated-JNK, phosphorylated-ERK, and phosphorylated-*p38* are preferred substrates for MKP-1 in vivo [117, 118].

Previously, we observed that AP-1 expression was up-regulated in the lungs of smoke-exposed ferrets with or without β-carotene supplementation (fig. 2) [48]. This over-expression of AP-1 was positively associated with increased levels of cyclin D1 protein and squamous metaplasia in the lungs of animals exposed to tobacco smoke [48]. It is conceivable that chronic excess β-carotene intake may modulate MAPK signaling and cause abnormal cell cycle regulation, promoting carcinogenesis. This hypothesis is supported by our recent observation that smoke exposure, high dose β-carotene, and their combination activated the phosphorylation of JNK and *p38*, and significantly reduced lung MKP-1 protein levels [119]. In contrast, low dose β-carotene attenuated smoke-induced JNK phosphorylation by preventing down-regulation of MKP-1 due to smoke exposure [119]. The mechanism behind this effect could be related to increased lung retinoic acid levels with low dose β-carotene supplementation, since it has been reported that retinoic acid can inhibit phosphorylation of MAPKs, such as JNK and ERK, by upregulation of MKP-1 [110, 120, 121]. These data may help to explain the conflicting results of the negative human β-carotene intervention trials (which used high doses of β-carotene [122, 123]) vs. the positive observational epidemiological studies showing that diets high in fruits and vegetables containing β-carotene (but at much lower concentrations than in the intervention studies and with other antioxidants present) are associated with a decreased risk for lung cancer [124–126].

Chronic alcohol intake leads to increased cellular proliferation in various tissues, such as liver, colon, and rectum [64, 66, 127]. Such hyperproliferation predisposes development of genetic instability and cancer development by increasing the number of cellular divisions. Several mechanisms contribute to increased cellular turnover after acute and chronic alcohol intake. One effect of alcohol intake on proliferative signaling pathways within the cells involves alteration of the MAPK pathway and its downstream cascades (e.g. activation of JNK and ERK increases expression and activity of c-Jun and c-Fos, fig. 2). Evidence has accumulated supporting a role for alcohol in the regulation of AP-1 gene expression. We have shown that chronic alcohol intake in rats significantly increases hepatic c-Jun and c-Fos protein levels [61]. Alcohol also activates AP-1 in HepG2 cells [128] and rat pancreatic stellate cells [129]. In transformed hepatocytes, alcohol administration leads to activation of ERK and increased DNA synthesis, and it enhances the MAPK activation after G protein signaling [130].

Retinoid receptors and AP-1 can interfere with each others' activities [131]. For example, all three RAR subtypes (RARα, RARβ, and RARγ) can effectively inhibit phorbol ester-induced AP-1 activity in either a retinoic acid-dependent or retinoic acid-independent manner [132]. Retinoic acid treatment in alcohol-fed rats has been shown to dramatically inhibit the alcohol-induced over-expression of c-Jun and cyclin D1, AP-1 DNA-binding activities, and phosphorylation of JNK [64, 110]. Because the transactivation of AP-1–dependent genes is required for tumor promotion [116] and cyclin D1 plays an important role in tumorigenesis and tumor progression in hepatocellular carcinoma [133], the identification of c-Jun and cyclin D1 as two potential targets of retinoic acid action in alcohol-fed rats indicates that retinoids plays an important role in the prevention of certain types of alcohol-promoted cancer. In addition, the down-regulation of MKP-1, which allows for increased levels of phosphorylated-JNK in alcohol-fed rats, can be reversed by retinoic acid supplementation [110]. This further supports the notion that JNK signaling may mediate alcohol-promoted hepatocyte proliferation and oncogenic transformation, owing to alcohol-impaired retinoic acid action and cross-talk with the JNK signaling pathway. It is also possible that the inhibition of JNK activation by retinoic acid may help to 'rescue' the functions of RARs. It has been recently reported that activation of JNK contributes to RAR dysfunction by phosphorylating RARα and inducing degradation through the ubiquitin-proteasomal pathway [134].

Conclusions

Disruption in retinoid metabolism and signaling may play a key role in the process of carcinogenesis. An understanding of the metabolic and molecular details behind the altered homeostasis of retinoids by smoke and alcohol may yield insights into pathophysiologic processes in related cancers. In animal studies, it seems that restoration of retinoic acid homeostasis could help to maintain normal cell proliferation and apoptosis and may be useful in the prevention and treatment of neoplastic transformations. However, a better understanding of the interactions between retinoids, cigarette smoke, and alcohol and the molecular mechanisms that underlie these interactions is needed before retinoids can be pursued in the prevention of carcinogenesis in humans. While no studies have specifically addressed the combined effects of smoke exposure and chronic, excessive ethanol consumption on retinoid homeostasis, there is reason to believe that these two conditions may interact additively or synergistically to inhibit the biosynthesis of retinoic acid by increasing acetaldehyde concentrations and to increase the catabolism of retinoic acid by inducing cytochrome P450 enzymes, thus compounding cancer risk at multiple sites. Future research should focus on the combined effects of alcohol and smoke and the ability of supplemental retinoids to delay or reverse these effects. However, special attention should be paid to the potential detrimental effects of polar metabolites of retinoids generated during smoke exposure and alcohol drinking. In considering the efficacy and complex biological functions of carotenoids and retinoids in human cancer prevention, nutritional intervention using combined agents that target different signaling pathways to protect tissues against smoke-borne chemical carcinogens, oxidative free radical damage, and alcohol-related insult would be an effective chemopreventive strategy to modify cancer risk.

References

1   Bagnardi V, Blangiardo M, La Vecchia C, Corrao G: Alcohol consumption and the risk of cancer: a meta-analysis. Alcohol Res Health 2001;25:263–270.
2   Donato F, Tagger A, Gelatti U, Parrinello G, Boffetta P, Albertini A, Decarli A, Trevisi P, Ribero ML, Martelli C, Porru S, Nardi G: Alcohol and hepatocellular carcinoma: the effect of lifetime intake and hepatitis virus infections in men and women. Am J Epidemiol 2002;155:323–331.
3   Kuper H, Tzonou A, Kaklamani E, Hsieh CC, Lagiou P, Adami HO, Trichopoulos D, Stuver SO: Tobacco smoking, alcohol consumption and their interaction in the causation of hepatocellular carcinoma. Int J Cancer 2000;85:498–502.
4   Freudenheim JL, Graham S, Byers TE, Marshall JR, Haughey BP, Swanson MK, Wilkinson G: Diet, smoking, and alcohol in cancer of the larynx: a case-control study. Nutr Cancer 1992;17:33–45.
5   Korte JE, Brennan P, Henley SJ, Boffetta P: Dose-specific meta-analysis and sensitivity analysis of the relation between alcohol consumption and lung cancer risk. Am J Epidemiol 2002;155: 496–506.
6   Freudenheim JL, Ritz J, Smith-Warner SA, Albanes D, Bandera EV, van den Brandt PA, Colditz G, Feskanich D, Goldbohm RA, Harnack L, Miller AB, Rimm E, Rohan TE, Sellers TA, Virtamo J,

Willett WC, Hunter DJ: Alcohol consumption and risk of lung cancer: a pooled analysis of cohort studies. Am J Clin Nutr 2005;82:657–667.

7   Smith-Warner SA, Spiegelman D, Yaun SS, van den Brandt PA, Folsom AR, Goldbohm RA, Graham S, Holmberg L, Howe GR, Marshall JR, Miller AB, Potter JD, Speizer FE, Willett WC, Wolk A, Hunter DJ: Alcohol and breast cancer in women: a pooled analysis of cohort studies. JAMA 1998;279:535–540.

8   Bandera EV, Freudenheim JL, Graham S, Marshall JR, Haughey BP, Swanson M, Brasure J, Wilkinson G: Alcohol consumption and lung cancer in white males. Cancer Causes Control 1992;3:361–369.

9   Stein CJ, Colditz GA: Modifiable risk factors for cancer. Br J Cancer 2004;90:299–303.

10 Sporn MB, Dunlop NM, Newton DL, Smith JM: Prevention of chemical carcinogenesis by vitamin A and its synthetic analogs (retinoids). Fed Proc 1976;35:1332–1338.

11 Hong W, Itri L: Retinoids and human cancer; in Sporn MB, Roberts AB, Goodman DS (eds): The Retinoids. New York, Raven Press, 1994, pp 597–658.

12 Moon RC, Mehta RG, Rao KVN: Retinoids and cancer in experimental animals; in Sporn MB, Roberts AB, Goodman DS (eds): The Retinoids. New York, Raven Press, 1994, pp 573–575.

13 Sun SY, Lotan R: Retinoids and their receptors in cancer development and chemoprevention. Crit Rev Oncol Hematol 2002;41:41–55.

14 O'Byrne SM, Blaner WS: Introduction to retinoids; in Packer L, Kraemer K, Obermuller-Jevic U, Sies H (eds): Carotenoids and Retinoids: Molecular Aspects and Health Issues. Champaign, IL, AOCS Press, 2005, pp 1–22.

15 Olson JA, Hayaishi O: The enzymatic cleavage of beta-carotene into vitamin A by soluble enzymes of rat liver and intestine. Proc Natl Acad Sci USA 1965;54:1364–1370.

16 Goodman DS, Blomstrand R, Werner B, Huang HS, Shiratori T: The intestinal absorption and metabolism of vitamin A and beta-carotene in man. J Clin Invest 1966;45:1615–1623.

17 Wang XD, Tang GW, Fox JG, Krinsky NI, Russell RM: Enzymatic conversion of beta-carotene into beta-apo-carotenals and retinoids by human, monkey, ferret, and rat tissues. Arch Biochem Biophys 1991;285:8–16.

18 Tang GW, Wang XD, Russell RM, Krinsky NI: Characterization of beta-apo-13-carotenone and beta-apo-14′-carotenal as enzymatic products of the excentric cleavage of beta-carotene. Biochem 1991;30:9829–9834.

19 Wang XD, Russell RM, Liu C, Stickel F, Smith DE, Krinsky NI: Beta-oxidation in rabbit liver in vitro and in the perfused ferret liver contributes to retinoic acid biosynthesis from beta-apocarotenoic acids. J Biol Chem 1996;271:26490–26498.

20 Napoli JL: Interactions of retinoid binding proteins and enzymes in retinoid metabolism. Biochim Biophys Acta 1999;1440:139–162.

21 Duester G: Families of retinoid dehydrogenases regulating vitamin A function: production of visual pigment and retinoic acid. Eur J Biochem 2000;267:4315–4324.

22 Duester G: Genetic dissection of retinoid dehydrogenases. Chem Biol Interact 2001;130–132:469–480.

23 White JA, Guo YD, Baetz K, Beckett-Jones B, Bonasoro J, Hsu KE, Dilworth FJ, Jones G, Petkovich M: Identification of the retinoic acid-inducible all-trans-retinoic acid 4-hydroxylase. J Biol Chem 1996;271:29922–29927.

24 Taimi M, Helvig C, Wisniewski J, Ramshaw H, White J, Amad M, Korczak B, Petkovich M: A novel human cytochrome P450, CYP26C1, involved in metabolism of 9-cis and all-trans isomers of retinoic acid. J Biol Chem 2004;279:77–85.

25 Fujii H, Sato T, Kaneko S, Gotoh O, Fujii-Kuriyama Y, Osawa K, Kato S, Hamada H: Metabolic inactivation of retinoic acid by a novel P450 differentially expressed in developing mouse embryos. EMBO J 1997;16:4163–4173.

26 Ray WJ, Bain G, Yao M, Gottlieb DI: CYP26, a novel mammalian cytochrome P450, is induced by retinoic acid and defines a new family. J Biol Chem 1997;272:18702–18708.

27 Gudas LJ, Sporn MB, Roberts AB: Cellular biology and biochemistry of the retinoids; in Sporn MB, Roberts AB, Goodman DS (eds): The Retinoids: Biology, Chemistry, and Medicine. New York, Raven Press, Ltd., 1994, pp 443–520.

28 Hansen LA, Sigman CC, Andreola F, Ross SA, Kelloff GJ, De Luca LM: Retinoids in chemoprevention and differentiation therapy. Carcinogenesis 2000;21:1271–1279.

29   De Luca L: Retinoids and their receptors in differentiation, embryogenesis, and neoplasia. FASEB J 1991;5:2924–2933.

30   Sun S-Y, Hail N Jr, Lotan R: Apoptosis as a novel target for cancer chemoprevention. J Natl Cancer Inst 2004;96:662–672.

31   Nagy L, Thomazy VA, Heyman RA, Davies PJ: Retinoid-induced apoptosis in normal and neoplastic tissues. Cell Death Differ 1998;5:11–19.

32   Soprano DR, Qin P, Soprano KJ: Retinoic acid receptors and cancers. Annu Rev Nutr 2004;24: 201–221.

33   Chambon P: A decade of molecular biology of retinoic acid receptors. FASEB J 1996;10:940–954.

34   Altucci L, Gronemeyer H: The promise of retinoids to fight against cancer. Nat Rev Cancer 2001;1:181–193.

35   Allenby G, Bocquel MT, Saunders M, Kazmer S, Speck J, Rosenberger M, Lovey A, Kastner P, Grippo JF, Chambon P: Retinoic acid receptors and retinoid X receptors: interactions with endogenous retinoic acids. Proc Natl Acad Sci USA 1993;90:30–34.

36   Kelloff GJ, Boone CW, Crowell JA, Steele VE, Lubet RA, Doody LA, Malone WF, Hawk ET, Sigman CC: New agents for cancer chemoprevention. J Cell Biochem Suppl 1996;26:1–28.

37   Lotan R: Retinoids in cancer chemoprevention. FASEB J 1996;10:1031–1039.

38   Langenfeld J, Kiyokawa H, Sekula D, Boyle J, Dmitrovsky E: Posttranslational regulation of cyclin D1 by retinoic acid: a chemoprevention mechanism. Proc Natl Acad Sci USA 1997;94:12070–12074.

39   Boyle JO, Langenfeld J, Lonardo F, Sekula D, Reczek P, Rusch V, Dawson MI, Dmitrovsky E: Cyclin D1 proteolysis: a retinoid chemoprevention signal in normal, immortalized, and transformed human bronchial epithelial cells. J Natl Cancer Inst 1999;91:373–379.

40   Spinella MJ, Freemantle SJ, Sekula D, Chang JH, Christie AJ, Dmitrovsky E: Retinoic acid promotes ubiquitination and proteolysis of cyclin D1 during induced tumor cell differentiation. J Biol Chem 1999;274:22013–22018.

41   Naderi S, Blomhoff HK: Retinoic acid prevents phosphorylation of pRB in normal human B lymphocytes: regulation of cyclin E, cyclin A, and p21(Cip1). Blood 1999;94:1348–1358.

42   Hsu S, Hsu J, Liu M, Chen L, Chang C: Retinoic acid-mediated G1 arrest is associated with induction of p27(Kip1) and inhibition of cyclin-dependent kinase 3 in human lung squamous carcinoma CH27 cells. Exp Cell Res 2000;258:322–331.

43   Fukao A, Tsubono Y, Kawamura M, Ido T, Akazawa N, Tsuji I, Komatsu S, Minami Y, Hisamichi S: The independent association of smoking and drinking with serum beta-carotene levels among males in Miyagi, Japan. Int J Epidemiol 1996;25:300–306.

44   Wei W, Kim Y, Boudreau N: Association of smoking with serum and dietary levels of antioxidants in adults: NHANES III, 1988–1994. Am J Public Health 2001;91:258–264.

45   Ross MA, Crosley LK, Brown KM, Duthie SJ, Collins AC, Arthur JR, Duthie GG: Plasma concentrations of carotenoids and antioxidant vitamins in Scottish males: influences of smoking. Eur J Clin Nutr 1995;49:861–865.

46   Bolton-Smith C, Casey CE, Gey KF, Smith WC, Tunstall-Pedoe H: Antioxidant vitamin intakes assessed using a food-frequency questionnaire: correlation with biochemical status in smokers and non-smokers. Br J Nutr 1991;65:337–346.

47   Stryker WS, Kaplan LA, Stein EA, Stampfer MJ, Sober A, Willett WC: The relation of diet, cigarette smoking, and alcohol consumption to plasma beta-carotene and alpha-tocopherol levels. Am J Epidemiol 1988;127:283–296.

48   Wang XD, Liu C, Bronson RT, Smith DE, Krinsky NI, Russell M: Retinoid signaling and activator protein-1 expression in ferrets given beta-carotene supplements and exposed to tobacco smoke. J Natl Cancer Inst 1999;91:60–66.

49   Wang XD, Russell RM: Procarcinogenic and anticarcinogenic effects of beta-carotene. Nutr Rev 1999;57:263–272.

50   Liu C, Wang XD, Bronson RT, Smith DE, Krinsky NI, Russell RM: Effects of physiological versus pharmacological beta-carotene supplementation on cell proliferation and histopathological changes in the lungs of cigarette smoke-exposed ferrets. Carcinogenesis 2000;21: 2245–2253.

51   Salgo MG, Cueto R, Winston GW, Pryor WA: Beta carotene and its oxidation products have different effects on microsome mediated binding of benzo[a]pyrene to DNA. Free Radic Biol Med 1999;26:162–173.

52   Paolini M, Antelli A, Pozzetti L, Spetlova D, Perocco P, Valgimigli L, Pedulli GF, Cantelli-Forti G: Induction of cytochrome P450 enzymes and over-generation of oxygen radicals in beta-carotene supplemented rats. Carcinogenesis 2001;22:1483–1495.

53   Perocco P, Paolini M, Mazzullo M, Biagi GL, Cantelli-Forti G: Beta-carotene as enhancer of cell transforming activity of powerful carcinogens and cigarette-smoke condensate on BALB/c 3T3 cells in vitro. Mutat Res 1999;440:83–90.

54   Yeh SL, Hu ML: Oxidized beta-carotene inhibits gap junction intercellular communication in the human lung adenocarcinoma cell line A549. Food Chem Toxicol 2003;41:1677–1684.

55   Siems W, Wiswedel I, Salerno C, Crifo C, Augustin W, Schild L, Langhans CD, Sommerburg O: Beta-carotene breakdown products may impair mitochondrial functions – potential side effects of high-dose beta-carotene supplementation. J Nutr Biochem 2005;16:385–397.

56   Liu C, Russell RM, Wang XD: Exposing ferrets to cigarette smoke and a pharmacological dose of beta-carotene supplementation enhance in vitro retinoic acid catabolism in lungs via induction of cytochrome P450 enzymes. J Nutr 2003;133:173–179.

57   Gradelet S, Leclerc J, Siess MH, Astorg PO: Beta-Apo-8'-carotenal, but not Beta-carotene, is a strong inducer of liver cytochromes P4501A1 and 1A2 in rat. Xenobiotica 1996;26:909–919.

58   Liu C, Russell RM, Wang X-D: Alpha-Tocopherol and ascorbic acid decrease the production of beta-apo-carotenals and increase the formation of retinoids from beta-carotene in the lung tissues of cigarette smoke-exposed ferrets in vitro. J Nutr 2004;134:426–430.

59   Kim Y, Chongviriyaphan N, Liu C, Russell RM, Wang XD: Combined antioxidant (beta-carotene, alpha-tocopherol and ascorbic acid) supplementation increases the levels of lung retinoic acid and inhibits activation of mitogen-activated protein kinase in the ferret lung cancer model. Carcinogenesis 2006;doi: 10.1093/carci

60   Wang XD, Seitz HK: Alcohol and retinoid interaction; in Watson RR, Preedy VR (eds): Nutrition and Alcohol: Linking Nutrient Interactions and Dietary Intake. Boca Raton, FL, CRC Press LLC, 2004, pp 313–322.

61   Wang XD, Liu C, Chung J, Stickel F, Seitz HK, Russell RM: Chronic alcohol intake reduces retinoic acid concentration and enhances AP-1 (c-Jun and c-Fos) expression in rat liver. Hepatology 1998;28:744–750.

62   Wang XD: Alcohol, vitamin A, and cancer. Alcohol 2005;35:251–258.

63   Molotkov A, Duester G: Retinol/ethanol drug interaction during acute alcohol intoxication in mice involves inhibition of retinol metabolism to retinoic acid by alcohol dehydrogenase. J Biol Chem 2002;277:22553–22557.

64   Chung J, Liu C, Smith DE, Seitz HK, Russell RM, Wang XD: Restoration of retinoic acid concentration suppresses ethanol-enhanced c-Jun expression and hepatocyte proliferation in rat liver. Carcinogenesis 2001;22:1213–1219.

65   Crabb DW, Pinairs J, Hasanadka R, Fang M, Leo MA, Lieber CS, Tsukamoto H, Motomura K, Miyahara T, Ohata M, Bosron W, Sanghani S, Kedishvili N, Shiraishi H, Yokoyama H, Miyagi M, Ishii H, Bergheim I, Menzl I, Parlesak A, Bode C: Alcohol and retinoids. Alcohol Clin Exp Res 2001;25:207S–217S.

66   Halsted CH, Villanueva J, Chandler CJ, Stabler SP, Allen RH, Muskhelishvili L, James SJ, Poirier L: Ethanol feeding of micropigs alters methionine metabolism and increases hepatocellular apoptosis and proliferation. Hepatology 1996;23:497–505.

67   Parlesak A, Menzl I, Feuchter A, Bode JC, Bode C: Inhibition of retinol oxidation by ethanol in the rat liver and colon. Gut 2000;47:825–831.

68   Seitz HK, Matsuzaki S, Yokoyama A, Homann N, Vakevainen S, Wang XD: Alcohol and cancer. Alcohol Clin Exp Res 2001;25:137S–143S.

69   Pöschl G, Seitz HK: Alcohol and cancer. Alcohol Alcohol 2004;39:155–165.

70   Salaspuro V, Salaspuro M: Synergistic effect of alcohol drinking and smoking on in vivo acetaldehyde concentration in saliva. Int J Cancer 2004;111:480–483.

71   Homann N, Tillonen J, Meurman JH, Rintamaki H, Lindqvist C, Rautio M, Jousimies-Somer H, Salaspuro M: Increased salivary acetaldehyde levels in heavy drinkers and smokers: a microbiological approach to oral cavity cancer. Carcinogenesis 2000;21:663–668.

72   Khalighi M, Brzezinski MR, Chen H, Juchau MR: Inhibition of human prenatal biosynthesis of all-trans-retinoic acid by ethanol, ethanol metabolites, and products of lipid peroxidation reactions: a possible role for CYP2E1. Biochem Pharmacol 1999;57:811–821.

73   Yokoyama A, Muramatsu T, Ohmori T, Higuchi S, Hayashida M, Ishii H: Esophageal cancer and aldehyde dehydrogenase-2 genotypes in Japanese males. Cancer Epidemiol Biomarkers Prev 1996;5:99–102.

74   Yokoyama A, Muramatsu T, Ohmori T, Yokoyama T, Okuyama K, Takahashi H, Hasegawa Y, Higuchi S, Maruyama K, Shirakura K, Ishii H: Alcohol-related cancers and aldehyde dehydrogenase-2 in Japanese alcoholics. Carcinogenesis 1998;19:1383–1387.

75   Yokoyama A, Kato H, Yokoyama T, Tsujinaka T, Muto M, Omori T, Haneda T, Kumagai Y, Igaki H, Yokoyama M, Watanabe H, Fukuda H, Yoshimizu H: Genetic polymorphisms of alcohol and aldehyde dehydrogenases and glutathione S-transferase M1 and drinking, smoking, and diet in Japanese men with esophageal squamous cell carcinoma. Carcinogenesis 2002;23:1851–1859.

76   Visapää JP, Gotte K, Benesova M, Li J, Homann N, Conradt C, Inoue H, Tisch M, Horrmann K, Vakevainen S, Salaspuro M, Seitz HK: Increased cancer risk in heavy drinkers with the alcohol dehydrogenase 1C*1 allele, possibly due to salivary acetaldehyde. Gut 2004;53:871–876.

77   Seitz HK, Poschl G, Simanowski UA: Alcohol and cancer. Recent Dev Alcohol 1998;14:67–95.

78   Sato M, Lieber CS: Increased metabolism of retinoic acid after chronic ethanol consumption in rat liver microsomes. Arch Biochem Biophys 1982;213:557–564.

79   Liu C, Russell RM, Seitz HK, Wang XD: Ethanol enhances retinoic acid metabolism into polar metabolites in rat liver via induction of cytochrome P4502E1. Gastroenterology 2001;120: 179–189.

80   Liu C, Chung J, Seitz HK, Russell RM, Wang XD: Chlormethiazole treatment prevents reduced hepatic vitamin A levels in ethanol-fed rats. Alcohol Clin Exp Res 2002;26:1703–1709.

81   Shimizu M, Lasker JM, Tsutsumi M, Lieber CS: Immunohistochemical localization of ethanol-inducible P450IIE1 in the rat alimentary tract. Gastroenterology 1990;99:1044–1053.

82   Gouillon Z, Lucas D, Li J, Hagbjork AL, French BA, Fu P, Fang C, Ingelman-Sundberg M, Donohue TM Jr, French SW: Inhibition of ethanol-induced liver disease in the intragastric feeding rat model by chlormethiazole. Proc Soc Exp Biol Med 2000;224:302–308.

83   Morimoto M, Reitz RC, Morin RJ, Nguyen K, Ingelman-Sundberg M, French SW: CYP-2E1 inhibitors partially ameliorate the changes in hepatic fatty acid composition induced in rats by chronic administration of ethanol and a high fat diet. J Nutr 1995;125:2953–2964.

84   Morimoto M, Hagbjork AL, Wan YJ, Fu PC, Clot P, Albano E, Ingelman-Sundberg M, French SW: Modulation of experimental alcohol-induced liver disease by cytochrome P450 2E1 inhibitors. Hepatology 1995;21:1610–1617.

85   Leo MA, Lieber CS: Alcohol, vitamin A, and beta-carotene: adverse interactions, including hepatotoxicity and carcinogenicity. Am J Clin Nutr 1999;69:1071–1085.

86   Dan Z, Popov Y, Patsenker E, Preimel D, Liu C, Wang X-D, Seitz HK, Schuppan D, Stickel F: Hepatotoxicity of alcohol-induced polar retinol metabolites involves apoptosis via loss of mitochondrial membrane potential. FASEB J 2005;19:845–847.

87   Leo MA, Kim C, Lieber CS: Increased vitamin A in esophagus and other extrahepatic tissues after chronic ethanol consumption in the rat. Alcohol Clin Exp Res 1986;10:487–492.

88   Mobarhan S, Seitz HK, Russell RM, Mehta R, Hupert J, Friedman H, Layden TJ, Meydani M, Langenberg P: Age-related effects of chronic ethanol intake on vitamin A status in Fisher 344 rats. J Nutr 1991;121:510–517.

89   Zachman RD, Grummer MA: Prenatal ethanol consumption increases retinol and cellular retinol-binding protein expression in the rat fetal snout. Biol Neonate 2001;80:152–157.

90   Friedman H, Mobarhan S, Hupert J, Lucchesi D, Henderson C, Langenberg P, Layden TJ: In vitro stimulation of rat liver retinyl ester hydrolase by ethanol. Arch Biochem Biophys 1989;269: 69–74.

91   Wang XD, Krinsky NI, Russell RM: Retinoic acid regulates retinol metabolism via feedback inhibition of retinol oxidation and stimulation of retinol esterification in ferret liver. J Nutr 1993;123:1277–1285.

92   Zolfaghari R, Ross AC: Lecithin:retinol acyltransferase from mouse and rat liver. cDNA cloning and liver-specific regulation by dietary vitamin A and retinoic acid. J Lipid Res 2000;41: 2024–2034.

93 Brabender J, Metzger R, Salonga D, Danenberg KD, Danenberg PV, Holscher AH, Schneider PM: Comprehensive expression analysis of retinoic acid receptors and retinoid X receptors in non-small cell lung cancer: implications for tumor development and prognosis. Carcinogenesis 2005; 26:525–530.

94 Gebert JF, Moghal N, Frangioni JV, Sugarbaker DJ, Neel BG: High frequency of retinoic acid receptor beta abnormalities in human lung cancer. Oncogene 1991;6:1859–1868.

95 Qiu H, Zhang W, El-Naggar AK, Lippman SM, Lin P, Lotan R, Xu XC: Loss of retinoic acid receptor-beta expression is an early event during esophageal carcinogenesis. Am J Pathol 1999;155:1519–1523.

96 Picard E, Seguin C, Monhoven N, Rochette-Egly C, Siat J, Borrelly J, Martinet Y, Martinet N, Vignaud JM: Expression of retinoid receptor genes and proteins in non-small-cell lung cancer. J Natl Cancer Inst 1999;91:1059–1066.

97 Xu X, Sozzi G, Lee J, Lee J, Pastorino U, Pilotti S, Kurie J, Hong W, Lotan R: Suppression of retinoic acid receptor beta in non-small-cell lung cancer in vivo: implications for lung cancer development. J Natl Cancer Inst 1997;89:624–629.

98 Xu XC, Lee JJ, Wu TT, Hoque A, Ajani JA, Lippman SM: Increased retinoic acid receptor-beta4 correlates in vivo with reduced retinoic acid receptor-beta2 in esophageal squamous cell carcinoma. Cancer Epidemiol Biomarkers Prev 2005;14:826–829.

99 Houle B, Rochette-Egly C, Bradley WE: Tumor-suppressive effect of the retinoic acid receptor beta in human epidermoid lung cancer cells. Proc Natl Acad Sci USA 1993;90: 985–989.

100 Song S, Xu XC: Effect of benzo[a]pyrene diol epoxide on expression of retinoic acid receptor-beta in immortalized esophageal epithelial cells and esophageal cancer cells. Biochem Biophys Res Commun 2001;281:872–877.

101 Chen GQ, Lin B, Dawson MI, Zhang XK: Nicotine modulates the effects of retinoids on growth inhibition and RAR beta expression in lung cancer cells. Int J Cancer 2002;99: 171–178.

102 Kurie JM, Lotan R, Lee JJ, Lee JS, Morice RC, Liu DD, Xu XC, Khuri FR, Ro JY, Hittelman WN, Walsh GL, Roth JA, Minna JD, Hong WK: Treatment of former smokers with 9-cis-retinoic acid reverses loss of retinoic acid receptor-beta expression in the bronchial epithelium: results from a randomized placebo-controlled trial. J Natl Cancer Inst 2003;95:206–214.

103 Napoli JL, Race KR: Biogenesis of retinoic acid from beta-carotene. Differences between the metabolism of beta-carotene and retinal. J Biol Chem 1988;263:17372–17377.

104 Witschi H: Carcinogenic activity of cigarette smoke gas phase and its modulation by beta-carotene and N-acetylcysteine. Toxicol Sci 2005;84:81–87.

105 Ponnamperuma RM, Shimizu Y, Kirchhof SM, De Luca LM: beta-Carotene fails to act as a tumor promoter, induces RAR expression, and prevents carcinoma formation in a two-stage model of skin carcinogenesis in male Sencar mice. Nutr Cancer 2000;37:82–88.

106 Prakash P, Liu C, Hu KQ, Krinsky NI, Russell RM, Wang XD: Beta-carotene and beta-apo-14′-carotenoic acid prevent the reduction of retinoic acid receptor beta in benzo[a]pyrene-treated normal human bronchial epithelial cells. J Nutr 2004;134:667–673.

107 Grummer MA, Zachman RD: Interaction of ethanol with retinol and retinoic acid in RAR beta and GAP-43 expression. Neurotoxicol Teratol 2000;22:829–836.

108 Li C, Wan YJ: Differentiation and antiproliferation effects of retinoic acid receptor beta in hepatoma cells. Cancer Lett 1998;124:205–211.

109 Wan YJ, Cai Y, Magee TR: Retinoic acid differentially regulates retinoic acid receptor-mediated pathways in the Hep3B cell line. Exp Cell Res 1998;238:241–247.

110 Chung J, Chavez PR, Russell RM, Wang XD: Retinoic acid inhibits hepatic Jun N-terminal kinase-dependent signaling pathway in ethanol-fed rats. Oncogene 2002;21:1539–1547.

111 Chen N, Nomura M, She QB, Ma WY, Bode AM, Wang L, Flavell RA, Dong Z: Suppression of skin tumorigenesis in c-Jun NH(2)-terminal kinase-2-deficient mice. Cancer Res 2001;61: 3908–3912.

112 Davis RJ: Signal transduction by the JNK group of MAP kinases. Cell 2000;103:239–252.

113 Karin M, Liu Z, Zandi E: AP-1 function and regulation. Curr Opin Cell Biol 1997;9: 240–246.

114   Yoshii S, Tanaka M, Otsuki Y, Fujiyama T, Kataoka H, Arai H, Hanai H, Sugimura H: Involvement of alpha-PAK-interacting exchange factor in the PAK1-c-Jun NH(2)-terminal kinase 1 activation and apoptosis induced by benzo[a]pyrene. Mol Cell Biol 2001;21:6796–6807.

115   Solhaug A, Ovrebo S, Mollerup S, Lag M, Schwarze PE, Nesnow S, Holme JA: Role of cell signaling in B[a]P-induced apoptosis: characterization of unspecific effects of cell signaling inhibitors and apoptotic effects of B[a]P metabolites. Chem Biol Interact 2005;151:101–119.

116   Young MR, Li JJ, Rincon M, Flavell RA, Sathyanarayana BK, Hunziker R, Colburn N: Transgenic mice demonstrate AP-1 (activator protein-1) transactivation is required for tumor promotion. Proc Natl Acad Sci USA 1999;96:9827–9832.

117   Liu Y, Gorospe M, Yang C, Holbrook NJ: Role of mitogen-activated protein kinase phosphatase during the cellular response to genotoxic stress. Inhibition of c-Jun N-terminal kinase activity and AP-1-dependent gene activation. J Biol Chem 1995;270:8377–8380.

118   Slack DN, Seternes OM, Gabrielsen M, Keyse SM: Distinct binding determinants for ERK2/p38alpha and JNK map kinases mediate catalytic activation and substrate selectivity of map kinase phosphatase-1. J Biol Chem 2001;276:16491–16500.

119   Liu C, Russell RM, Wang X-D: Low dose beta-carotene supplementation of ferrets attenuates smoke-induced lung phosphorylation of JNK, p38 MAPK, and p53 proteins. J Nutr 2004;134: 2705–2710.

120   Hirsch DD, Stork PJ: Mitogen-activated protein kinase phosphatases inactivate stress-activated protein kinase pathways in vivo. J Biol Chem 1997;272:4568–4575.

121   Furukawa M, Zhang YQ, Nie L, Shibata H, Kojima I: Role of mitogen-activated protein kinase and phosphoinositide 3-kinase in the differentiation of rat pancreatic AR42J cells induced by hepatocyte growth factor. Diabetologia 1999;42:450–456.

122   Omenn GS, Goodman GE, Thornquist MD, Balmes J, Cullen MR, Glass A, Keogh JP, Meyskens FL, Valanis B, Williams JH, Barnhart S, Hammar S: Effects of a combination of beta carotene and vitamin A on lung cancer and cardiovascular disease. N Engl J Med 1996;334:1150–1155.

123   The Alpha-Tocopherol, Beta Carotene Cancer Prevention Study Group: The effect of vitamin E and beta carotene on the incidence of lung cancer and other cancers in male smokers. N Engl J Med 1994;330:1029–1035.

124   Ziegler RG, Mason TJ, Stemhagen A, Hoover R, Schoenberg JB, Gridley G, Virgo PW, Fraumeni JF Jr: Carotenoid intake, vegetables, and the risk of lung cancer among white men in New Jersey. Am J Epidemiol 1986;123:1080–1093.

125   Speizer FE, Colditz GA, Hunter DJ, Rosner B, Hennekens C: Prospective study of smoking, antioxidant intake, and lung cancer in middle-aged women (USA). Cancer Causes Control 1999;10:475–482.

126   Swanson CA, Mao BL, Li JY, Lubin JH, Yao SX, Wang JZ, Cai SK, Hou Y, Luo QS, Blot WJ: Dietary determinants of lung-cancer risk: results from a case-control study in Yunnan Province, China. Int J Cancer 1992;50:876–880.

127   Simanowski UA, Homann N, Knuhl M, Arce L, Waldherr R, Conradt C, Bosch FX, Seitz HK: Increased rectal cell proliferation following alcohol abuse. Gut 2001;49:418–422.

128   Román J, Colell A, Blasco C, Caballeria J, Pares A, Rodes J, Fernandez-Checa JC: Differential role of ethanol and acetaldehyde in the induction of oxidative stress in HEP G2 cells: effect on transcription factors AP-1 and NF-kappaB. Hepatology 1999;30:1473–1480.

129   Masamune A, Kikuta K, Satoh M, Satoh A, Shimosegawa T: Alcohol activates activator protein-1 and mitogen-activated protein kinases in rat pancreatic stellate cells. J Pharmacol Exp Ther 2002;302:36–42.

130   Aroor AR, Shukla SD: MAP kinase signaling in diverse effects of ethanol. Life Sci 2004;74: 2339–2364.

131   Göttlicher M, Heck S, Herrlich P: Transcriptional cross-talk, the second mode of steroid hormone receptor action. J Mol Med 1998;76:480–489.

132   Lin F, Xiao D, Kolluri SK, Zhang X: Unique anti-activator protein-1 activity of retinoic acid receptor beta. Cancer Res 2000;60:3271–3280.

133   Uto H, Ido A, Moriuchi A, Onaga Y, Nagata K, Onaga M, Tahara Y, Hori T, Hirono S, Hayashi K, Tsubouchi H: Transduction of antisense cyclin D1 using two-step gene transfer inhibits the growth of rat hepatoma cells. Cancer Res 2001;61:4779–4783.

134 Srinivas H, Juroske DM, Kalyankrishna S, Cody DD, Price RE, Xu XC, Narayanan R, Weigel NL, Kurie JM: c-Jun N-terminal kinase contributes to aberrant retinoid signaling in lung cancer cells by phosphorylating and inducing proteasomal degradation of retinoic acid receptor alpha. Mol Cell Biol 2005;25:1054–1069.

Xiang-Dong Wang
Nutrition and Cancer Biology Laboratory, Jean Mayer United States Department of
Agriculture Human Nutrition Research
Center on Aging at Tufts University 711
Washington St., Boston, MA 02111 (USA)
Tel. +1 617 556 3130, Fax +1 617 556 3344
E-Mail xiang-dong.wang@tufts.edu

Cho CH, Purohit V (eds): Alcohol, Tobacco and Cancer.
Basel, Karger, 2006, pp 160–174

....

Role of *S*-Adenosyl-ʟ-Methionine in Alcohol-Associated Liver Cancer

Shelly C. Lu[a]*, José M. Mato*[b]

[a]USC Research Center for Liver Diseases, USC-UCLA Research Center for Alcoholic
Liver and Pancreatic Diseases, Keck School of Medicine USC, Los Angeles, Calif.,
USA; [b]CIC Biogune, Center for Cooperative Research in Biosciences, Parque
Tecnológico de Bizkaia, Derio, Spain

Abstract

S-adenosylmethionine (SAMe), widely available as a nutritional supplement, has
rapidly moved from being a methyl donor to a key metabolite that regulates hepatocyte
growth, death and differentiation. Biosynthesis of SAMe occurs in all mammalian cells as
the first step in methionine catabolism in a reaction catalyzed by methionine adenosyltransferase (MAT). Two genes (*MAT1A* and *MAT2A*) encode for the catalytic subunits of MAT
isoenzymes. A third gene (*MAT2β*) encodes for the regulatory subunit that regulates the
MAT2A encoded isoenzyme. *MAT1A* is expressed mostly in the liver whereas *MAT2A* is
widely distributed. In the liver, *MAT1A* expression correlates with a differentiated phenotype
while *MAT2A* expression correlates with increased growth and de-differentiation. In normal
liver, *MAT2A* and *MAT2β* expression are very low or undetectable. *MAT2A* and *MAT2β* are
induced in human hepatocellular carcinoma and both can offer a growth advantage.
Decreased hepatic SAMe biosynthesis is a consequence of all forms of chronic liver injury,
including those associated with alcohol consumption. This occurs due to decreased *MAT1A*
expression and inactivation of the *MAT1A* encoded isoenzymes. The *MAT1A* knockout
mouse model has provided important insights on the consequences of chronic hepatic SAMe
deficiency. In this model, the liver is prone to injury, develops spontaneous steatohepatitis by
8 months and hepatocellular carcinoma by 18 months. Recent data show SAMe can regulate
hepatocyte growth and death, independent of its role as a methyl donor. Importantly, SAMe
is anti-apoptotic in normal hepatocytes but pro-apoptotic in liver cancer cells, which might
make SAMe an attractive agent in the chemoprevention and treatment of human hepatocellular carcinoma.

Since *S*-adenosylmethionine (SAMe, also abbreviated as SAM and
AdoMet)'s discovery more than half a century ago, it has enjoyed steady interest as

a methyl donor and used therapeutically in cholestatic and alcoholic liver injury, depression and arthritis. SAMe biosynthesis occurs in all mammalian cells in a reaction catalyzed by the essential enzyme methionine adenosyltransferase (MAT). SAMe is particularly important in liver as all forms of chronic liver injury result in reduced SAMe biosynthesis due to decreased hepatic MAT activity. Our recent data indicate SAMe is a key regulator of hepatocyte growth and death by mechanisms that are independent of methylation. Chronic hepatic SAMe deficiency occurs in *MAT1A* knockout mice and leads to livers that are more susceptible to further injury, development of steatohepatitis and hepatocellular carcinoma (HCC). In this regard, chronic alcoholics are predisposed to HCC but the molecular mechanisms are unknown. Many of these patients have decreased *MAT1A* expression, some with only liver fibrosis. This chapter reviews SAMe metabolism, regulation of MAT genes, changes that occur in alcoholic liver disease and how they may contribute to the development of HCC. Possible therapeutic role of SAMe in chemoprevention and treatment of HCC is also proposed.

SAMe Metabolism

SAMe is synthesized in the cytosol of every cell but the liver plays a central role in the homeostasis of SAMe as the major site of its synthesis and degradation [1] (fig. 1). Liver is where up to half of the daily intake of methionine is converted to SAMe and up to 85% of all methylation reactions takes place [2]. SAMe is the principal biological methyl donor, the precursor for polyamine biosynthesis and in liver, SAMe is also a precursor of glutathione (GSH) via the transsulfuration pathway [1]. MAT is the only enzyme responsible for SAMe biosynthesis [1]. Under normal physiological conditions, most of the 6–8 g of SAMe generated per day is used in transmethylation reactions in which methyl groups are added to compounds and SAMe is converted to *S*-adenosylhomocysteine (SAH) [3]. SAH is a potent competitive inhibitor of transmethylation reactions; both an increase in SAH level as well as a decrease in the SAMe level or SAMe to SAH ratio are known to inhibit transmethylation reactions [4]. For this reason, SAH needs to be rapidly catabolized. The reaction that converts SAH to homocysteine and adenosine is reversible and catalyzed by SAH hydrolase [4]. Interestingly, the thermodynamics favor synthesis of SAH [4]. In vivo, the reaction proceeds in the direction of hydrolysis only if the products, adenosine and homocysteine, are rapidly removed [5]. In liver, there are three pathways that metabolize homocysteine. One is the transsulfuration pathway, which converts homocysteine to cysteine, the rate-limiting precursor for GSH synthesis [6]. This pathway is very active in the liver, allowing methionine to serve as a precursor for cysteine and GSH [6]. The other two pathways that metabolize homocysteine resynthesize

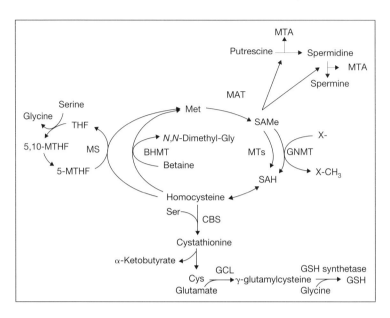

Fig. 1. Hepatic SAMe metabolis. The first step in methionine (met) catabolism is the generation of S-adenosylmethionine (SAMe) in a reaction catalyzed by methionine adenosyl-transferase (MAT). SAMe is the link to three key metabolic pathways, polyamine synthesis, transmethylation and transsulfuration. In the polyamine synthesis pathway, SAMe is decarboxylated and the remaining propylamino moiety attached to its sulfonium ion is donated to putrescine to form spermidine and methylthioadenosine (MTA) and to spermidine to form spermine and a second molecule of MTA. In transmethylation, SAMe donates its methyl group to a large variety of acceptor molecules in reactions catalyzed by methyltransferase (MTs), the most abundant in the liver being glycine-*N*-methyltransferase (GNMT). S-adeno-sylhomocysteine (SAH) is generated as a product of transmethylation and is hydrolyzed to form homocysteine (Hcy) and adenosine through a reversible reaction catalyzed by SAH hydrolase. SAH is a potent competitive inhibitor of methylation reactions and prompt removal of adenosine and Hcy is required to prevent accumulation of SAH. Hcy can be remethylated to form methionine by two enzymes: methionine synthase (MS), which requires normal levels of folate and vitamin B_{12}; and betaine methyltransferase (BHMT), which requires betaine, a metabolite of choline. Remethylation of homocysteine via MS requires 5-methyltetrahydrofo-late (5-MTHF), which is derived from 5,10-methylenetetrahydrofolate (5,10-MTHF) in a reaction catalyzed by methylenetetrahydrofolate reductase. 5-MTHF is then converted to tetrahydrofolate (THF) as it donates its methyl group and THF is converted to 5,10-MTHF to complete the folate cycle. Hcy can also undergo the transsulfuration pathway to form cysteine (the rate-limiting precursor for GSH) via a two-step enzymatic process catalyzed by cys-tathionine β-synthetase (CBS) and cystathionase, both requiring vitamin B_6. However, the transsulfuration pathway is limited to the liver, kidney and pancreas, as the rest of the organs lack expression of one or more enzymes of the transsulfuration pathway. All mammalian tissues express MAT and MS, whereas BHMT expression is limited to the liver and kidney.

methionine from homocysteine. One is catalyzed by methionine synthase (MS) and the other is catalyzed by betaine-homocysteine methyltransferase (BHMT) [4]. If homocysteine metabolism is impaired, the ratio of SAMe to SAH may decrease leading to inhibition of transmethylation reactions. SAMe plays a regulatory role on methionine metabolism by inhibiting BHMT and MS and activating cystathionine β-synthase (CBS), a key enzyme in the trans-sulfuration pathway [3]. Thus, when SAMe is depleted, homocysteine is channeled to remethylation to regenerate SAMe; whereas when SAMe level is high, homocysteine is channeled to the trans-sulfuration pathway. SAMe is also a precursor of polyamines as the propylamino moiety of SAMe is converted via a series of enzymatic steps to spermidine and spermine. In the biosynthesis of polyamines, 5'-methylthioadenosine (MTA) is an end product. Thus, in the liver SAMe provides the link to three key metabolic pathways, transmethylation, trans-sulfuration and polyamine synthesis.

MAT Genes and Isoenzymes

MAT is an essential enzyme because it catalyzes the only reaction that generates SAMe [1]. In mammals, *MAT1A* and *MAT2A* are two genes that encode for two homologous MAT catalytic subunits, α1 and α2 [7]. *MAT1A* is expressed mostly in the liver and it encodes the α1 subunit found in two native MAT isozymes, which are either a dimer (MAT III) or tetramer (MAT I) of this subunit [7]. *MAT2A* encodes for the α2 catalytic subunit found in a native MAT isozyme (MAT II), which is widely distributed [7, 8]. Fetal liver expresses only *MAT2A* and as the liver matures, MAT expression switches from *MAT2A* to *MAT1A* shortly after birth [9, 10]. The composition of MAT II varies depending on the tissue [11–13]. In lymphocytes, a third gene, *MAT2β*, encodes for the regulatory β subunit which controls the activity of MAT II by lowering the K_m and K_i of MAT II for methionine and SAMe, respectively, thereby making the enzyme work more efficiently but also more susceptible to feedback inhibition by SAMe [13–15]. The β subunit also plays a similar regulatory role on MAT activity in liver cancer cells [16].

MAT1A and *MAT2A* share a great deal of similarity, both coding for a 3.4 kb mRNA product, and may have originated from a common ancestral gene. Despite the similarities, MAT isoenzymes differ in kinetic and regulatory properties [1]. MAT II has the lowest K_m ($\sim$4–10 μM), MAT I has intermediate K_m (23 μM–1 mM), and MAT III has the highest K_m (215 μM–7 mM) for methionine [17–21]. MAT activity is modulated by SAMe. SAMe strongly inhibits MAT II ($IC_{50} = 60 \mu M$, which is close to the normal intracellular SAMe concentration [4]; whereas it minimally inhibits MAT I ($IC_{50} = 400 \mu M$) and stimulates MAT III (up to 8-fold at 500 μM SAMe concentration) [19]. Thus, the type of MAT expressed by the cell should have a significant influence on the steady state SAMe level. Indeed, we have verified this using an in vitro cell line model that expresses different MAT genes by showing that cells that express *MAT1A* had the highest SAMe levels while cells that express *MAT2A* had the opposite [22]. While MAT expression can influence SAMe level, SAMe level can in turn affect MAT expression. When rat hepatocytes are placed in primary culture, a switch in MAT expression occurs from *MAT1A* to *MAT2A*, likely due to de-differentiation of the cells [23]. This change can be prevented by adding SAMe to the medium. The human *MAT2A* gene also appears to be similarly regulated by SAMe so that its expression is rapidly induced when SAMe level falls and down-regulated when SAMe is added [24, 25]. Thus, a fall in hepatic SAMe level may feed into a vicious cycle that favors a switch in MAT expression and liver de-differentiation.

A switch in MAT expression from *MAT1A* to *MAT2A* occurs in human HCC at the transcriptional level [25, 26]. To examine whether this switch in MAT expression is important pathogenetically, we established a cell line model that differs only in the MAT gene expressed [22]. We found that cells that express *MAT1A* had higher SAMe levels and DNA methylation and grew much slower than cells that express *MAT2A* [22]. SAMe treatment of liver cancer cells raised intracellular SAMe levels and inhibited cell growth, supporting a causal role of SAMe and growth [22]. Thus, the switch in MAT expression in HCC is important because it facilitates liver cancer growth and also provides a target for the design of therapy against HCC. Indeed, eliminating *MAT2A* expression in liver cancer cells led to cell death but cells transfected with *MAT1A* were spared [22]. However, anti-*MAT2A* treatment must be restricted to the liver to avoid systemic toxicity.

Recently Martínez-Chantar et al. [16] examined the role of the β subunit in liver disease. *MAT2β* is expressed in rat extrahepatic tissues but not in normal rat liver. In human liver, β subunit expression was associated with cirrhosis and HCC. Expression of the β subunit was observed in most (HepG2, PLC and Hep3B) but not all (HuH7) human hepatoma cell lines. Transfection of β subunit reduced SAMe content and stimulated DNA synthesis in HuH7 cells, whereas down-regulation of β subunit expression diminished DNA synthesis in HepG2 cells [16]. Thus, in the liver, β subunit is associated with cirrhosis and cancer and provides another proliferative advantage in human hepatoma cell lines. Whether SAMe also regulates *MAT2β* expression is unknown.

Changes in MAT Expression and SAMe Levels in Alcoholic Liver Injury

Kinsell and coworkers described over 50 years ago that patients with cirrhosis often have hypermethioninemia and delayed plasma clearance of methionine after intravenous injection [27, 28]. Subsequent studies showed that this could be attributed to a 50–60% decrease in the activity of MAT I/III [29], which may contribute to decreased hepatic GSH level in these patients as SAMe administration normalized GSH levels [30]. The decrease in MAT activity occurs by two mechanisms. First is decreased *MAT1A* expression in the majority of cirrhotic patients [31]; second is inactivation of MAT I/III, likely due to covalent modification of critical cysteine residues [1]. In particular, the cysteine at position 121, conserved in rat, mice and human MAT I/III, but absent in MAT II, is a target of covalent modification by oxidation (i.e. by hydroxyl radical) or by the formation of a nitrosothiol [1]. Although this cysteine is not essential for activity, when it is covalently modified, the enzyme is inactivated [1]. The inactivation could be reversed by GSH and other thiol-reducing agents. Nitrosylation of Cys 121 of MAT I/III and its inactivation have been demonstrated both in vitro and in vivo in animals treated with lipopolysaccharide [32, 33]. Since oxidative and nitrosative stress occur in the majority of liver diseases, MAT I/III inactivation would be expected. Recently, we found decreased hepatic *MAT1A* expression and reduced SAMe levels in patients hospitalized for alcoholic hepatitis [34]. Some of these patients have only liver fibrosis. Thus, decreased hepatic *MAT1A* expression and SAMe levels should not be considered as complications seen only with end-stage cirrhosis.

Abnormal methionine metabolism is well documented in several animal models of alcoholic liver injury including baboons, micropigs, rats and mice [29, 35–37]. We showed in rats fed intragastric ethanol for 9 weeks that *MAT1A* and *MAT2A* mRNA levels doubled but only the protein level of *MAT2A* was comparably increased [37]. This represents an example of dissociation between the *MAT1A* mRNA and protein levels and the possibility of increased MAT I/III degradation in this experimental model is worthy of further investigation. Hepatic levels of methionine, SAMe and DNA methylation fell by ~40%. Reduced SAMe level was believed to contribute to abnormal mitochondrial membrane fluidity in rats fed ethanol [38] and decreased liver GSH level in baboons fed ethanol [35]. We examined what impact a change in DNA methylation may have in ethanol-induced liver injury and predisposition to cancer. In many types of cancer global DNA hypomethylation is common, although there is often regional hypermethylation [39]. A common hypothesis is that hypomethylation of growth-promoting protooncogenes and/or hypermethylation of tumor suppressor genes will alter the expression of these genes to promote a selective growth advantage for the initiated cell [40]. In addition, DNA hypomethylation may promote malignant transformation by inducing regional alterations in DNA conformation and chromatin structure rendering affected regions more accessible to DNA-damaging agents [40]. In our animal model of alcoholic injury, we found hypomethylation and increased expression of c-*myc*, and increased genome-wide DNA strand break accumulation [37]. These changes may predispose the liver to malignant degeneration.

Consequences of Chronic Hepatic SAMe Depletion

The *MAT1A* knockout mouse model has provided important insights with regards to the consequences of chronic hepatic SAMe depletion. This model is highly relevant to human liver disease as both *MAT1A* expression and hepatic SAMe levels are markedly reduced in the majority of cirrhotic patients. *MAT1A* knockout mice have markedly increased serum methionine levels, reduced hepatic SAMe (76% lower) and GSH (40% lower) levels [41]. This confirms the importance of *MAT1A* in systemic methionine catabolism and the influence of MAT expression on hepatic SAMe level. *MAT1A* knockout mice display a phenotype that resembles that observed in liver injury or stress with a vast array of genes involved in growth, de-differentiation and acute phase response upregulated [41]. At three months of age, *MAT1A* knockout mice have histologically normal liver but the livers are heavier and more susceptible to choline-deficient diet-induced fatty liver and CCl_4-induced liver injury [41, 42]. Following 2/3 partial hepatectomy, *MAT1A* knockout mice have impaired liver regeneration due to an inability to upregulate cyclin D1 and abnormal extracellular signal-regulated kinases signaling [43]. In addition, there is loss of responsiveness of the *MAT1A* knockout hepatocytes to mitogenic signals [43]. The molecular mechanisms of these abnormalities remain to be elucidated. By eight months, *MAT1A* knockout mice develop spontaneous non-alcoholic steatohepatitis (NASH) on a normal diet [41]; and by 18 months, the majority of the knockout mice develop HCC [42].

We used genomics and proteomics to better understand the phenotype of the *MAT1A* knockout mice. In addition to genes involved in growth and acute phase response, a number of genes involved in lipid and glucose metabolism are also altered [42]. In addition, both cytochrome P450 2E1 (CYP2E1) and uncoupling protein 2 are upregulated, which may have played a role in the propensity to develop NASH and predisposition to CCl_4-induced liver injury [42]. Another SAMe target important in the phenotype of the *MAT1A* knockout mice is prohibitin, which was identified using the proteomics approach [44]. We found prohibitin 1 to be downregulated at the protein level (but not mRNA) in the livers of *MAT1A* knockout mice from the time of birth to the development of NASH [44]. Prohibitin 1 is the product of a nuclear gene that is targeted to the inner mitochondrial membrane [45], and act as a chaperone-like protein that participates in the correct folding and assembly of some of the components of the mitochondrial respiratory chain [46, 47]. Consistent with this, we found both cytochrome C oxidase I and II protein levels to be downregulated [44]. These changes correlated with impaired mitochondrial function and likely also contribute to NASH development and propensity to liver injury. Indeed, prohibitin 1 protein levels are also diminished in obese patients and ob/ob mice [44]. In addition to mitochondrial targeting, prohibitin 1 has also been shown to target to the nucleus and act like a tumor suppressor by recruiting histone deacetylase and other co-repressors to silence the E2F genes, which are important for cell cycle progression [48]. This may be one mechanism to explain why a chronic reduction in SAMe favors HCC. Future research should elucidate the mechanism by which SAMe modulates prohibitin 1 protein level.

Liver Cancer in *MAT1A* Knockout Mice and Animals Fed Lipotrope-Deficient Diets

In mammals, choline is synthesized from phosphatidylethanolamine (PE). In this process, which takes place mainly in the liver [49], PE is converted into phosphatidylcholine by three successive *N*-methylations of the ethanolamine moiety of PE. In this reaction, which is catalyzed by the enzyme PE-*N*-methyltransferase (PEMT), SAMe is the donor of the three methyl groups. This pathway, however, is not sufficient to cover the daily requirements of choline. Thus, *PEMT* knockout mice fed a choline deficient diet develop fatal liver damage [50]. In humans and experimental cirrhosis in baboons, PEMT activity is markedly reduced [51, 52]. These results suggest that, although choline requirements in humans, especially in adults, are not well known, the maintenance of adequate dietary choline in cirrhosis may be particularly important to prevent the progression of the liver disease.

It is well known that when rats and mice are fed a diet deficient in lipotropes (choline, methionine, folate, and vitamin B12), the liver develops steatosis within a few days [53]. If the deficient diet continues, the liver develops NASH, fibrosis and cirrhosis, with some animals ultimately developing HCC [54]. Numerous nutritional studies have shown that dietary methyl insufficiency causes a decrease in hepatic levels of SAMe and an increase in SAH content [55, 56]. Data from the *MAT1A* null mice would suggest that the lipotrope deficiency exert its pathogenic effect in the liver through a decreased availability of SAMe. Furthermore, these results suggest that the deficiency in MAT activity observed in human liver cirrhosis may contribute to the pathogenesis and progression of the disease as well as predisposition to HCC.

SAMe Regulation of Hepatocyte Growth

How does chronic SAMe deficiency lead to HCC? One mechanism may be related to decreased prohibitin 1 level but recent data clearly show SAMe can regulate hepatocyte growth directly. SAMe level is related to the differentiation status of the hepatocyte. In this regard, quiescent and proliferating hepatocytes have high and low SAMe levels, respectively [22]. Following 2/3 partial hepatectomy, SAMe levels fall dramatically, coinciding with the onset of DNA synthesis and the induction of early response genes [57]. When this fall in SAMe was prevented by exogenous SAMe administration, hepatocyte DNA synthesis was inhibited [58, 59]. Additionally, exogenous SAMe inhibits the growth of hepatoma cells in culture [22] and prevents development of HCC in rats treated with hepatocarcinogen [60, 61]. The question is how SAMe exerts this growth modulatory response in hepatocytes.

Hepatocyte growth factor (HGF) is the most potent liver mitogen that can activate *MAT2A* transcriptionally and induce hepatocyte DNA synthesis [62]. *MAT2A* induction has been shown to be required for liver cell proliferation [63]. SAMe inhibited the mitogenic activity of HGF and prevented *MAT2A* induction [62]. Although HGF is a potent mitogen for cultured hepatocytes, it elicits a poor mitogenic response when administered in vivo to normal rats [64]. These observations suggest that hepatic parenchymal cells need to be primed in order to respond to proliferative signals. Tumor necrosis factor α (TNFα) and interleukin 6 are cytokines that may be part of this priming event [65–67]. These cytokines can induce the expression of the inducible nitric oxide synthase (iNOS) gene [68], which also occurs in the liver after partial hepatectomy and before the onset of DNA synthesis [69]. NO production seems to be important because inhibition of iNOS decreases hepatocyte proliferation and iNOS knockout mice have impaired liver regeneration [70]. Since NO can inactivate MAT I/III [1], the relationship between NO, SAMe and HGF's mitogenic response was examined [71]. The mitogenic activity of HGF was dependent on iNOS activity and a fall in SAMe level. Consistent with this, *MAT1A* knockout mice do not exhibit a fall in SAMe level following partial hepatectomy (since MAT II is not inactivated by NO) and *MAT1A* knockout hepatocytes are resistant to the mitogenic effect of HGF [43]. Taken together, these observations suggest that the fall in liver SAMe contents after partial hepatectomy, due to the NO-mediated inhibition of MAT I/III, releases the inhibitory tone that SAMe exerts on growth factors such as HGF and allows hepatocytes to proliferate. Although a transient fall in SAMe appears necessary for the liver to regenerate, chronic depletion would favor a malignant degeneration.

SAMe Regulation of Hepatocyte Death

Apoptotic cell death contributes to the development of many liver injuries that are ameliorated by SAMe treatment. Hence, it was important to examine the effect of SAMe on apoptosis. We found that while SAMe protected against okadaic acid-induced apoptosis in normal hepatocytes by blocking cytochrome c release, it induced apoptosis in liver cancer cell lines HepG2 and HuH-7 [72]. The same anti-apoptotic action in normal hepatocytes and pro-apoptotic action in liver cancer cells were observed with MTA. MTA can be derived from SAMe enzymatically and non-enzymatically. MTA is a product of SAMe metabolism in the polyamine pathway (fig. 1). Exogenous SAMe can also undergo non-enzymatic hydrolysis into MTA [73]. In contrast to SAMe, MTA does not contribute to GSH synthesis, is not a methyl donor and inhibits methyltransferases [74]. Thus, the death regulatory effects of SAMe are GSH-independent and may be mediated in part through its conversion to MTA. These observations are in agreement with the reported chemopreventive action of SAMe and MTA in an animal model of chemical hepatocarcinogenesis in rats, which was accompanied by an increase of apoptotic bodies in atypical nodules and HCC foci in SAMe treated animals [61, 75].

One mechanism of SAMe and MTA's differential effect on apoptosis in normal hepatocytes and liver cancer cells was recently identified [76]. In microarray studies, SAMe treatment induced Bcl-x expression. Bcl-x is alternatively spliced to produce two distinct mRNAs and proteins, Bcl-x_L and Bcl-x_S. Bcl-x_L is anti-apoptotic while Bcl-x_S is pro-apoptotic. SAMe and MTA induced selectively Bcl-x_S in a time- and dose-dependent manner in HepG2 cells by increasing alternative splicing of Bcl-x. Alternative splicing requires dephosphorylation of SR proteins, which are specific substrates of protein phosphatase 1. Interestingly, both SAMe and MTA increased phosphatase 1 catalytic subunit mRNA and protein levels and de-phosphorylation of SR proteins. Furthermore, inhibitors of protein phosphatase 1 blocked the ability of SAMe and MTA to induce Bcl-x_S expression and apoptosis in HepG2 cells. Importantly, SAMe and MTA treatment of normal human hepatocytes had no effect on protein phosphatase 1 catalytic subunit or Bcl-x_S expression. This can explain the differential effect these agents have on apoptosis in normal versus cancerous hepatocytes. Thus, in liver cancer cells SAMe and MTA have the ability to affect cellular phosphorylation state and alternative splicing of genes, in this case resulting in the induction of Bcl-x_S leading to apoptosis.

SAMe Treatment in Liver Disease Including Cancer

The importance of methyl groups in general, and SAMe in particular, to normal hepatic physiology, coupled with the convincing body of evidence linking abnormal SAMe content with the development of experimental and human liver disease, led to the examination of the effect of SAMe supplementation in a variety of animal models of liver disease. SAMe administration to alcohol-fed rats and baboons reduced GSH depletion and liver damage [29]. SAMe improved survival in animal models of galactosamine-, acetaminophen- and thioacetamide-induced hepatotoxicity, and in ischemia-reperfusion induced liver injury [3]. SAMe treatment also diminished liver fibrosis in rats treated with carbon tetrachloride [3], and reduced neoplastic hepatic nodules in animal models of HCC [58, 60, 61]. In patients with alcoholic cirrhosis, SAMe is the only treatment shown to improve survival in those with less advanced cirrhosis [77]. Various mechanisms including increased GSH levels, a change in DNA methylation, improved membrane fluidity, and decreased TNFα expression have been proposed for SAMe's protective action [35, 38, 60, 78, 79]. Our recent studies further illustrate the complexity of SAMe's actions on growth and death that are independent of its role as a methyl donor or GSH precursor. Given SAMe's differential effect on apoptosis in normal versus cancerous hepatocytes, SAMe is an attractive agent to consider both in chemoprevention and treatment of HCC. Finally, it would also be worthwhile to examine whether MTA can be used in the same settings since it is much more potent in vitro as compared to SAMe. Future studies should address the molecular mechanisms of the growth and death regulatory effects of SAMe and MTA, as well as their efficacy in chemoprevention and treatment of HCC.

Acknowledgements

This work was supported by NIH grants DK51719 (to S.C. Lu), AA12677, AA013847 and AT1576 (to S.C. Lu and J.M. Mato), P50 AA11999 (USC-UCLA Research Center for Alcoholic Liver and Pancreatic Diseases), and P30 DK48522 (USC Research Center for Liver Diseases), and Plan Nacional of I+D SAF2002–00168 of the Ministerio de Educación y Ciencia (to J.M. Mato).

References

1 Mato JM, Corrales FJ, Lu SC, Avila MA: S-Adenosylmethionine: a control switch that regulates liver function. FASEB J 2002;16:15–26.
2 Mudd SH, Poole JR: Labile methyl balances for normal humans of various dietary regimens, Metabolism 1975;24:721–735.
3 Mato JM, Alvarez L, Ortiz P, Pajares MA: S-adenosylmethionine synthesis: molecular mechanisms and clinical implications. Pharmacol Ther 1997;73:265–280.
4 Finkelstein JD: Methionine metabolism in mammals. J Nutr Biochem 1990;1:228–237.
5 Hoffman DR, Marion DW, Cornatzer WE, Duerre JA: S-adenosylmethionine and S-adenosylhomocysteine metabolism in isolated rat liver. Effects of L-methionine, L-homocysteine and adenosine. J Biol Chem 1980;255:10822–10827.

6   Lu SC: Regulation of hepatic glutathione synthesis: Current concept and controversies. FASEB J 1999;13:1169–1183.

7   Kotb M, Mudd SH, Mato JM, Geller AM, Kredich NM, Chou JY, Cantoni GL: Consensus nomenclature for the mammalian methionine adenosyltransferase genes and gene products. Trends Genet 1997;13:51–52.

8   Horikawa S, Tsukada K: Molecular cloning and developmental expression of a human kidney S-adenosylmethionine synthetase. FEBS Lett 1992;312:37–41.

9   Horikawa S, Ozasa H, Ota K, Tsukada K: Immunohistochemical analysis of rat S-adenosylmethionine synthetase isozymes in developmental liver. FEBS Lett 1993;330:307–311.

10   Gil B, Casado M, Pajares M, Boscá L, Mato JM, Martín-Sanz P, Alvarez L: Differential expression pattern of methionine adenosyltransferase isoenzymes during rat liver development. Hepatology 1996;24:876–881.

11   Horikawa S, Sasuga J, Shimizu K, Ozasa H, Tsukada K: Molecular cloning and nucleotide sequence of cDNA encoding the rat kidney Methionine adenosyltransferase. J Biol Chem 1990;265:13683–13686.

12   Mitsui K, Teraoka H, Tsukada K: Complete purification and immunochemical analysis of Methionine adenosyltransferase from bovine brain. J Biol Chem 1988;263:12211–12216.

13   Kotb M, Kredich NM: Methionine adenosyltransferase from human lymphocytes. Purification and characterization. J Biol Chem 1985;260:3923–3930.

14   Halim A, Leighton L, Geller A, Kotb M: Expression and functional interaction of the catalytic and regulatory subunits of human methionine adenosyltransferase in mammalian cells. J Biol Chem 1999;274:29720–29725.

15   LeGros HL Jr, Halim AB, Geller AM, Kotb M: Cloning, expression, and functional characterization of the β regulatory subunit of human methionine adenosyltransferase (MAT II). J Biol Chem 2000;275:2359–2366.

16   Martínez-Chantar ML, Garcia-Trevijano ER, Latasa MU, Martin-Duce A, Fortes P, Caballeria J, Avila MA, Mato JM: Methionine adenosyltransferase II β subunit gene expression provides a proliferative advantage in human hepatoma. Gastroenterology 2003;124:940–948.

17   Liau MC, Chang CF, Becker FF: Alteration of methionine adenosyltransferases during chemical hepatocarcinogenesis and in resulting carcinomas. Cancer Res 1979;39:2113–2119.

18   Okada G, Teraoka H, Tsukada K: Multiple species of mammalian S-adenosylmethionine synthetase. Partial purification and characterization. Biochemistry 1981;20:934–940.

19   Sullivan DM, Hoffman J: Fractionation and kinetic properties of rat liver and kidney methionine adenosyltransferase isozymes. Biochemistry 1983;22:1636–1641.

20   Cabrero C, Puerta J, Alemany S: Purification and comparison of two forms of S-adenoysl-L-methionine synthetase from rat liver. Eur J Biochem 1987;170:299–304.

21   Pajares MA, Duran C, Corrales F, Pliego M, Mato JM: Modulation of rat liver S-adnosylmethionine synthetase activity by glutathione. J Biol Chem 1992;267:17598–17605.

22   Cai J, Mao Z, Hwang JJ, Lu SC: Differential expression of methionine adenosyltransferase genes influences the rate of growth of human hepatocellular carcinoma cells. Cancer Res 1998;58:1444–1450.

23   Garcia-Trevijano ER, Latasa MU, Carretero MV, Berasain C, Mato JM, Avila MA: S-Adenosylmethionine regulates MAT1A and MAT2A gene expression in cultured rat hepatocytes: a new role for S-adenosylmethionine in the maintenance of the differentiated status of the liver. FASEB J 2000;14:2511–2518.

24   Martínez-Chantar ML, Latasa MU, Varela-Rey M, Lu SC, García-Trevijano E, Mato JM, Avila MA: L-Methionine availability regulates the expression of methionine adenosyltransferase 2A gene in human hepatocarcinoma cells. J Biol Chem 2003;278:19885–19890.

25   Yang HP, Huang ZZ, Zeng ZH, Chen CJ, Selby RR, Lu SC: Role of promoter methylation in increased methionine adenosyltransferase 2A expression in human liver cancer. Am J Physiol 2001;280:G184–G190.

26   Cai J, Sun WM, Hwang JJ, Stain S, Lu SC: Changes in S-adenosylmethionine synthetase in human liver cancer: molecular characterization and significance. Hepatology 1996;24:1090–1097.

27   Kinsell LW, Harper HA, Barton HC, Michaels GD, Weiss HA: Rate of disappearance from plasma of intravenously administered methionine in patients with liver damage. Science 1947;106:589–594.

28   Horowitz JH, Rypins EB, Henderson JM, Heymsfield SB, Moffitt SD, Bain RP, Chawla RK, Bleier JC, Rudman D: Evidence for impairment of transsulfuration pathway in cirrhosis. Gastroenterology 1981;81:668–675.

29   Tsukamoto HC, Lu SC: Current concepts in the pathogenesis of alcoholic liver injury. FASEB J 2001;15:1335–1349.

30   Vendemiale G, Altomare E, Trizio T, Le Grazie C, Di Padova C, Salerno MT, Carrieri V, Albano O: Effects of oral S-adenosyl-L-methionine on hepatic glutathione in patients with liver disease. Scand J Gastroenterol 1989;24:407–415.

31   Avila MA, Berasain C, Torres L, Martin-Duce A, Yang HP, Prieto J, Lu SC, Caballería J, Rodés J, Mato JM: Reduced mRNA abundance of the main enzymes involved in methionine metabolism in human liver cirrhosis and hepatocellular carcinoma. J Hepatol 2000;33:907–914.

32   Avila MA, Mingorance J, Martínez-Chantar ML, Casado M, Martín-Sanz P, Bascá L, Mato JM: Regulation of rat liver S-adenosylmethionine synthetase during septic shock: role of nitric oxide. Hepatology 1997;25:391–396.

33   Ruiz F, Corrales FJ, Miqueo C, Mato JM: Nitric oxide inactivates rat hepatic methionine adenosyltransferase in vivo by S-nitrosylation. Hepatology 1998;28:1051–1057.

34   Lee TD, Sadda ME, Mendler MH, Bottiglieri T, Kanel G, Mato JM, Lu SC: Abnormal hepatic methionine and GSH metabolism in patients with alcoholic hepatitis. Alcohol Clin Exp Res 2004;28:173–181.

35   Lieber CS, Casini A, DeCarli LM, Kim C, Lowe N, Sasaki R, Leo MA: S-adenosyl-L-methionine attenuates alcohol-induced liver injury in the Baboon. Hepatology 1990;11:165–172.

36   Villanueva JA, Halsted CH: Hepatic transmethylation reactions in micropigs with alcoholic liver disease. Hepatology 2004;39:1303–1310.

37   Lu SC, Huang ZZ, Yang HP, Mato JM, Avila MA, Tsukamoto H: Changes in methionine adenosyltransferase and S-adenosylmethionine homeostasis in the alcoholic rat liver. Am J Physiol 2000;279:G178–G185.

38   Colell A, Garcia-Ruiz C, Morales A, Ballesta A, Ookhtens M, Rodés J, Kaplowitz N, Fernandez-Checa JC: Transport of reduced glutathione in hepatic mitochondria and mitoplasts from ethanol-treated rats: effect of membrane physical properties and S-adenosyl-L-methionine. Hepatology 1997;26:699–708.

39   Baylin SB, Herman JG, Graff JR, Vertino PM, Issa JP: Alterations in DNA methylation: a fundamental aspect of neoplasia. Adv Cancer Res 1998;72:141–196.

40   Pogribny IP, Basnakian AG, Miller BJ, Lopatina NG, Poirier LA, James SJ: Breaks in genomic DNA and within the p53 gene are associated with hypomethylation in livers of folate/methyl-deficient rats. Cancer Res 1995;55:1894–1901.

41   Lu SC, Alvarez L, Huang ZZ, Chen LX, An W, Corrales FJ, Avila MA, Kanel G, Mato JM: Methionine adenosyltransferase 1A knockout mice are predisposed to liver injury and exhibit increased expression of genes involved in proliferation. Proc Natl Acad Sci USA 2001;98: 5560–5565.

42   Martínez-Chantar ML, Corrales FJ, Martínez-Cruz A, García-Trevijano ER, Huang ZZ, Chen LX, Kanel G, Avila MA, Mato JM, Lu SC: Spontaneous oxidative stress and liver tumors in mice lacking methionine adenosyltransferase 1A. FASEB J 2002;16:1292–1294.

43   Chen L, Zeng Y, Yang HP, Lee TD, French SW, Corrales FJ, García-Trevijano ER, Avila MA, Mato JM, Lu SC: Impaired liver regeneration in mice lacking methionine adenosyltransferase 1A. FASEB J 2004;18:914–916.

44   Santamaría E, Avila MA, Latasa MU, Rubio A, Martín-Duce A, Lu SC, Mato JM, Corrales FJ: Functional proteomics of non-alcoholic steatohepatitis: mitochondrial proteins as targets of S-adenosylmethionine. Proc Natl Acad Sci USA 2003;100:3065–3070.

45   Ikonen E, Fiedler K, Parton RG, Simons K: Prohibitin, an antiproliferative protein, is localized to mitochondria. FEBS Lett 1995;358:273–277.

46   Nijtmans LG, de Jong L, Artal Sanz M, Coates PJ, Berden JA, Back JW, Muijsers AO, van der Spek H, Grivell LA: Prohibitins act as a membrane-bound chaperone for the stabilization of mitochondrial proteins. EMBO J 2000;19:2444–2451.

47 Nijtmans LG, Artal SM, Grivell LA, Coates PJ: The mitochondrial PHB complex: roles in mitochondrial respiratory complex assembly, aging and degenerative disease. Cell Mol Life Sci 2002;59:143–155.

48 Wang S, Fusaro G, Padmanabhan J, Chellappan SP: Prohibitin co-localizes with Rb in the nucleus and recruits N-CoR and HDAC1 for transcriptional repression. Oncogene 2002;21:8388–8396.

49 Alemany S, Varela I, Mato JM: Inhibition of phosphatidylcholine synthesis by vasopressin and angiotensin in rat hepatocytes. Biochem J 1982;208:453–457.

50 Waite KA, Cabilio NR, Vance DE: Choline deficiency – induced liver damage is reversible in Pemt⁻/⁻ mice. J Nutr 2002;132:68–71.

51 Duce AM, Ortiz P, Cabrero C, Mato JM: S-adenosyl-L-methionine synthetase and phospholipid methyltransferase are inhibited in human cirrhosis. Hepatology 1988;8:65–68.

52 Lieber CS, Robins SJ, Leo MA: Hepatic phosphatidylethanolamine methyltransferase activity is decreased by ethanol and increased by phosphatidylcholine. Alcohol Clin Exp Res 1994;18: 592–595.

53 Best CH, Hershey JM, Huntsman AG: The effect of lecithin on fat deposition in the liver of the normal rat. J Physiol 1932;75:56–66.

54 Newberne PM: Lipotropic factors and oncogenesis. Adv Exp Med Biol 1986;206:223–251.

55 Shivapurkar N, Poirier LA: Tissue levels of S-adenosylmethionine and S-adenosylhomocysteine in rats fed methyl-deficient, amino acid-defined diets for one to five weeks. Carcinogenesis 1983;4:1051–1057.

56 Cook RJ, Horne DW, Wagner C: Effect of dietary methyl group deficiency on one-carbon metabolism in rats. J Nutr 1989;119:612–617.

57 Huang ZZ, Mao Z, Cai J, Lu SC: Changes in methionine adenosyltransferase during liver regeneration in the rat. Am J Physiol 1998;38:G14–G21.

58 Pascale RM, Simile MM, Satta G, Seddaiu MA, Daino L, Pinna G, Vinci MA, Gaspa L, Feo F: Comparative effects of L-methionine, S-adenosyl-L-methionine and 5'-methylthioadenosine on the growth of preneoplastic lesions and DNA methylation in rat liver during the early stages of hepatocarcinogenesis. Anticancer Res 1991;11:1617–1624.

59 Shivapurkar N, Hoover KL, Poirier LA: Effect of methionine and choline on liver tumor promotion by phenobarbital and DDT in diethylnitrosamine-initiated rats. Carcinogenesis 1986;8: 615–617.

60 Pascale RM, Marras V, Simile MM, Daino L, Pinna G, Bennati S, Carta M, Seddaiu MA, Massarelli G, Feo F: Chemoprevention of rat liver carcinogenesis by S-adenosyl-L-methionine: a long-term study. Cancer Res 1992;52:4979–4986.

61 Pascale RM, Simile MM, De Miglio MR, Nufris A, Daino L, Seddaiu MA, Rao PM, Rajalakshmi S, Sarma DS, Feo F: Chemoprevention by S-adenosyl-L-methionine of rat liver carcinogenesis initiated by 1,2-dimethylhydrazine and promoted by orotic acid. Carcinogenesis 1995;16:427–430.

62 Latasa MU, Boukaba A, García-Trevijano ER, Torres L, Rodríguez J, Caballería L, Lu SC, Lüpez-Rodas G, Franco L, Mato JM, Avila MA: Hepatocyte growth factor induces MAT2A expression and histone acetylation in rat hepatocytes. Role in liver regeneration. FASEB J 2001;15: 1248–1250.

63 Pañeda C, Gorospe I, Herrera B, Nakamura T, Fabregat I, Varela-Nieto I: Liver cell proliferation requires methionine adenosyltransferase 2A mRNA up-regulation. Hepatology 2002;35:1381–1391.

64 Michalopoulos GK, DeFrances MC: Liver regeneration. Science 1997;276:60–66.

65 Akerman P, Cote P, Yang SQ, McClain C, Nelson S, Bagby GJ, Diehl AM: Antibodies to tumor necrosis factor-alpha inhibit liver regeneration after partial hepatectomy. Am J Physiol 1992;263:G579–G585.

66 Yamada Y, Kirillova I, Peschon JJ, Fausto N: Initiation of liver growth by tumor necrosis factor: deficient liver regeneration in mice lacking type I tumor necrosis factor receptor. Proc Natl Acad Sci USA 1997;94:1441–1446.

67 Cressman DE, Greenbaum LE, DeAngelis RA, Ciliberto G, Furth EE, Poli V, Taub R: Liver failure and defective hepatocyte regeneration in interleukin-6-deficient mice. Science 1996;274:1369–1383.

68 Li J, Billiar T: Nitric oxide. IV. Determinants of nitric oxide protection and toxicity in liver. Am J Physiol 1999;276:G1069–G1073.

69 Obolenskaya M, Schulze-Specking A, Plaumann B, Frenzer K, Freudenberg N, Decker K: Nitric oxide production by cells isolated from regenerating rat liver. Biochem Biophys Res Commun 1994;204:1305–1311.

70 Rai RM, Lee FYJ, Rosen A, Yang SQ, Lin HZ, Koteish A, Liew FY, Zaragoza C, Lowenstein C, Diehl AM: Impaired liver regeneration in inducible nitric oxide synthase-deficient mice. Proc Natl Acad Sci USA 1998;95:13829–13834.

71 García-Trevijano ER, Martínez-Chantar ML, Latasa MU, Mato JM, Avila MA: NO sensitizes rat hepatocytes to proliferation by modifying S-adenosylmethionine levels. Gastroenterology 2002; 122:1355–1363.

72 Ansorena E, García-Trevijano ER, Martínez-Chantar ML, Huang ZZ, Chen LX, Mato JM, Iraburu M, Lu SC, Avila MA: S-adenosylmethionine and methylthioadenosine are anti-apoptotic in cultured rat hepatocytes but pro-apoptotic in human hepatoma cells. Hepatology 2002;35:274–280.

73 Simile MM, Banni S, Angioni E, Carta G, De Miglio MR, Muroni MR, Calvisi DF, Carru A, Pascale RM, Feo F: 5'-Methylthioadenosine administration prevents lipid peroxidation and fibrogenesis induced in rat liver by carbon-tetrachloride intoxication. J Hepatol 2001;34:386–394.

74 Dante R, Anaud M, Niveleau A: Effects of 5'-deoxy-5'-methylthioadenosine on the metabolism of S-adenosylmethionine. Biochem Biophys Res Commun 1983;114:214–221.

75 Garcea R, Daino L, Pascale R, Simile MM, Puddu M, Frasseto S, Cozzolino P, Seddaiu MA, Feo F: Inhibition of promotion and persistent nodule growth by S-adenosyl-L-methionine in rat liver carcinogenesis: role of remodeling and apoptosis. Cancer Res 1989;49:1850–1856.

76 Yang HP, Sadda MR, Li M, Zeng Y, Chen LX, Ou XP, Runnegar MT, Mato JM, Lu SC: S-Adenosylmethionine and its metabolite induce apoptosis in HepG2 cells: role of protein phosphatase 1 and Bcl-x_S. Hepatology 2004;40:221–231.

77 Mato JM, Cámara J, Ortiz P, Rodés J, Spanish Collaborative Group for the Study of Alcoholic Liver Cirrhosis: S-adenosylmethionine in the treatment of alcoholic liver cirrhosis: a randomized, placebo-controlled, double-blind multicentre clinical trial. J Hepatol 1999;30:1081–1089.

78 Chawla RK, Watson WH, Eastin CE, Lee EY, Schmidt J, McClain CJ: S-adenosylmethione deficiency and TNF-α in lipopolysaccharide-induced hepatic injury. Am J Physiol 1998;38:G125–G129.

79 Watson WH, Chawla RK: S-adenosylmethionine attenuates the lipopolysaccharide-induced expression of the gene for tumour necrosis factor alpha. Biochem J 1999;342:21–25.

Shelly C. Lu
HMR 415, Keck School of Medicine USC
2011 Zonal Ave.
Los Angeles, CA 90033 (USA)
Tel. +1 323 442 2441, Fax +1 323 442 3234
E Mail shellylu@usc.edu

Cho CH, Purohit V (eds): Alcohol, Tobacco and Cancer.
Basel, Karger, 2006, pp 175–188

....

Alcohol and Iron as Cofactors in Hepatocellular Carcinoma

Dennis R. Petersen

Department of Pharmaceutical Sciences, University of Colorado Health Sciences
Center, Denver, Colo., USA

Abstract

Oxidative stress is recognized as a central mechanism in the initiation and promotion
stages of carcinogenesis. The combined pro-oxidant potential of ethanol and iron ranges
from additive to synergistic with respect to initiating hepatocellular oxidative stress through
antioxidant depletion. Disorders such as hemochromatosis and alcoholic liver disease are
associated with significant oxidative stress and the hepatic accumulation of iron, both of
which are predisposing factors for development of hepatocellular carcinoma (HCC).
Sustained oxidative stress and redox imbalance associated with exposure to alcohol and iron
results in DNA damage and the activation of cellular transcription factors which, in turn, initiate or contribute to neoplastic progression of the liver. Evidence is now emerging that activation of the transcription factor nuclear factor-κB in hepatic macrophage plays a significant
role in initiation and propagation of early malignant clones that eventually lead to the development of HCC. Consequently, there is persuasive evidence that the potential of ethanol and
iron to induce oxidative stress is an important pathomechanism underlying development of
HCC. This review summarizes recently published studies, which describe the cellular events
and mechanisms potentially involved in initiation and progression of HCC in individuals
experiencing oxidative stress as a consequence of hepatic iron accumulation and chronic
alcohol ingestion.

Organisms and cells existing in the presence of oxygen are continually
confronted with toxic concentrations of reactive oxygen species (ROS) and
reactive nitrogen species (RNS). In addition, mammalian cells sequester various transition metals such as iron that are essential for their existence but also
have the potential to function as catalysts of autocatalytic reactions producing
cytotoxic concentrations of ROS and RNS. Thus, mammalian cells are faced
with a complex problem in that oxygen and iron are absolute requirements for

sustaining essential cellular processes but also contribute significantly to oxidative stress. Accordingly, mammalian cells have a battery of diverse enzymatic and antioxidant systems capable of neutralizing ROS as well as precisely regulated systems capable of storing iron in protein-associated complexes which limit its reactivity. The sustained overproduction of ROS/RNS results in diminished antioxidant systems, which are ultimately manifested in oxidative stress and cellular injury. This is consistent with the observation that oxidative stress is a central pathophysiologic element in a number of diseases associated with chronic inflammation including cardiovascular disease [1], iron storage diseases [2], cancer [3], alcoholic liver disease [4] and neurodegenerative diseases [5].

The observations that the prolonged ingestion of ethanol or accumulation of iron are manifested in organ dysfunction are significant. Therefore, it is reasonable to assume that, when present together, the interaction of these pro-oxidants would establish a cellular environment conducive to DNA damage and dysregulation of gene expression, both of which are significant events in carcinogenesis. This overview will examine the growing body of evidence suggesting that the combined pro-oxidative effects of alcohol and iron are involved in hepatocarcinogenesis. Particularly, the molecular mechanisms augmenting oxidative stress and potential cellular responses proposed to be involved in development of hepatocellular carcinoma (HCC) will be examined.

The Properties and Role of Iron as a co-Morbid Factor in Carcinogenesis

The role of iron, alone or in combination with other co-morbid variables, as a predisposing factor for various cancers has been problematic to establish. This is attributable to the fact that iron is among the most abundant transition elements on earth. Further, the bioactivity of iron is indisputably complex in that iron deficiency is manifested in anemia whereas excess stores of body iron results in cellular and organ damage.

The biological importance, as well as toxicity, of iron is attributable to the chemical properties of this transition metal, which facilitate its involvement in one-electron oxidation/reduction reactions involving the interconversion of ferric (Fe^{3+}) and ferrous (Fe^{2+}) states. Ferrous iron (Fe^{2+}) is characterized by the presence of 4 unpaired electrons whereas 5 unpaired electrons exist with Fe^{3+}. Using the definition that a free radical is any species containing one or more unpaired electrons and is capable of independent existence [6], both Fe^{2+} and Fe^{3+} can be classified as free radicals. As illustrated below, of the two forms of iron, Fe^{2+} is the most capable of redox cycling, a chemical property that explains the ability of iron to participate in Fenton redox chemistry.

$$Fe^{2+} + H_2O_2 \rightarrow Fe^{3+} + \,^{\bullet}OH + OH \tag{1}$$

$$Fe^{2+} + O_2 \rightarrow Fe^{3+} + O_2^{\bullet} \tag{2}$$

$$Fe^{3+} + O_2^{\bullet-} \rightarrow Fe^{2+} + O_2 \tag{3}$$

Reaction (1) illustrates the well-known Fenton reaction that produces the highly reactive hydroxyl radical ($^{\bullet}OH$) which rapidly interacts with cellular protein, DNA or lipids. Reaction (2) depicts the production of superoxide ($O_2^{\bullet-}$) from the oxidation of iron by molecular oxygen. As with all chemical reactions, an important predictor of the predominance of redox cycling reactions is the concentration of reactants [7]. Relevant to biological systems are the estimates that steady-state levels of H_2O_2 in liver cells are in the range of $10^{-8}\,M$ [8] whereas cellular concentrations of O_2 approximate $10^{-5}\,M$ [9]. Presuming equivalence of Fe^{2+} reaction rates with H_2O_2 or O_2, and an oxidizable substrate concentration which approaches $1\,M$ in biologic systems, it has been estimated that Reaction (2) would predominate over Reaction (1) by a factor of 10^8 [7, 10]. These calculations suggest the interaction of Fe^{2+} with dioxygen is potentially a significant source of ROS in biologic system. It is important to consider the reactivity of the species of reactive oxygen produced by Reactions (1) and (2). In the case of Reaction (1), the hydroxyl radical produced is the most reactive ROS known as evidenced by a highly positive reduction potential (2.31 V) which exceeds that of $O_2^{\bullet-}$ by a factor of 2.5. Indeed, it has been estimated that $^{\bullet}OH$ oxidizes molecular targets within 5 molecular diameters of its site of generation [11]. However, even though there is a significant difference in the reactivity of $O_2^{\bullet-}$ and $^{\bullet}OH$, the reduction of Fe^{2+} by $O_2^{\bullet-}$ (Reaction (3)) results in acceleration or a net increase in production of $^{\bullet}OH$ through the superoxide-assisted the Fenton reaction. Thus, the production of $O_2^{\bullet-}$ in the iron-catalyzed reactions presented in Reactions (1)–(3) propagates formation of the highly reactive $^{\bullet}OH$.

Hydroxyl radical reacts readily with all cellular constituents including proteins, lipids, nucleic acids, carbohydrates and a spectrum of organic acids. The reactions include, electron transfer, addition or, as demonstrated below in Reaction (4), hydrogen abstraction from molecules such as polyunsaturated fatty acids resulting in formation of carbon-centered lipid radical ($L^{\bullet}$).

$$^{\bullet}OH + LH \rightarrow L^{\bullet} + H_2O \tag{4}$$

$$Fe^{2+} + LOOH \rightarrow Fe^{3+} + LO^{\bullet} + OH^+ \tag{5}$$

As shown in Reaction (5), Fe^{2+} also readily interacts with lipid hydroperoxides to form a lipid alkoxyl radical ($LO^{\bullet}$). Progression of the Reactions (1)–(3) has well-documented detrimental effects on biological systems. The oxidation of iron in Reaction (1) yields $^{\bullet}OH$ which can extract a hydrogen atom from a lipid giving rise to a L• that can, in turn, extract a hydrogen atom from anther lipid molecule. Consequently, Fe^{2+} can mediate production of •OH which readily functions in the initiation stage of cellular lipid peroxidation. Under aerobic conditions, O_2 adds to the lipid radical producing the corresponding lipid peroxyl radical (LOO•). This peroxyl radical is readily reduced, by abstraction of a hydrogen atom from another lipid, to form a lipid hydroperoxide (LOOH) which undergoes decomposition via Fe^{2+} (Reaction (5)). Thus, Fe^{2+} also plays an important role in the autocatalytic propagation step of lipid peroxidation.

As illustrated in Scheme 1, the ability of iron to readily accept or donate electrons has significant impact on the peroxidative decomposition of membrane polyunsaturated fatty acids such as linoleic acid. It must be considered that this and other polyunsaturated fatty acids in biological systems are bound to membrane phospholipids, triglycerides and cholesterol. β-cleavage of the C–C double bonds can occur on either side hydroperoxy group which will yield a triglyceride- or phospholipids-bound aldehyde or a free aldehyde with a methyl terminus (i.e. 4-hydroxy-2-nonenal (4-HNE)). It is evident from Scheme 1 that, in addition to participating in the initiation and propagation steps of lipid peroxidation, iron is also involved in β-cleavage of the lipid hydroperoxide. This latter reaction also gives rise to numerous alkanes, alkanals and alkenals as well as several very electrophilic α, β-unsaturated hydroxyalkenals including malondialdehyde, acrolein and crotonaldehyde. The resulting biogenic aldehydes such 4-HNE, acrolein, crotonaldehyde and malondialdehyde have cellular half-lives upwards to 2 or 3 min and therefore are capable of diffusing from their site of production to more distant sites within the cell to interact with DNA or protein nucleophiles. It is the diffusible and electrophilic nature of these lipid peroxidative products that is central to their proposed involvement in carcinogenesis. Similarly, the ability of iron to enhance formation of •OH documented to attack sugars, purines and pyrimidines has important implications in DNA damage and mutations associated with cancer. Collectively, the reactions presented in Reactions (1)–(5) and outlined in Scheme 1 illustrate the iron-mediated pathways proposed to be involved in oxidative stress associated with chemical carcinogenesis [3].

Iron is the most abundant transition metal in the human body with estimates of total body iron concentrations of up to 6 g, of which 20–30% is stored in hepatocytes [12]. Therefore, given the abundance of iron in human body and the potential of this transition metal to participate in one-electron oxidation/reduction reactions, it is understandable that mammals have elaborate systems to regulate its accumulation and storage. New components and concepts involved in the molecular control of mammalian iron metabolism are emerging and discussed in two recent reviews [13, 14]. Dietary iron, existing primarily in the Fe^{3+} state is reduced

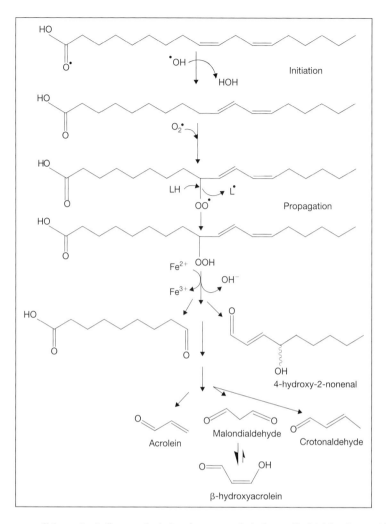

Scheme 1. A diagram depicting the autocatalytic free radical-initiated peroxidation of the polyunsaturated fatty acid, linolenate. As outlined in the text, chronic alcohol ingestion and hepatic iron overload result in the production of ROS and RNS that function in the initiation step to produce •OH. Fe^{2+} readily interacts with the lipid hydroperoxide (LOOH) to produce the corresponding lipid alkoxyl radical (LO•), one intermediate involved in the propagation step of lipid peroxidation. Further, Fe^{2+} is involved in degradation of the lipid hydroperoxide through β-cleavage of the carbon–carbon bonds adjacent to the hydroperoxide group, a reaction resulting in production of α, β-unsaturated hydroxyl aldehydes such as 4-HNE proposed to be involved in the codon-specific mutation of *p53*. Thus, catalytic iron is potentially involved in the initiation, propagation and degradation steps of lipid peroxidation. The peroxidation of polyunsaturated fatty acids also results in production a variety of other end products of lipid peroxidation including crotonaldehyde, acrolein and malondialdehyde which have a variety of biologic effects.

to the Fe^{2+} and absorbed at the level of the enterocyte by means of a divalent metal transporter, (DMT)-1. It is widely accepted that iron homeostasis is primarily regulated at the level of intestinal absorption. Once inside the enterocyte, iron is bound to the abundant cellular protein ferritin which has the capacity to sequester up to 4,500 molecules of iron in the Fe^{3+} state. In the instance that iron is needed for erythrocytes, or for other purposes, an activation of transport mechanisms, primarily the iron transport protein ferroportin also known as IRGE1, initiates transfer of iron from the enterocyte to the serum iron binding protein, transferrin (Tf). Each molecule of Tf avidly binds, in fully ligated form, one or two molecules of iron in the Fe^{3+} state thus decreasing the occurrence of participation in one-electron reduction/oxidation reactions. The uptake of circulating iron by hepatic and extrahepatic cells occurs via the interaction of Tf with Tf receptors (TfR) initiating cellular uptake of the TfR–Tf–iron complex that is subsequently internalized by endosomes. The acidic endosomal environment assists release of Fe^{3+} from the TfR–Tf complex, which is then processed for cellular release and recycling of Tf. Fe^{3+} is enzymatically reduced and sequestered by ferritin or enters other intracellular pools of iron characterized by iron bound to low molecular weight ligands such as ATP, ADP, GTP or citrate. Currently, approximations of the cellular concentrations of low molecular weight pools of iron range from $0.2\,\mu M$ in erythroid cells to $10\,\mu M$ in hepatocytes [15, 16]. Iron present in these pools is exchangeable, chelatable for uptake by ferritin or bioavailable for detection by iron responsive element-binding proteins that function as cellular regulators of iron homeostasis. These low molecular weight chelates of iron are also catalytically reactive in oxidant formation and are responsible for iron toxicity [17]. Similarly, the catalytic nature of these low molecular weight iron chelates very likely plays important roles in iron co-carcinogenicity. Regulation of intracellular iron concentrations occurs via modulation of TfR mRNA stability, which is increased when intracellular iron concentrations are low whereas stability is decreased when cellular iron concentrations are sufficient.

As noted in a series of recent reviews, discovery of the protein hepcidin has led to the conclusion that the liver plays a major role in the regulation of iron absorption, sequestration and mobilization [14, 18, 19]. This regulatory hormone produced predominantly in the liver as the 84 amino-acid prohepcidin is present in blood as the cysteine-rich, hepcidin-derived peptides consisting of 25, 22, and 20 amino acids. The concentration of hepcidin in blood generally reflects the body stores of iron. Enhanced iron uptake by enterocytes and the release of iron from reticuloendothelial macrophage occurs in response to low levels of hepcidin. Conversely, increased secretion of hepcidin results in decreased iron absorption and the retention of iron by reticuloendothelial macrophage. Thus, in instances of iron deficiency, hypoxia or anemia, hepcidin expression/secretion is decreased. Replete iron storage or inflammation is associated with decreased dietary iron absorption and sequestration of iron by reticuloendothelial macrophage. Iron Tf saturation is one potential signal for stimulation or inhibition of hepatic hepcidin. As discussed in the following sections, the physiological accumulation of iron by the liver is linked to the development of HCC, suggesting that dysregulation in secretion of hepcidin could be an important variable is this and other diseases associated with iron overload.

Several metals, including arsenic, nickel and chromium are recognized carcinogens while the carcinogenicity or co-carcinogenicity of iron remains controversial. The questionable cause-effect association between iron exposure and cancer is confirmed in recent comprehensive reviews [7, 20, 21]. These reviews, as well as others [22, 23] provide compelling evidence of associations between body iron stores of iron and colorectal cancer as well as liver cancer. It is significant that the colon is exposed to high concentrations of iron derived from the diet while significant concentrations of iron are stored in the liver. The high incidence of human HCC in individuals with the hereditary iron storage disease hemochromatosis suggests that iron may function as a co-carcinogen in the liver [7, 20]. The essential role of iron in cell cycle progression and proliferation of neoplastic cells has recently been reviewed [24]. Likewise, the use of various therapeutic approaches to sequester iron and inhibit the proliferation of neoplastic cells underscores the importance of iron in tumor cell growth [21]. However, the use of human subjects often necessitates indirect measurements of body stores of iron based on Tf or ferritin saturation and do not provide information concerning the cellular concentrations of free or 'catalytic' iron that may be of greatest mechanistic importance in the carcinogenic or co-carcinogenic process.

A recent review [25] summarizes a series of studies strongly suggesting that oxidative stress is mechanistically associated with liver cancer. Relevant to the potential role of transition metals such as iron, is a recent report presents evidence of an increased frequency of *p53* mutation in non-tumorous liver obtained from patients with the Wilson disease, a syndrome associated with copper accumulation, and patients with hemochromatosis [26]. This communication reports significantly higher frequencies of G:C to T:A transversions and at codon 249 and C:G to A:T transversions in codon 250 in patients with Wilson disease whereas the C:G to A:T transversions were most prevalent in patients with hemochromatosis. In addition, parallel in vitro studies using TK-6 lymphoblastoid cells treated with 4-HNE revealed an increase in G–T transversions which also occurred at codon 249 of *p53*. These experimental results indicate that subjects with cooper or iron storage display oxidative stress leading to generation of 4-HNE that specifically results in mutations of the *p53* tumor suppressor gene. These observations are supported by a series of well-designed mechanistic studies characterizing the formation of DNA base adducts with various aldehydic products of lipid peroxidation. These experiments evaluated the iron catalyzed peroxidation of ω-3 and ω-6 lipid substrates in the presence of deoxyguanosine reported the formation of cyclic adducts with acrolein, crotonaldehyde and 4-HNE [27]. Further, these same investigators [28] have used nuclease excision repair to establish that 4-HNE preferentially forms DNA adducts at codon 249 of the human *p53* gene suggesting this is a targeted mutational site for HCC. Subsequent studies have demonstrated that glutathione depletion enhances formation of these adducts further establishing oxidative stress as an integral mechanistic component [29]. Collectively, these studies establish a mechanistic basis for the association between iron accumulation, oxidative stress and production of reactive lipid aldehyde capable of forming a site-specific mutation potentially linked to HCC.

The Combined Pro-Oxidant Actions of Ethanol and Iron in Hepatocellular Carcinogenesis

As reviewed elsewhere, different experimental approaches point to oxidative stress as a central factor involved in alcohol-induced liver injury [4]. Recognition of specific free radical species which function as initiators of oxidative stress in biological systems have proven difficult since they are short-lived due to their chemical reactivity. For instance, although superoxide anion ($O_2^{\bullet-}$) is reasonably stable in aqueous solutions at pH 7.4, and its reaction rates with lipids, proteins or DNA are quite slow, the reaction of this radical with Fe^{3+} is significant proceeding at a rate of $1.5 \times 10^8 \ M^{-1}s^{-1}$. Therefore, the predicated half-life of $O_2^{\bullet-}$ in biological systems which contain iron is extremely limited. An example of a biologically relevant reactions include reduction of $O_2^{\bullet-}$ to H_2O_2 which can serve as a source for the reactive hydroxyl anion ($^{\bullet}OH$). In addition, the rapid reaction of $O_2^{\bullet-}$ with nitric oxide $NO^{\bullet}$ ($1.9 \times 10^{-10} \ M^{-1}s^{-1}$) results in formation of the very potent nitrating species, peroxynitrite ($ONOO^-$). Accordingly, $O_2^{\bullet-}$ likely serves as a precursor for an assortment of reactions which generate molecules with marked cytotoxic potential.

The liver is the primary organ involved in the oxidation of ethanol. During this process, the production of the ethanol-derived α-hydroxyethyl radical ($CH3C^{\bullet}HOH$) has been detected in the livers of mice and rats treated chronically with alcohol [30, 31]. This reactive intermediate, derived from the complex catalytic interactions of ethanol with cytochrome P4502E1, is proposed to mediate its hepatotoxic effects through adduct formation with cellular proteins [32]. Given the high reactivity of $O_2^{\bullet-}$, $^{\bullet}OH$, Fe^{2+}, $ONOO^-$ and $CH3C^{\bullet}HOH$ it is predictable that biomolecules (i.e. proteins, lipids and nucleic acids) within close proximity of cellular microenvironments producing these radicals would be most susceptible to damage. Given this scenario, it is likely that the ongoing production of such free radicals, as a consequence of oxidative stress, contributes to the cocarcinogenic actions of alcohol and iron. This proposition is consistent with a recent, comprehensive evaluation of the relevant literature [33] linking this complex panel of pro-oxidants to oxidant-induced DNA damage leading to mutations, alterations of gene expression and cellular signaling cascades manifested in uncontrolled cell proliferation.

Two recent reviews present convincing evidence that oxidative stress is a common pathway by which chronic alcohol ingestion and iron induce liver injury [4, 34]. Additionally, hepatic iron overload develops in a significant proportion of individuals who consume alcohol on a chronic basis [35–37]. These observations are consistent with reports of a significant increase in hepatic iron concentrations in animal models of chronic alcohol consumption [38, 39]. Consequently, the presence of both pro-oxidants would noticeably increase oxidative stress resulting in the production of aldehydic products of lipid peroxidation such as 4-HNE that have been implicated in targeting codon 249 of *p53*. Therefore, it is predictable that the occurrence of HCC in individuals who consume alcohol on a regular basis should be higher as compared to those subjects who do not drink alcohol and the risk of developing HCC in subjects who have hereditary hemochromatosis and consume alcohol should be considerable. For instance results of a recent report indicate that heavy alcohol consumption, viral hepatitis and diabetes exert independent and synergistic effects on the risk of developing HCC [40]. When data from this study were adjusted for confounding factors, subjects consuming more than 4 alcohol-containing drinks per day displayed a 3-fold increased risk of HCC. As noted, individuals with hereditary hemochromatosis who consume alcohol would be expected to be at increased risk of developing HCC. This proposition is supported by a recent study documenting a high degree of heterozygosity for the *HFE* C282Y mutation in patients with alcoholic cirrhosis and HCC [41]. While cirrhosis is a recognized risk factor for development of HCC, these results suggest this *HFE* mutation could be a predisposing factor for HCC. Evidence from a study investigating the incidence of cirrhosis in *HFE* C282Y homozygous hemochromatosis subjects strongly suggest that alcohol consumption contributes to the development of hepatic fibrosis and cirrhosis [36]. Assessment of subjects with respect to hepatic iron accumulation and alcohol consumption indicated that subjects with hemochromatosis consuming more than 60 g of alcohol per day were nearly 10 times more likely to develop cirrhosis. This study, which involved 224 subjects, reported the incidence of HCC was 10%. A striking feature of the individuals documented to have HCC was that they all consumed more than 60 g of alcohol per day further supporting a role for alcohol and iron in development of HCC. Taken together, these studies imply a potential mechanistic association of ethanol, iron and the hemochromatosis gene (*HFE*) C282Y mutation in development of HCC.

Hepatic Macrophage (Kupffer cells) and Nuclear Factor-kB (NF-kB) as Targets of Pro-Oxidant-Induced Hepatocarcinogenesis

Admittedly, the pathomechanisms responsible for the synergistic actions of ethanol and iron in the liver are multiple and predictably complex involving oxidative stress, depletion of antioxidant systems, activation of Kupffer cells, the modulation of redox-sensitive transcription factors and activation of cellular stress-activated signaling cascades such as the mitogen-activated protein kinase cascade [42]. However, data obtained from studies using specific animal models of alcohol-induced liver injury indicate that Kupffer cells play an important role in cellular responses associated with inflammation and cell death. In one pivotal study, the histochemical analysis of liver sections obtained from rats fed an alcohol-containing diet for 8 weeks revealed the deposition of ferric iron primarily in the centrilobular region with comparable deposition of the iron in both hepatocytes and Kupffer cells [38]. These results were confirmed through the observation that the non-heme iron content of Kupffer cells prepared from livers of rats treated chronically with ethanol was elevated 1.8-fold above that of Kupffer cells isolated pair-fed animals [43]. Interestingly, the elevated non-heme iron concentrations in Kupffer cells isolated from alcohol-treated rats was associated with significant elevations in electron paramagnetic resonance-radical generation, NF-kB binding, and increased mRNA for tumor necrosis factor alpha (TNF-α) and macrophage inflammatory protein. Exposure of isolated Kupffer cells to the iron chelator, deferiprone, restored all of these parameters to the levels observed in Kupffer cells isolated from control rats, clearly implicating a mechanistic role iron in these cellular responses. These results have important implications in terms of the potential role of iron as a signaling molecule in modulating the redox-sensitive transcription factor, NF-kB, and, in turn, hepatic TNF-alpha. In addition to its well-documented proinflammatory properties [44], initiation of TNF-alpha mediated hepatotoxicity has been linked to the actions of lipopolysacchardie (*LPS*) [45], oxidative stress [46] and perturbation of the mitochondrial permeability transition pore [47]. Therefore, it is likely that accumulation of iron by Kupffer cells is a key event for activation or sensitization of these cells to stimuli such as LPS or other extracellular effectors.

The responsiveness of the redox-sensitive transcription factor, NF-kB, to iron was evaluated in a recent study which demonstrated that physiologically relevant concentrations of ferrous iron ($10 \mu M$) increased I-kappa B kinase activity,

NF-kB binding and TNF-N release by cultured rat Kupffer cells [48]. Again, iron chelation blocked these responses confirming that Fe^{2+} is a central signaling molecule in this cascade. Cultured hepatic Kupffer cells subjected to Fe^{59} followed by stimulation with LPS were used as a cellular model system to evaluate the intracellular pools of iron involved in NF-kB activation [49]. Exposure of Kupffer cells to LPS or TNF-alpha resulted in a transient sequestration of Fe^{59} into intracellular low-molecular weight iron complexes (LMW-Fe) which was followed by a decrease of the labeled intracellular LMW-Fe and subsequent activation of I-kappa B kinase activity and NF-kB. The effectiveness of iron chelation in blocking these responses suggest the involvement of LMW-Fe complexes which are thought to be the cellular pools of iron involved in cytotoxicity [17]. Further, these investigators used specific inhibitors to implicate involvement of superoxide and peroxynitrite in modulation of signaling events mediated by iron in Kupffer cells suggesting the involvement ROS and RNS in mobilization of iron from intracellular iron storage proteins such as ferritin for translocation to and from the LMW-Fe complexes. These systematic studies reveal the intricate cellular mechanisms responding to oxidative stress that in turn modulate intracellular iron pools. Of central importance is the association of iron mobilization with activation of proinflammatory cellular signaling cascades involving NF-kB. Whereas the responses of Kupffer cells to various stimuli are usually beneficial, it is conceivable that sustained oxidative stress associated with chronic alcohol ingestion and/or hepatic iron overload could play a significant role in the initiation and promotion stages of carcinogenesis.

The potential role of NF-kB activation in liver carcinogenesis has been the subject of a number of recent investigations. Whereas it is beyond the scope of this chapter to summarize these investigations, a recent insightful review of this redox-sensitive transcription factor in liver carcinogenesis has been published [50]. This comprehensive review summarizes numerous independent investigations implicating NF-kB activation as an early event in the multistep processes involved in initiation and progression of HCC. Especially compelling is the evidence suggesting the involvement of NF-kB in transitions involving the modulation of growth factors, inflammation, compensatory cellular proliferation, cell survival and neoplastic progression. This growing body of literature point to the possibility of developing pharmacologic agents which modulate NF-kB activation to selectively destroy malignant liver cells.

Perspectives and Future Directions

Although not supported by extensive epidemiological data, the results of a number of independent studies suggest that the interaction of alcohol and iron at the cellular level could play significant roles in the initiation and promotion stages of carcinogenesis, especially in the liver. Certainly, the evidence that ethanol or iron function independently as relatively potent pro-oxidants is very convincing. When present together, the pro-oxidant actions of these agents are at least additive and possibly synergistic resulting in profound cellular oxidative stress which is a critical event in carcinogenesis. This is evident from the mounting evidence pointing to the potential of end products of lipid peroxidation such as 4-HNE that target the *p53* tumor suppressor gene as an important mechanistic event in HCC. More recent experimental results also indicate that hepatic macrophages are targets of iron mobilization in response to ROS and RNS, the result of which is modulation of intricate signaling cascades which activate NF-kB. One result is production of proinflammatory cytokines such as TNF-alpha which further create a cellular environment conducive to the molecular events of carcinogenesis. It is clear that oxidative stress is a consequence of continued exposure to ethanol and iron. However, the mechanisms by which oxidative stress resulting from these two agents is linked to HCC are complex. Future studies directed at maintaining the oxidant balance in hepatocytes and regulating iron homeostasis in Kupffer cells will provide additional mechanistic information as well as insight into unique therapeutic approaches to prevent HCC.

Acknowledgement

This work was supported by National Institutes of Health grants NIH/AA RO1 AA09300.

References

1   Molavi B, Mehta JL: Oxidative stress in cardiovascular disease: molecular basis of its deleterious effects, its detection, and therapeutic considerations. Curr Opin Cardiol 2004;19:488–493.
2   Wang XW, Hussain SP, Huo T-I, Wu C-G, Forgues M, Hofseth LJ, Brechot C, Harris CC: Molecular pathogenesis of human hepatocellular carcinoma. Toxicology 2002;181–182:43–47.
3   Klaunig JE, Kamendulis LM: The role of oxidative stress in carcinogenesis. Annu Rev Pharmacol Toxicol 2004;44:239–267.
4   Arteel GE: Oxidants and antioxidants in alcohol-induced liver disease. Gastroenterology 2003;124:778–790.
5   Shelly ML: (*E*)-4hydroxynonenal may be involved in the pathogenesis of Parkinson's disease. Free Radic Boil Med 1998;25:169–174.
6   Halliwell B, Gutteridge JMC: *Free radicals in biology and medicine*, ed 3. India; Thompson Press, 2003.
7   Huang X: Iron overload and its association with cancer risk in humans: evidence of iron as a carcinogenic metal. Mutat Res 2003;533:153–171.
8   Boveris A, Cadenas E: Cellular Sources and Steady-State Levels of Reactive Oxygen Species. New York, Marcel Dekker, Inc., 1997.
9   Jones DP: Intracellular diffusion of O_2 and ATP. Am J Physiol 1986;250:C663–C675.

10   Qian SY, Buettner GR: Iron and dioxygen chemistry is an important route to initiation of biological free radical oxidations: an electron paramagnetic resonance spin trapping study. Free Radic Biol Med 1999;26:1447–1456.

11   Pryor WA: Oxy-radicals and related species: their formation, lifetimes, and reactions. Annu Rev Physiol 1986;48:657–667.

12   Wriggleworth JM, Baum H: The biochemical functions of iron; in Jacobs A, Woodward M (eds): Iron in Biochemistry and Medicine. London, Academic Press, 1980, pp 29–86.

13   Hentze MW, Muckenthaler MU, Andrews NC: Balancing acts: control of mammalian iron metabolism. Cell 2004;117:285–297.

14   Papanikolaou G, Pantopoulos K: Iron metabolism and toxicity. Toxicol Appl Pharmacol 2005;202:199–211.

15   Cabantchik ZI, Glickstein H, Milgram P, Breuer W: A fluorescence assay for assessing chelation of intracellular iron in a membrane model system of mammalian cells. Anal Biochem 1996;233: 221–227.

16   Petrat F, Rauen H, deGroot H: Determination of the chelatable iron pool of isolated rat hepatocytes by digital fluorescence microscopy using the fluorescent probe, phengreen SK. Hepatology 1999;29:1171–1179.

17   Kruszewski M: Labile iron pool: the main determinant of cellular response to oxidative stress. Mutat Res 2003;531:81–92.

18   Frazer DM, Anderson GJ: Iron imports: intestinal iron absorption and its regulation. Am J Physiol 2005;289:631–635.

19   Vyoral D, Petrak J: Hepcidin: a direct link between iron metabolism and immunity. Int J Biochem Cell Biol 2005;37:1768–1773.

20   Deugnier Y: Iron and liver cancer. Alcohol 2003;30:145–150.

21   Deugnier YM, Loreal O: Iron as a carcinogen; in Barton J, Edwards C (eds): Hemochromatosis. Cambridge, Cambridge University Press 2000, pp 239–249.

22   Toyokuni S: Iron and carcinogenesis; from Fenton reaction to target genes. Redox Report 2002;7: 189–191.

23   Toyokuni S: Iron-induced carcinogenesis: the role of redox regulation. Free Radic Biol Med 1996;20:553–566.

24   Le NTV, Richardson DR: The role of iron in cell cycle progression and proliferation of neoplastic cells. Biochim Biophys Acta 2002;1063:31–46.

25   Chung FL, Choudhury S, Roy R, Hu W, Tang MS: Formation of trans-4-hydroxy-2-nonenal- and other enal-derived cyclic DNA adducts from omega-3 and omega-6 polyunsaturated fatty acids and their roles in DNA repair and human p53 gene mutation. Mutat Res 2003;531:25–26.

26   Hussain SP, Raja K, Amstad PA, Sawyer M, Trudel LJ, Wogan GN, Hofseth LJ, Shields PG, Billiar TR, Trautwein C, Hohler T, Galle PR, Phillips DH, Markin R, Marrogi AJ, Harris CC: Increased p53 mutation load in nontumorous human liver of Wilson disease and hemochromatosis: oxyradical overload disease. Proc Natl Acad Sci USA 2000;97:12770–12775.

27   Pan J, Chung FL: Formation of cyclic deoxyguanosine adducts from omega-3 and omega-6 polyunsaturated fatty acids oxidative conditions. Chem Res Toxicol 2002;15:367–372.

28   Hu W, Feng Z, Eveleigh J, Iyer G, Pan J, Amin S, Chung F-L, Tang M-S: The major lipid product of trans-4-hydroxy-2-nonenal, preferentially forms DNA adducts at codon 249 of human p53 gene, a unique mutational hotspot in hepatocellular carcinoma. Carcinogenesis 2002;23: 1781–1789.

29   Chung FL, Komninou D, Zhang L, Nath R, Pan J, Amin S, Richie J: Glutathione depletion enhances the formation of endogenous cyclic DNA adducts derived from t-4-hydroxy-2-nonenal in rat liver. Chem Res Toxicol 2005;18:24–27.

30   Knecht KT, Bradford BU, Mason RP, Thurman RG: In vivo formation of a free radical metabolite of ethanol. Mol Pharmacol 1990;38:26–30.

31   Kono H, Rusyn I, Yin M, Gabele E, Yamashina S, Dikalova A, Kadiiska MB, Connor HD, Mason RP, Segal BH, Bradford BU, Holland SM, Thurman RG: NADPH oxidase-derived free radicals are key oxidants in alcohol-induced liver disease. J Clin Invest 2000;106:867–872.

32   Stewart SF, Vidali M, Day CP, Albano E, Jones DE: Oxidative stress as a trigger for cellular immune responses in patients with alcoholic liver disease. Hepatology 2004;39:197–203.

33 Klaunig JE, Kamendulis LM: The role of oxidative stress in carcinogenesis. Annu Rev Pharmacol Toxicol 2004;44:239–267.

34 Tavill AS, Qadri AM: Alcohol and iron. Semin Liver Dis 2004;24:317–325.

35 Suzuki M, Fujimoto Suzuki Y, Hoski Y, Saito H, Sakurai S, Kohgo Y: Up-regulation of transferrin receptor expression in hepatocytes by habitual alcohol drinking is implicated in hepatic iron overload in alcoholic liver disease. Alcohol Clin Exp Res 2002;26:26S–31S.

36 Fletcher LM, Dixon JL, Purdie DM, Powell LW, Crawford DHG: Excess alcohol greatly increases the prevalence of cirrhosis in hereditary hemochromatosis. Gastroenterol 2002;122:281–289.

37 Fletcher IM, Halliday JW, Powell IW: Interrelationships of alcohol and iron in liver disease with particular reference to the iron-binding proteins, ferritin and transferrin. J Gastroenterol Hepatol 1999;14:202–214.

38 Valerio LG, Parks T, Petersen DR: Alcohol mediates increases in hepatic and serum non-heme iron in a rodent model for alcohol-induced liver injury. Alcohol Clin Exp Res 1996;20:1352–1361.

39 Tsukamoto H, Horne W, Kamimura S, Niemela O, Parkkila S, Yla-Hert S, Brittenham GM: Experimental liver cirrhosis induced by alcohol and iron. J Clin Invest 1995;96:620–630.

40 Yuan J-M, Govindarajan S, Kazuko A, Yu M-C: Synergism of alcohol, diabetes and viral hepatitis on the risk of hepatocellular carcinoma in blacks and whites in the US. Cancer 2004;101:1009–1017.

41 Lauret E, Rodriguez M, Gonzalez S, Linares A, Lopez-Vazquez A, Martinez-Borra J, Rodrigo L: *HFE* gene mutations in alcoholic and virus-related cirrhotic patients with hepatocellular carcinoma. Am J Gastroenterol 2002;97:1016–1021.

42 Hoek JB, Pastorino JG: Cellular signaling mechanisms in alcohol-induced liver damage. Semin Liver Dis 2004;24:257–271.

43 Tsukomoto H, Lin M, Ohata M, Giulivi C, French SW, Brittenham G: Iron primes hepatic macrophages for NFkappaB activation in alcoholic liver injury. Am J Physiol 1999;277:G1240–G1250.

44 Baeuerle PA: The inducible transcription activator NF-kB; regulation by distinct protein subunits. Biochim Biophys Acta 1991;1072:63–80.

45 McClain CJ, Barve S, Deaciuc I, Kugelmas M, Hill D: Tumor necrosis factor and alcoholic liver disease. Semin Liver Dis 1999;19:205–219.

46 Zhou Z, Wang L, Song Z, Lambert JC, McClain CJ, Kang J: A critical involvement of oxidative stress in acute alcohol-induced hepatic TNF-A production. Am J Pathol 2003;163:1137–1146.

47 Soriano ME, Nicolosi L, Bernardi P: Desensitization of the permeability transition pore by cyclosporine A prevents activation of the mitochondrial apoptotic pathway and liver damage by tumor necrosis factor-α. J Biol Chem 2004;279:36803–36808.

48 She H, Xiong S, Lin M, Zandi E, Giullivi C, Tsukamoto H: Iron activates NF-kappaB in Kupffer cells. Am J Physiol Gastrointest Liver Physiol 2002;283:G719–G726.

49 Xiong S, She H, Takeuchi H, Han B, Engelhardt JF, Barton CH, Zandi E, Giulivi C, Tsukamoto H: Signaling role of intracellular iron in NF-kB activation. J Biol Chem 2003;278:17646–17654.

50 Arsura M, Cavin LG: Nuclear factor-κB and liver carcinogenesis. Cancer Lett 2005;229:157–169.

Dennis R. Petersen, PhD
Department of Pharmaceutical Sciences, Box C238
University of Colorado Health Sciences Center, 4200 East 9th Avenue
Denver, CO 80262 (USA)
Tel. +1 303 315 6159, E-Mail dennis.petersen@uchsc.edu

Cho CH, Purohit V (eds): Alcohol, Tobacco and Cancer.
Basel, Karger, 2006, pp 189–204

········· ············

Tobacco Use and Cancer: An Epidemiologic Perspective

Michael J. Thun

American Cancer Society, Atlanta, Ga., USA

Abstract

Over 1.3 billion people use tobacco worldwide. Tobacco products currently account for more cancer deaths than can be attributed to all other known causes of the disease. Although lung cancer is the most common type of cancer caused by smoking, tobacco products are causally related to at least 14 other types of cancer, and cause even more deaths from cardiovascular and respiratory diseases than from cancer. This review considers the carcinogenic substances in tobacco and tobacco smoke, the multitude of cancers caused by tobacco use, the carcinogenicity of second-hand smoke, the health benefits of cessation, and the 'environmental' factors (social, economic, technological, and political) that continue to promote the global spread of manufactured cigarettes. Even as the wealthiest countries of the world have made progress in reducing tobacco use, the multinational tobacco companies have expanded their markets in economically developing countries and among women.

Over 1.3 billion people use tobacco worldwide [1, 2]. Tobacco products currently account for more cancer deaths than can be attributed to all other known causes of the disease [3]. Cigarette smoking alone causes approximately 20% of cancer deaths globally [4], about 30% in economically developed countries, and as much as 50% among men in some populations [5]. The proportion in women is smaller, but is increasing because of aggressive cigarette marketing.

Cancer is only one of many detrimental effects of tobacco use. About half of all long-term smokers (those who start in adolescence or young adulthood and cannot quit) will eventually be killed by tobacco [6]. Although lung cancer is the most common cancer caused by smoking, and has become the most common type of cancer in the world [7], it is rarely the predominant disease by which smoking kills people. Analyses of deaths caused by smoking show that vascular deaths (stroke and coronary heart disease) outnumber smoking-attributable lung cancer deaths in North America [8]; chronic obstructive pulmonary disease predominates in China, and tuberculosis in India [6].

The geographic distribution of cancers and other diseases caused by smoking has also begun to shift from economically developed to developing countries. About 80% of the 1,142 million current smokers now live in low and middle-income countries [9]. Consequently, the number of deaths caused by smoking in low and middle-income countries will soon surpass than in the high income countries [10].

Carcinogens in Tobacco and Tobacco Smoke

All currently available tobacco products deliver a complex mixture of toxic and carcinogenic chemicals to the user [3]. Tobacco smoke is a heterogeneous mixture containing approximately 4,000 compounds, more than 50 of which are known to cause cancer in humans and/or animals [3]. Some carcinogens, such as arsenic, cadmium, chromium, nickel, and polonium 210 are taken up by the tobacco plant from soil, pesticides, and/or phosphate fertilizers. Others, such as polycyclic aromatic hydrocarbons are produced during pyrolysis; still others accumulate during processing and curing. While nicotine itself is not carcinogenic, addiction to nicotine is a major obstacle to smoking cessation.

The levels of 'tar' and nicotine in cigarettes, as measured by machine smoking, bear little relationship to the actual amount of carcinogens or nicotine delivered to the smoker. Unlike smoking machines, humans can modify their manner of smoking to maintain a desired nicotine intake [11, 12]. Cigarettes that are marketed as 'Light', 'Ultralight', and 'Low Tar' do not reliably deliver less 'tar', nor do they result in less lung cancer risk than cigarettes labeled 'regular' or 'full flavored' [3, 13].

The concentration of specific carcinogens in tobacco and/or tobacco smoke has also fluctuated over time due to changes in processing or manufacturing tobacco products. For example, the introduction of reconstituted tobacco in the 1950s enabled manufactures to include ribs and stems as well as tobacco leaf in cigarettes [14], greatly increasing exposure to tobacco specific nitrosamines (TSNAs). Exposure to TSNAs is also increased by the use of gas heaters to cure tobacco [15] and the use of fermentation to produce cigars and moist snuff [14]. The concentration of TSNAs in moist snuff sold in America is currently as much as 38 times higher than the concentration in Swedish snus [16].

The Rise of Manufactured Cigarettes

Most tobacco use prior to the twentieth century involved snuff, chewing tobacco, pipes, and/or cigars rather than cigarette smoking [17]. Not until the

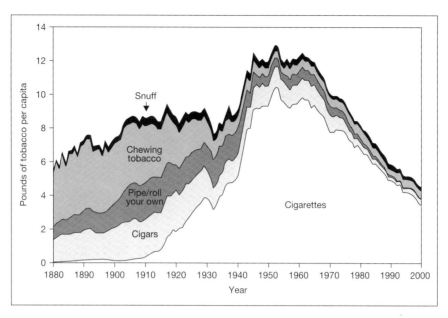

Fig. 1. Adult (age ≥18 years) per capita consumption of different forms of tobacco in the USA, 1880–2000. Adapted from [68].

early 20th century did cigarette smoking begin to displace these products in North America. Two inventions that facilitated the rise of manufactured cigarettes were the development of portable safety matches [18] and a cigarette rolling machine that could mass produce low cost cigarettes [19]. Figure 1 depicts the rapid increase in per capita consumption of cigarettes in the United States, beginning in the early 20th century. The rate of increase was accelerated by mass advertising campaigns that glamorized particular brands and by the distribution of free cigarettes to soldiers in World Wars I and II. By the mid-1940s, sales of manufactured cigarettes had largely displaced other forms of tobacco in most Western countries.

Within countries, the uptake of manufactured cigarettes and the ensuing epidemic of cancer and other diseases caused by smoking follows a remarkably predictable pattern. Lopez et al. [20] have developed a conceptual model that illustrates the usual progression of events (fig. 2). Initially (stage 1), cigarette smoking is largely confined to young and middle-aged men. No more than 20% of adults smoke and there is as yet no apparent increase in smoking-attributable diseases. In stage 2, the prevalence of cigarette smoking among adults increases to over 50% in men and to 30–40% in women. Smoking initiation occurs at progressively younger ages, and there is a small increase in deaths from lung

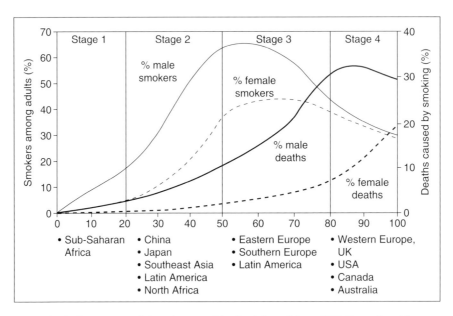

Fig. 2. Four stages of the tobacco epidemic. Adapted from [20]. Reproduced by permission of the BMJ Publishing Group.

cancer and other smoking-attributable disease in men. Stage 3 involves a substantial downturn in smoking prevalence among men, a more gradual decline in women, and a convergence of male and female smoking prevalence. The burden of diseases caused by smoking continues to increase in this stage. Smoking attributable deaths comprise 10–30% of all deaths, about three quarters of these being in men. Stage 4 of the epidemic is characterized by a marked downturn in smoking prevalence in both men and women. Deaths attributable to smoking among men peak at 30–35% of all deaths (40–45% of deaths in middle aged men) and subsequently decline. Among women, smoking-attributable deaths rise to about 20–25% of all deaths. This model is used by the World Health Organization to allow virtually any country to find itself in relation to the larger pandemic. It illustrates that the full consequences of lifetime smoking are not manifest until manufactured cigarettes have been used extensively for 80–100 years.

Cancers Caused by Tobacco Use

Cigarette smoking causes at least 15 different types of cancer (table 1) [3]. The strongest associations with both current and former smoking are with

Table 1. Smoking attributable mortality in the United States, 1995–1999, by disease, smoking status, and sex[a]

Disease	Relative risk estimates by smoking status							
	men		women		attributable percent of deaths		number of deaths	
	current	former	current	former	men (%)	women (%)	men	women
Cancers								
Lip, oral cavity, pharynx	10.9	3.4	5.1	2.3	75	50	3,900	1,300
Esophagus	6.8	4.5	7.8	2.8	73	57	6,300	1,600
Stomach	2.0	1.5	1.4	1.3	29	11	2,200	600
Pancreas	2.3	1.2	2.3	1.6	23	24	3,100	3,400
Larynx	14.6	6.3	13.0	5.2	83	75	2,500	600
Trachea, lung, bronchus	23.3	8.7	12.7	4.5	88	72	80,600	44,200
Cervix uteri	na	na	1.6	1.1	na	12	na	500
Urinary bladder	3.3	2.1	2.2	1.9	47	29	3,700	1,100
Kidney, other urinary tract	2.7	1.7	1.3	1.1	39	5	2,800	200
Acute myeloid leukemia	1.9	1.3	1.1	1.4	25	11	800	300
All vascular diseases[b]	–	–	–	–	21	12	87,600	55,000
All respiratory diseases[b]	–	–	–	–	58	46	53,700	44,300

[a]Source: Adapted from [8].
[b]Relative risk estimates vary by disease and (for coronary heart disease and stroke) by age.

cancers of the lung and larynx. However, the toxic and carcinogenic constituents of tobacco smoke reach virtually every organ in the body, so it is not surprising that cigarette smoking causes malignancies at multiple sites throughout the respiratory, gastrointestinal, genitourinary, and hematopoeitic systems.

Trachea, Bronchus, and Lung

All smoked tobacco products (cigarettes, pipes, cigars, bidis etc.) are known to cause lung cancer in humans. In Western countries, approximately 90% of lung cancer deaths in men and over 70% in women could be prevented by the avoidance of cigarette smoking [3, 17, 21]. Cigarette smoking is also an established cause of all histological types of lung cancer. Studies conducted in the 1950s linked smoking primarily to Kreyberg Type 1 lung cancers (squamous and small cell carcinomas). More recent studies have also established an association with Kreyberg Type 2 cancers (adenocarcinoma or large cell carcinoma) [22–24], although not as strong as that with Type 1 cancers [8, 25]. Two changes in cigarette design may have strengthened the association between smoking and adenocarcinoma. First, the introduction of filter-tip and lower nicotine cigarettes motivates smokers to inhale more deeply in order to maintain their accustomed intake of nicotine. This transports carcinogens in tobacco smoke to the periphery of the lung, where most pulmonary adenocarcinomas occur. A second factor that potentially contributed to the rise in adenocarcinoma was the introduction of reconstituted tobacco in the 1950s, as mentioned above. TSNAs released from combustion of tobacco ribs and stems are potent inducers of pulmonary adenocarcinoma in rodents, even when injected [25].

Lung cancer risk is more strongly influenced by the duration of smoking than by any other parameter of smoking [26, 27]. Hence, the relative risk (RR) of lung cancer death is approximately 50 among men who have smoked 40 cigarettes per day for 40 or more years [28]. Within countries, the average RR associated with current cigarette smoking continues to increase with the time period that cigarette smoking has been entrenched in that country, due to progressively earlier age of initiating regular smoking. In the large American Cancer Society cohort, Cancer Prevention Study II, the median delay between the initiation of smoking and death from lung cancer is 50 years [29]. This protracted delay may help to explain national lung cancer rates in countries such as Japan, where the current incidence and death rates among Japanese men appear unexpectedly low, given the current prevalence of cigarette smoking. It is known that the limited availability of cigarettes in Japan after World War II delayed the resumption of regular cigarette smoking for much longer in Japan than in the West. It is possible that this, rather than differences in genetic susceptibity, accounts for the disproportionately low lung cancer rates.

The lifetime probability that a smoker will develop lung cancer is also increasing over time. It is often said that 'only 10% of cigarette smokers develop lung cancer' [30, 31]. However, the actual probability is conditional on the duration and intensity of smoking, the manner in which cigarettes are smoked, and on competing causes of death. Based on a study of U.S. veterans conducted in the 1950s [32], Mattson et al. [33] estimated that 9.3% of men who smoked <25 cigarettes per day at age 35 would develop lung cancer by age 85, and 17.9% of those who smoked ≥25 cigarettes per day [33]. More recent analyses of the large American Cancer Society cohort indicate that the cumulative probability of death from lung cancer, not conditioned on surviving other causes of death, is 14.6% in male smokers and 8.3% in female smokers through age ≥85 years [29]. These compare with cumulative probabilities of 1.1% and 0.9% among male and female lifelong non-smokers respectively. If the impact of competing causes of death were excluded from this calculation, the lifetime probabilities would be 24.1% and 11.0% in male and female smokers respectively, and 1.6% and 1.1% in men and women never smokers. The latter estimates are more relevant for estimating the fraction of genetically susceptible persons in the population than are the unconditional percentages, but even these substantially underestimate the probability of premature death from smoking, because they do not consider all of the other adverse effects on survival.

Upper Aerodigestive Tract Cancers

All forms of tobacco use cause cancer throughout the upper respiratory and digestive tracts (upper aerodigestive tract) of humans. Combustion products in tobacco smoke come into direct contact with the oral and nasal cavities, pharynx, larynx, and respiratory epithelium. Tobacco leaf and extracts of tobacco dissolved in saliva reach tissues of the oropharynx, larynx, and digestive tract. For many aerodigestive tract cancers, the combination of tobacco smoking and heavy alcohol consumption results in a higher risk than would be expected from the sum of the risk from either exposure alone, as discussed below. Abstinence from either tobacco or alcohol would prevent many cancers of the oropharynx, larynx, and esophagus.

Lip, Oral Cavity and Pharynx

All forms of tobacco use cause dysplasia and squamous cell carcinoma of the oral cavity and pharynx. The magnitude of risk from cigar and pipe smoking is similar to that from cigarettes. Use of chewing tobacco, snuff, and/or betel also causes dysplasia and cancer, often at the location that directly contacts the tobacco product. On average, the RR of death from oropharyngeal cancer among persons who currently and exclusively smoke cigarettes is approximately eleven in men and five in women compared to lifelong non-smokers (table 1) [3, 8]. The extent to which tobacco use combined with heavy alcohol consumption magnifies the risk of either exposure alone is illustrated in figure 3. In this large population-based case-control study, the RR associated with smoking ≥40 cigarettes per day for ≥20 years is 7.4 in men who drank <1 alcoholic drink per week (and undefined among women); however, in those who drink >30 alcoholic drinks per week, the RR associated with this level of smoking is 37.7 in men and 107.9 in women [34]. After cessation of smoking, the RR of oropharyngeal cancer decreases substantially within the first 10 years after quitting [35]. Premalignant oral lesions such as leukoplakia and erythroplasia regress after cessation of tobacco use [36].

Nasal Cavity and Paranasal Sinuses

Cancer of the nasal cavity and paranasal sinuses are rare and were not designated as causally related to smoking until 2004 [3]. Case-control studies in the U.S., Europe, Hong Kong, and Japan report RR estimates of about two in

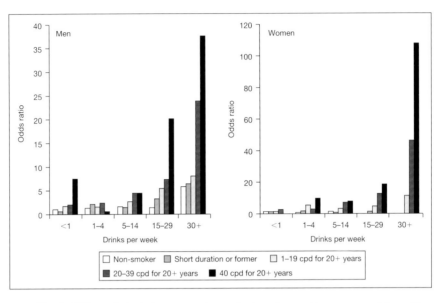

Fig. 3. Effect of alcohol consumption and cigarette smoking compared to neither drinking nor smoking on oropharyngeal cancer incidence. Adapted from [34].

current smokers compared to lifelong non-smokers for squamous cell carcinoma [3]. The association is weaker than that between cigarette smoking and cancers of the lung and larynx. However, tissues in the nasal passages are only exposed to the constituents of mainstream cigarette smoke during exhalation, or through the bloodstream.

Larynx

Cancer of the larynx is second only to lung cancer in the strength of its association with cigarette smoking (table 1). The risk of death from laryngeal cancer among current smokers, compared to lifelong non-smokers is similar for men (RR = 14.6) and women (RR = 13.0) (table 1). Several population-based case control studies report RRs of 15 and above in men who smoke more than one pack of cigarettes per day [8]. Cancer of the larynx is extremely rare among lifelong non-smokers. Risk increases with the duration and intensity of smoking and decreases rapidly after the cessation of smoking [37]. The combination of tobacco smoking with heavy alcohol consumption greatly increases the RR for laryngeal cancer [38–42], although most studies have not been formally evaluated for statistical interaction.

Esophagus

Tobacco smoking has been designated a cause of esophageal cancer for over 20 years [43, 44]. The risk of esophageal cancer among cigar and pipe smokers is similar to that of cigarette smokers [3]. The two major subtypes of esophageal cancer are squamous cell carcinoma, usually located in the upper two-thirds of the esophagus, and adenocarcinoma that occurs in the lower third of the esophagus or gastroesophageal junction. Studies that do not separate these histologic subtypes report RRs of 4–7 associated with current cigarette smoking [3]. This risk increases with the amount and duration of smoking and decreases with earlier age at cessation [3, 8]. Strong interaction is seen with combined exposure to tobacco and alcohol [3]. In case control studies that can separate by histological subtype, smoking is associated with both squamous cell and adenocarcinoma of the esophagus (studies reviewed in [3, 8]). However, the association is stronger for squamous cell carcinoma. Several studies suggest that black (air cured) tobacco may confer a greater risk of esophageal cancer than blond (flue cured) tobacco [45, 46].

Other Gastrointestinal

Stomach

Over 20 cohort studies and nearly 40 case-control studies have reported an association between tobacco smoking and stomach cancer, with RR estimates averaging approximately 1.6 in current cigarette smokers compared to never smokers [3]. Until recently, stomach cancer was not classified as smoking-related because of uncertainty about potential confounding by *Helicobacter pylori* (*H. pylori*) infection and diet [44, 47]. However, several case-control studies have stratified the analysis based on *H. pylori* seropositivity and reported substantially stronger associations between smoking and stomach cancer among persons who are seropositive for *H. pylori* than among uninfected individuals (reviewed in [3, 8]). There is some evidence that tobacco smoking may adversely affect the progression of intestinal metaplasia to dysplasia in *H. pylori*-affected people. In case control studies, smoking appears to be associated with cancers of both the gastric cardia and elsewhere in the stomach [3]. It has been estimated [48] that worldwide, the proportion of stomach cancer attributable to smoking is 11% in men and 4% in women in economically developing countries and 17% among men and 11% among women in developed countries.

Pancreas

Numerous cohort and case-control studies report an association between cigarette smoking and cancer of the exocrine pancreas (reviewed in [3]). The risk of pancreatic cancer is also increased among persons who smoke cigars and/or pipes in most large studies [49, 50]. The evidence that tobacco smoke is carcinogenic to the human pancreas was classified as sufficient by IARC in 1986 [44] and by the U.S. Surgeon General in 1982 [51]. The RR of death from pancreatic cancer among male and female current cigarette smokers, compared to never smokers, is 2.3 for both men and women in the ACS Cancer Prevention Study II [8]. Risk becomes lower in persons who stop smoking compared to those who continue in both cohort and case-control studies (reviewed in [3]).

Liver

At least 22 cohort and 27 case-control studies have examined the relationship between tobacco smoking and hepatocellular carcinoma (reviewed in [3, 8]). Most of these studies report an association between smoking and liver cancer, with RR estimates of about 1.5–2.5 [3]. Liver cancer was not classified as smoking-related by IARC in 1986, however, because of uncertainty about potential confounding by hepatitis virus infection and heavy consumption of alcohol [44, 47]. Hepatitis B virus infection causes most liver cancer worldwide, and hepatitis C virus infection accounts for a large fraction of the disease in Japan, North Africa, and Southern Europe [52]. Heavy, but not moderate, consumption of alcohol also contributes to risk [53].

Recent studies have resolved these concerns about confounding. Several studies have reported a higher risk of liver cancer among non-drinking smokers compared to non-drinking non-smokers. Smoking is also associated with liver cancer among Chinese [54] and Japanese [55] women in whom heavy alcohol consumption is extremely rare [3]. Several studies have stratified on or adjusted for HbsAg and anti-hepatitis C virus and found little attenuation of the association between smoking and liver cancer. The risk of chronic infection with hepatitis viruses was not higher in smokers than non-smokers in one study [56]; however, compared to never-smokers, smokers did experience greater risk of progression from chronic hepatitis B virus and hepatitis C virus infection to liver cirrhosis [57] and/or liver cancer [58]. The designation of cancers of the liver and stomach as causally related to smoking has a much larger impact on smoking-attributable diseases in countries such as China and India than in the U.S.

Intestinal Tract

There is no evidence that tobacco increases the risk of cancer of the small intestine, although this cancer is so rare that most studies have little or no power to study it. Neither the U.S. Surgeon General report [8] nor IARC [3] designate colorectal cancer among the cancer sites causally related to smoking, although there is increasing evidence that it may be. The risk of colorectal adenomatous polyps is associated with current cigarette smoking in three prospective studies (reviewed in [59]). With one exception, the RR estimates associated with current smoking range between 1.5 and 3.8, adjusting for age and other covariates. Most prospective cohort studies of colon cancer report RR estimates ranging between 1.2 and 1.4 for current cigarette smoking. Most prospective studies of rectum cancer report RR estimates between 1.4 and 2.0. However, there has not yet been a systematic meta-analysis that evaluates this association across all studies, controlling for other factors known to increase or decrease the risk of colorectal cancer. Cancer of the anus, a malignancy with squamous or transitional cell histology, has repeatedly been found to be positively associated with cigarette smoking [60], although, confounding by human papillomas virus has not been excluded.

Urinary Tract

Tobacco smoking is strongly associated with transitional cell carcinomas of the urinary bladder, ureters, and renal pelvis, especially in men (table 1). Smoking is also associated with adenocarcinoma of the renal parenchyma (reviewed in [3, 8]). Renal adenocarcinomas constitute the majority of kidney cancers. The U.S. Surgeon General does not distinguish between adenocarcinoma and transitional cell carcinoma in estimating that cigarette smoking accounts for 26% of deaths from cancers of the kidney, ureter, and urethra in the U.S. (40% in men and 5% in women) [8].

The risk of bladder cancer among male current cigarettes smokers is approximately 2–3 times higher than that of never smokers in prospective studies from the U.S., Japan, and Europe. Metabolites of heterocyclic aromatic amines, polycyclic aromatic hydrocarbons, and other carcinogens in tobacco can be detected in urine (reviewed in [3]). Smokers have a higher prevalence of preneoplastic changes in the bladder. DNA adducts have been detected in exfoliated urothelial cells from cigarette smokers. There is limited evidence that the association between smoking and bladder cancer is stronger for smoking black than blond tobacco. Meta-analyses of case-control and case series studies of smoking in relation to bladder cancer have reported a stronger association among persons with the slow acetylator N-acetyltransferase phenotype [61]. The aromatic amines 4-aminobiphenyl and 2-napthylamine are known to cause bladder cancer in occupationally-exposed populations and are thought to contribute to the increased risk of bladder cancer in smokers.

Leukemia

Acute myeloid leukemia is the only hematopoeitic malignancy consistently associated with smoking (reviewed in [3, 8]). Tobacco smoke contains several leukemogens such as benzene [62] and radioactive isotopes of polonium and lead. Smokers have much higher levels of benzene in the blood than non-smokers. Based on linear extrapolation from the known effects at high doses, Korte et al. [63] have estimated that benzene in cigarettes may account for 12–58% of smoking-induced myeloid leukemia.

Uterine Cervix

Cancer of the uterine cervix, predominantly involving squamous cell carcinoma, is consistently associated with cigarette smoking in many studies. The association has now been designated as causal [3, 8]. The main uncertainty, now resolved, concerned the possibility of confounding by sexually transmitted diseases. Since the identification of human papillomavirus (HPV) infection as the main cause of cervical cancer [64], studies have examined whether tobacco smoking acts as a co-factor with HPV infection in causing progression from preneoplastic lesions to cancer. Analyses have either been restricted to study participants who are positive for HPV DNA or have tried to adjust for HPV infection. In the IARC multicenter, pooled analysis of invasive cervical cancer, restriction did not materially alter the association between smoking and risk [65]. Smoking is associated with a spectrum of cervical abnormalities from dysplasia to cervical intraepithelial neoplasia, cervical cancer in situ, and invasive squamous cell carcinoma. Risk increases with increasing intensity and duration of smoking. Whereas smoking is associated with increased risk of squamous cell carcinoma of the uterine cervix, it was associated with lower risk of adenocarcinoma of the cervix in one multicenter case-control study [66].

Secondhand Smoke

Many scientific consensus committees have concluded that exposure to environmental tobacco smoke causes lung cancer in humans [3]. Environmental tobacco smoke exposure (also called secondhand smoke, passive smoking, and involuntary exposure to tobacco smoke) is associated with increased lung cancer risk among non-smokers married to smokers in more than 50 studies (reviewed in [3]).

The Health Benefits of Cessation

The risk of developing or dying from cancer or other adverse effects of tobacco use is substantially lower in people who quit than in those who continue using tobacco products. This provides major opportunities for prevention of tobacco-related diseases. It also indicates that continuing exposure affects even the late stages of carcinogenesis. The health benefits from cessation are greatest when successful quitting occurs at an early age but are substantial even when cessation occurs by age 50 or 60 [67].

Conclusions

Epidemiologic studies over the last half-century have demonstrated the immense burden of disease caused by tobacco use. Population studies reveal the scope of the global pandemic caused by manufactured cigarettes, its temporal progression within and across countries, and the remarkable number of disease it causes. All of the major 'environmental' factors that promote tobacco use (economic, social, and political), are highly modifiable. The wealthiest countries of the world have made substantial progress in reducing tobacco use at the population level, whereas the opposite is true in many economically developing countries.

Acknowledgements

The author thanks Lindsey Hannan and Jane Henley for their help in preparing this manuscript.

References

1   Guindon GE, Boisclair D: Past, current, and future trends in tobacco use. HNP Discussion Paper 2003.
2   Thun MJ, Costa e Silva VL: Introduction and overview of global tobacco surveillance; in Shafey O, Guindon GE, Dolwick S (eds): Tobacco Control Country Profiles. Atlanta, GA, American Cancer Society, 2003.
3   IARC: Tobacco Smoke and Involuntary Smoking. IARC Monographs on the Evaluation of the Carcinogenic Risk of Chemicals to Humans. Lyon, International Agency for Research on Cancer, 2004, vol 83, p 1452.
4   Ezzati M, et al: Role of smoking in global and regional cancer epidemiology: current patterns and data needs. Int J Cancer 2005;116:963–971.
5   Peto R, et al: Mortality from smoking in developed countries 1950–2000. New York, NY, Oxford University Press, 1994.
6   Peto R, Lopez A: The future worldwide health effects of current smoking patterns; in Boyle P, et al (eds): Tobacco and Public Health: Science and Policy, Oxford, England, Oxford University Press, 2004, pp 281–286.
7   Parkin D, Pisani P, Ferlay J: Estimates of the worldwide incidence of 25 major cancers in 1990. Int J Cancer 1999;80:827–841.
8   U.S. Department of Health and Human Services, The Health Consequences of Smoking: A Report of the Surgeon General. Atlanta, Georgia, U.S., Department of Health and Human Services,

Centers for Disease Control and Prevention, National Center for Chronic Disease Prevention and Health Promotion, Office on Smoking and Health, 2004, p 941.

9   Jha P, Chaloupka F: The economics of global tobacco control. Br Med J 2000;321:358–361.

10   Ezzati M, Lopez AD: Estimates of global mortality attributable to smoking in 2000. Lancet 2003;362:847–852.

11   Jarvis M, et al: Nicotine yield from machine-smoked cigarettes and nicotine intakes in smokers: evidence from a representative population survey. J Natl Cancer Inst 2001;93:134–138.

12   Benowitz N: Compensatory smoking of low-yield cigarettes; in Shopland D, et al (eds): Risks Associated with Smoking Cigarettes with Low Machine-Measured Yields of Tar and Nicotine. Smoking and Tobacco Control Monograph No. 13, Bethesda, MD, U.S. Department of Health and Human Services, National Institutes of Health, National Cancer Institute, NIH Pub. No. 02–5074, 2001, pp 39–63.

13   Harris J, et al: Cigarette tar yields in relation to lung cancer mortality in the Cancer Prevention Study-II prospective cohort, 1982–8. Br Med J 2004;328:72–76.

14   Hoffmann D, Hoffmann I: The changing cigarette, 1950–1995. J Toxicol Environ Health 1997;50:307–364.

15   Peele DM, Riddick MG, Edwards ME: Formation of tobacco-specific nitrosamines in flue-cured tobacco. Recent Adv Tob Sci 2001;27:3–12.

16   Osterdahl BG, Jansson C, Paccou A: Decreased levels of tobacco-specific N-nitrosamines in moist snuff on the Swedish market. J Agric Food Chem 2004;52:5085–5088.

17   Doll R: Uncovering the effects of smoking: historical perspective. Stat Methods Med Res 1998;7:87–117.

18   Anonymous. Matches. 2001 [cited 12/5/2001].

19   Slade J: The tobacco epidemic: lessons from history. J Psychoactive Drugs 1989;21:281–291.

20   Lopez A, Collishaw N, Piha T: A descriptive model of the cigarette epidemic in developed countries. Tob Control 1994;3:242–247.

21   Medical Research Council: Tobacco smoking and cancer of the lung. Br Med J 1957;1:1523.

22   Doll R, Hill A, Kreyberg L: The significance of cell type in relation to the aetiology of lung cancer. Br J Cancer 1957;ii:43–48.

23   Kreyberg L: Histological lung cancr types: a morphological and biological correlation. Acta Pathol Microbiol Immunol Scand Suppl 1962;157:11–92.

24   Wynder E, Hoffmann D: Smoking and lung cancer: scientific challenges and opportunities. Cancer 1994;54:5284–5295.

25   Thun M, et al: Cigarette smoking and changes in the histopathology of lung cancer. J Natl Cancer Inst 1997;89:1580–1586.

26   Doll R, Peto R: Cigarette smoking and bronchial carcinoma: dose and time relationships among regular smokers and lifelong non-smokers. J Epidemiol Community Health 1978;32:303–313.

27   Flanders W, et al: Lung cancer mortality in relation to age, duration of smoking, and daily cigarette consumption: results from Cancer Prevention Study II. Cancer Res 2003;63:6556–6562.

28   Thun M, et al: Age and the exposure-response relationships between cigarette smoking and premature death in Cancer Prevention Study II; in Shopland D, et al (eds): Changes in Cigarette-related Disease Risks and Their Implication for Prevention and Control. Smoking and Tobacco Control Monograph No. 8. Bethesda, MD, U.S. Department of Health and Human Services, National Institutes of Health, National Cancer Institute, NIH Pub. No. 97–4213 1997, pp 383–475.

29   Thun M, Henley S, Calle E: Tobacco use and cancer: an epidemiologic perspective for geneticists. Oncogene 2002;21:7307–7325.

30   Mabry M, Nelkin B, Baylin S: Lung Cancer; in Vogelstein B, Kinzler K (eds): The Genetic Basis of Human Cancer. New York, McGraw-Hill, 1998, pp 671–679.

31   Spivack S, et al: The molecular epidemiology of lung cancer. Crit Rev Toxicol 1997;27:319–365.

32   Kahn H: The Dorn study of smoking and mortality among U.S. veterans: report on eight and one-half years of observation; in Haenszel W (ed): Epidemiological Study of Cancer and Other Chronic Diseases. Bethesda, MD, U.S. Department of Health, Education, and Welfare, Public Health Service, National Cancer Institute, 1966, pp 1–126.

33   Mattson M, Pollack E, Cullen J: What are the odds that smoking will kill you? Am J Public Health 1987;77:425–431.

34   Blot WJ, et al: Smoking and drinking in relation to oral and pharyngeal cancer. Cancer Res 1988;48:3282–3287.
35   US DHHS: The Health Benefits of Smoking Cessation. A Report of the Surgeon General. Rockville, MD, 1990, p 625.
36   Martin G, et al: Oral leukoplakia status six weeks after cessation of smokeless tobacco use. J Am Dent Assoc 1999;130:945–954.
37   U.S. Department of Health and Human Services, The Health Benefits of Smoking Cessation: A Report of the Surgeon General. Rockville, MD, U.S. Department of Health and Human Services, Public Health Service, Centers for Disease Control, Center for Chronic Disease Prevention and Health Promotion, Office on Smoking and Health, 1990.
38   Tuyns AJ, et al: Cancer of the larynx/hypopharynx, tobacco and alcohol: IARC international case-control study in Turin and Varese (Italy), Zaragoza and Navarra (Spain), Geneva (Switzerland) and Calvados (France). Int J Cancer 1988;41:483–491.
39   Falk RT, et al: Effect of smoking and alcohol consumption on laryngeal cancer risk in coastal Texas. Cancer Res 1989;49:4024–4029.
40   Franceschi S, et al: Smoking and drinking in relation to cancers of the oral cavity, pharynx, larynx, and esophagus in northern Italy. Cancer Res 1990;50:6502–6507.
41   Choi SY, Kahyo H: Effect of cigarette smoking and alcohol consumption in the etiology of cancers of the digestive tract. Int J Cancer 1991;49:381–386.
42   Freudenheim JL, et al: Diet, smoking, and alcohol in cancer of the larynx: a case-control study. Nutr Cancer 1992;17:33–45.
43   U.S. Department of Health and Human Services, The Health Consequences of Smoking: Cancer. A Report of the Surgeon General. Rockville, MD, U.S. Department of Health and Human Services, Public Health Service, Office on Smoking and Health, 1982.
44   IARC: Tobacco Smoking. IARC Monographs on the Evaluation of the Carcinogenic Risk of Chemicals to Humans. Lyon, International Agency for Reseach on Cancer, 1986, vol 38, p 421.
45   De Stefani E, Barrios E, Fierro L: Black (air-cured) and blond (flue-cured) tobacco and cancer risk. III. Oesophageal cancer. Eur J Cancer 1993;29A:763–766.
46   Castellsague X, et al: Independent and joint effects of tobacco smoking and alcohol drinking on the risk of esophageal cancer in men and women. Int J Cancer 1999;82:657–664.
47   U.S. Department of Health and Human Services, Reducing the Health Consequences of Smoking: 25 Years of Progress. A Report of the Surgeon General. Rockville, MD, U.S. Department of Health and Human Services, Public Health Service, Centers for Disease Control, Center for Chronic Disease Prevention and Health Promotion, Office on Smoking and Health, 1989.
48   Tredaniel J, et al: Tobacco smoking and gastric cancer: review and meta-analysis. Int J Cancer 1997;72:565–573.
49   Shanks T, Burns D: Disease consequences of cigar smoking; in Shopland D, et al (eds): Cigars-Health Effects and Trends. Smoking and Tobacco Control Monograph No. 9. Bethesda, MD, U.S. Department of Health and Human Services, National Institutes of Health, National Cancer Institute, NIH Pub. No. 98–4302, 1998, pp 105–160.
50   Shapiro J, Jacobs E, Thun M: Cigar smoking in men and risk of death from tobacco-related cancers. J Natl Cancer Inst 2000;92:333–337.
51   US DHHS: The Health Consequences of Smoking. A Report of the Surgeon General. DHHS(PHS) 82–50179. Washington, DC, 1982, p 302.
52   IARC: Hepatitis Viruses. IARC Monographs on the Evaluation of Carcinogenic Risks to Humans. Lyon, IARC Sci Publ., 1994, vol 59.
53   IARC: Alcohol drinking. IARC Monographs on the Evaluation of the Carcinogenic Risk of Chemicals to Humans. Lyon, International Agency for Reseach on Cancer, 1988, vol 59.
54   Liu BQ, et al: Emerging tobacco hazards in China: 1. Retrospective proportional mortality study of one million deaths. Br Med J 1998;317:1411–1422.
55   Tanaka K, et al: A long-term follow-up study on risk factors for hepatocellular carcinoma among Japanese patients with liver cirrhosis. Jpn J Cancer Res 1998;89:1241–1250.
56   Evans A, et al: Eight-year follow-up of the 90,000-person Haimen City cohort: I. Hepatocellular carcinoma mortality, risk factors, and gender differences. Cancer Epidemiol Biomarkers Prev 2002;11:369–376.

57 Yu GP, Hsieh CC: Risk factors for stomach cancer: a population-based case-control study in Shanghai. Cancer Causes Control 1991;2:169–174.

58 Tsukuma H, et al: Risk factors for hepatocellular carcinoma among patients with chronic liver disease. N Engl J Med 1993;328:1797–1801.

59 Giovanucci E: An updated review of the epidemiological evidence that cigarette smoking increases risk of colorectal cancer. Cancer Epidemiol Biol Prev 2001;10:725–731.

60 Daling JR, et al: Cigarette smoking and the risk of anogenital cancer. Am J Epidemiol 1992;135: 180–189.

61 Marcus PM, Vineis P, Rothman N: NAT2 slow acetylation and bladder cancer risk: a meta-analysis of 22 case-control studies conducted in the general population. Pharmacogenetics 2000;10: 115–122.

62 IARC: Overall Evaluations of Carcinogenicity: An Updating of IARC Monographs. IARC Monographs on the Evaluation of the Carcinogenic Risk of Chemicals to Humans. Lyon, International Agency for Reseach on Cancer, 1987, vol 7.

63 Korte JE, et al: The contribution of benzene to smoking-induced leukemia. Environ Health Perspect 2000;108:333–339.

64 IARC: Human Papillomaviruses. IARC Monographs on the Evaluation of Carcinogenic Risks to Humans. Lyon, IARC Sci Publ., 1995, vol 64.

65 Plummer M, et al: Smoking and cervical cancer: pooled analysis of a multicentric case-control study; in 19th International Pappilomavirus Conference. Florianopolis, Brazil, 2001.

66 Lacey JV Jr, et al: Associations between smoking and adenocarcinomas and squamous cell carcinomas of the uterine cervix (United States). Cancer Causes Control 2001;12:153–161.

67 Peto R, et al: Smoking, smoking cessation and lung cancer in the UK since 1950: combination of national statistics with two case-control studies. Br Med J 2000;321:323–329.

68 U.S. Department of Agriculture. NCI Smoking and Tobacco Control Monograph 8, 1997, p 13.

Michael J. Thun
American Cancer Society, 1599 Clifton Road
Atlanta, GA 303290–4251 (USA)
Tel. +1 404 329 5747, Fax +1 404 327 6450
E-Mail mthun@cancer.org

Cho CH, Purohit V (eds): Alcohol, Tobacco and Cancer.
Basel, Karger, 2006, pp 205–228

....

Tobacco Associated Lung Cancer

Hildegard M. Schuller

Experimental Oncology Laboratory, Department of Pathobiology, College
of Veterinary Medicine, University of Tennessee, Knoxville, Tenn., USA

Abstract

Lung cancer is the leading cause of cancer death, and about 90% of the cases are associated with exposure to cigarette smoke. While genotoxic effects of polycyclic aromatic
hydrocarbon and nitrosamine carcinogens contained in tobacco have been extensively
reported in the literature, only a handful of laboratories have addressed the potential interaction of tobacco-specific agents with neurotransmitter receptors of the autonomic nervous
system, their associated signaling cascades, and the role of these events for the development
of lung cancer. This review summarizes currently available information on this recently
emerging field. In vitro and in vivo data presented document that nicotine and the carcinogenic nitrosamines *N*-nitrosonornicotine and 4-(methylnitrosamino)-1-(3-pyridyl)-1-
butanone bind as agonists to nicotinic acetylcholine receptors, with each nitrosamine having
higher affinity than nicotine for a subset of nicotinic acetylcholine receptors. The main
signaling cascade downstream of these receptors includes Ca^{2+}, protein kinase C, Raf,
ERK1/2 and the AKT pathway with effects on cell proliferation, apoptosis, angiogenesis, cell
migration and invasion. 4-(methylnitrosamino)-1-(3-pyridyl-1-butanone additionally binds
as an agonist to β-adrenergic receptors. These receptors stimulate the proliferation of small
airway epithelial cells and the adenocarcinomas derived from them via a complex signaling
network that includes cAMP, protein kinase A, CREB as well as c-src, ras, Raf and ERK1/2
and transactivation of the epidermal growth factor receptor as well as the release of arachidonic acid. β-Carotene, dexamethasone, caffeine, theopylline and green tea stimulate this
pathway by increasing cAMP and promote the development of pulmonary adenocarcinoma
while inhibiting small cell lung carcinoma. Diagnostic tools are urgently needed to assess
which signaling cascade is hyperactive in individual patients before successful cancer prevention and therapy can be implemented.

Lung cancer is the leading cause of cancer death in industrialized countries,
and has increased in incidence by 600% in women during the last 20 years [1]
while slightly declining in men [2]. About 90% of all lung cancer cases are

attributed to smoking, and the observed changes in lung cancer incidence among men and women are tightly linked with changes in smoking patterns. The high mortality from lung cancer is not only caused by the high incidence of this disease but also by a lack of tests for early detection and the absence of effective therapies. The relative contribution to the overall lung cancer rate of the three major histological types of lung cancer, pulmonary adenocarcinoma (PAC), small cell lung carcinoma (SCLC) and squamous cell carcinoma (SQC) has changed significantly during the last 20 years. In most countries, the rate of SCLC and SQC has decreased in males whereas an increase was observed in females, reflecting an increase in the number of women who smoke. By contrast, the rate of PAC rose in both, men and women in virtually every country, a trend detected in smokers as well as non-smokers [2, 3]. From being a rare type of lung cancer, PAC has thus advanced to being the leading type of lung cancer in smokers and non-smokers today. This shift in the frequency of histological lung cancer types occurred at the time when at first, filtered cigarettes, and later, low nicotine cigarettes were introduced. The cigarette filters trap the tar fraction of smoke, which is rich in polycyclic aromatic hydrocarbons (PAHs). This family of chemicals causes the development of SQC when deposited into the upper airways of laboratory animals. On the other hand, changes in the manufacturing of cigarettes (e.g., use of different tobacco blends) increased the concentration of nitrate [4], enhancing the formation of the nicotine-derived nitrosamine 4-(methylnitrosamino)-1-(3-pyridyl)-1-butanone (NNK). NNK causes PAC in laboratory rodents irrespective of the route of administration [4, 5]. The observed increase in the rate of PAC in smokers may therefore be directly related to the increased concentration of this carcinogen in cigarettes.

The carcinogenic nitrosamines N-nitrosonornicotine (NNN) and NNK (fig. 1) are formed from nicotine in the presence of nitrosating agents during the processing of tobacco and in the mammalian organism. Each of these nitrosamines acts systemically, causing the development of PAC in laboratory rodents regardless of the route of administration. However, NNK is significantly more potent than NNN, causing the development of PAC at a lower cumulative dose and at a higher incidence [4, 5]. NNK is therefore thought to be responsible for the development of most PACs in smokers. Both nitrosamines are converted by oxidative enzymes in mammalian cells to reactive forms that methylate and pyridyloxobutylate DNA [4, 6]. In turn, this results in activating point mutations in the K-ras gene and in inactivation of the p53 gene [4, 7]. The predominating PAHs contained in tobacco smoke are benzo(a)pyrene (B[a]P) and benz[a]anthracene (B[a]A). Both agents are locally active carcinogens and cause the development of SQC when introduced directly into the airways of rodents [4]. PAHs undergo oxidative metabolism to reactive intermediates that form DNA-adducts associated with mutations in the p53 and p21 genes [8]. Several excellent recent

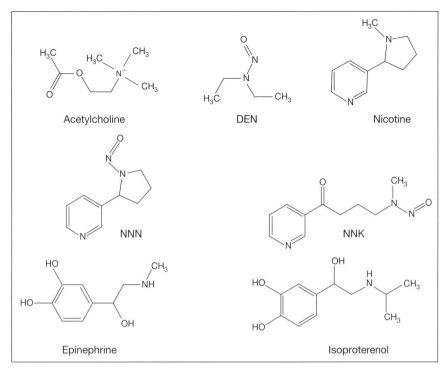

Fig. 1. Chemical structures of the physiological agonist for nAChRs, acetylcholine, the physiological agonist for β-ARs, epinephrine, the synthetic selective agonist for β-ARs and of the tobacco-agents DEN, nicotine, NNN and NNK.

reviews have summarized the widely accepted traditional views on these genotoxic effects of tobacco nitrosamines and PAHs [4, 9, 10].

Data generated during the last 10 years suggest that the tobacco nitrosamines NNK and NNN as well as nicotine additionally modulate the regulation of lung cell proliferation and apoptosis as well as angiogenesis via direct interaction with neurotransmitter receptors of the autonomic nervous system. The autonomic nervous system with its two branches, the sympathicus and the vagus, regulates a wide range of non-voluntary functions of the mammalian organism by the release of neurotransmitters from nerves that bind to cell surface receptors in the effector cells. The role of disturbed autonomic regulation in the genesis of tobacco-associated lung cancer has been a neglected area of study. Research in this novel field is still in its infancy but promises the potential for the development of novel strategies for the prevention and treatment of lung cancer. This review provides a summary of current knowledge in this emerging field.

The Role of Nicotinic Acetylcholine Receptors and Their Signaling Cascades in Smoking-Associated Lung Carcinogenesis

Nerves of the vagus regulate diverse organ and cell functions by releasing the neurotransmitter acetylcholine, which binds to acetylcholine receptors on the cell membrane. These receptors are comprised of two families, the nicotinic acetylcholine receptors (nAChRs), to which nicotine binds with high affinity, and the muscarinic acetylcholine receptors (mAChRs), to which nicotine does not bind. All nAChRs are comprised of a central ion channel with high permeability for Ca^{2+} surrounded by subunits termed alpha through delta. Neuronal nAChRs are further sub classified based on the expression of α_2 through α_9 and β_2 through β_4 subunits whereas the muscle nAChRs contain in addition gamma and delta subunits [11–13]. The chronic interaction of nicotine with neuronal nAChRs in the brain comprised of α_4/β_2 subunits has long been recognized as a major cause for nicotine addiction in smokers [14, 15]. On the other hand, nAChRs with the subunit compositions α_2–α_5 and β_2–β_4 regulate the release of catecholamines from the adrenal medulla and their chronic stimulation by nicotine contributes to the development of smoking-associated cardiovascular disease [16].

A potential role for neuronal nAChRs in smoking associated lung carcinogenesis was first suggested in 1989 when it was shown that nicotine and NNK stimulated the proliferation of SCLC cells and that this response was blocked by antagonists for nAChRs but not by an antagonist for mAChRs [17, 18]. These reports were followed by publications from another laboratory in 1990 and 1994 which documented a nicotine-induced reversal of apoptosis in response to opioids in a large panel of SCLC and NSCLC cell lines [19, 20]. A third independent laboratory described in 1993 that nicotine stimulated the proliferation of human SCLC cell lines via stimulation of a serotonergic autocrine loop [21]. The initial findings from these three laboratories suggested that nicotine itself may contribute to the development of smoking-associated lung cancer by interaction with nAChR-mediated proliferative and apoptotic signaling pathways. In addition, the data with NNK [17, 18] indicated that the extreme potency of NNK as a pulmonary carcinogen may be linked to an ability to function as an agonist for nAChRs. Radio receptor assays showed a high Bmax in saturation binding assays with the selective antagonist for nAChRs comprised of homomeric α_7 subunits, α-bungarotoxin (α-BTX), indicative of high levels of expression of this receptor in human SCLC cell lines as opposed to PAC cell lines which demonstrated low or non-detectable binding of α-BTX [22]. These findings were extended by recent investigations that showed expression of mRNA for the α_7nAChR in a large panel of cell lines derived from different types of human lung cancers and in immortalized human small airway epithelial cells while significant amounts of receptor protein were only detected in SCLC cell lines [23]. Taken together, these findings indicate that the α_7nAChR is expressed in many lung cell types but demonstrates particularly high levels of expression in SCLC cells. Radio receptor assays assessing the relative binding affinities of nicotine, NNN, and NNK in competition with α-BTX identified NNK as a ligand with unprecedented high affinity for the α_7nAChR [22]. Analysis of these binding data by non-linear regression in fact showed that the affinity of NNK for the α_7nAChR was about 1,300 times greater than that of nicotine. By contrast, NNN bound with significantly lower affinity to this receptor than nicotine. On the other hand, saturation binding assays with the selective ligand for nAChRs containing α_2–α_6 and β_2–β_4 nAChR subunits, epibatidine (EB), revealed a high Bmax suggestive of high receptor expression in cell lines from human PACs while binding of EB was below detectable levels in SCLC cell lines [22]. Radio receptor assays in which nicotine, NNK or NNN competed for nicotinic binding sites with EB revealed an exceptionally high affinity of NNN for this receptor (about 5,100 times higher than nicotine) while the affinity of NNK was very low (about 12 times lower than nicotine). Stimulation of the α_7nAChr by NNK (30 nM) significantly increased cell number and DNA synthesis in SCLC cells and this response was blocked by the α_7nAChR antagonist, α-BTX [22]. Flow cytometric analysis showed a significant increase in intracellular Ca^{2+} in response to 1 nM NNK and this effect was blocked by α-BTX [24, 25], thus confirming NNK as an agonist for the α_7nAChR. In conjunction with earlier reports, these findings clearly established an important regulatory role of the α_7nAChR in SCLC. On the other hand, binding of nicotine or NNN to the EB-sensitive nAChRs expressed in PAC cells did not modulate cell proliferation under the assay conditions used [22]. Recent studies have revealed the expression of the EB-sensitive nAChR family as well as α_7nAChRs in a wide variety of cell types in the monkey lung [26, 27]. In vitro studies with immortalized human bronchial epithelial cells and human small airway epithelial cells have shown that both receptor families participate in the activation of the serine/threonine kinase AKT in response to nicotine or NNK, an effect resulting in the attenuation of apoptosis induced by etoposide, radiation or hydrogen peroxide as well as the induction of a transformed phenotype [28]. However, the concentrations of nicotine (10–100 μM) or NNK (1 μM) required to elicit these effects were significantly higher than those reported in studies with pulmonary neuroendocrine cells (PNECs) or SCLC cell lines. Another laboratory additionally reported the activation of NF-κB and up regulation of cyclin D1 in human bronchial epithelial cells NHBE and human small airway epithelial cells exposed to NNK at concentrations ranging from 0.5–10 μM [29]. Again, these concentrations are considerably higher than those required to stimulate mitogenic signaling in SCLC cells. Additional support for an important role of α_7nAChR stimulation in the growth regulation of SCLC and the putative cell of origin of this cancer, PNEC, came from a number of in vitro studies showing that binding of nicotine (1 μM) or NNK (100 pM) to the α_7nAChR resulted in phosphorylation of protein kinase C (PKC), Raf-1, ERK1/2 and c-myc [25, 30, 31]. Another laboratory also reported activation of ERK1/2 in response to nicotine in human SCLC cell lines [21]. Recent investigations additionally have reported α_7nAChR-mediated NNK-induced functional cooperation between Bcl2 and c-myc that inhibited apoptosis while stimulating cell proliferation of human SCLC cell lines [32]. In addition, NNK phosphorylated μ- and m-calpains in human SCLC cells in an ERK1/2 and Ca^{2+}-dependent manner, resulting in the induction of cell migration and invasion and these effects were abrogated by the α_7nAChR antagonist αBTX [33]. It is of note that the concentration of NNK required to elicit these effects all three laboratories and were extremely low (100 pM), thus underlining the very high affinity of NNK for the α_7nAChR which is over expressed in SCLC cells. These findings are in accord with the frequent expression of amplified c-myc in SCLC [34]. In addition, it has been shown that the PKC/Raf-1/ERK1/2 signaling cascade is also stimulated by autocrine growth factors for SCLC, including the neuropeptide growth factors bradykinin, vasopressin, bombesin (MB), neurotensin and galananine as well as serotonin and acetylcholine [21, 35–37]. In turn, it was shown that the release of some of these growth factors was triggered by the influx of Ca^{2+} caused by stimulation of the α_7nAChR [21, 35]. Collectively, these data suggest that the major signaling pathway that regulates SCLC growth, apoptosis and invasiveness includes Ca^{2+} influx, activation of PKC, Raf-1, ERK1/2 Bcl2, c-myc as well as calpains and that binding of agonists to the α_7nAChR and to receptors for autocrine growth factors initiate the activation of this pathway (fig. 2). Additional nAChR types appear to contribute to the modulation of proliferation and apoptosis in bronchial epithelial cells and small airway epithelial cells upon exposure to relatively high concentrations of nicotine or NNK.

In addition to the summarized direct effects of nicotinic agonists on lung cancer cells, it has been recently discovered that nicotine stimulates angiogenesis and enhances the neovascularization of Lung tumors [38]. In fact, studies in xenographs from Lewis Lung cancer cells even showed that exposure to side stream smoke (the experimental equivalent of second hand smoke) increased tumor size and angiogenesis and that this effect was inhibited by the broad-spectrum antagonist for neuronal nAChRs, mecamylamine [39].

While the in vitro evidence for a critical role of α_7nAChR-mediated signaling in the genesis of smoking-associated SCLC is strong, proof for this hypothesis should come from animal experiments. However, nicotine does not

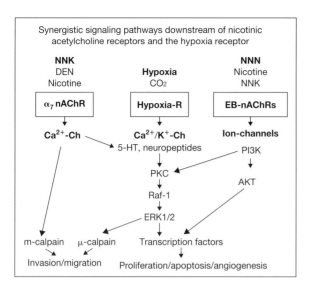

Fig. 2. Simplified scheme of signaling network downstream of nAChRs and hypoxia receptor leading to modulation of proliferation, apoptosis, angiogenesis, invasion and cell migration. The agonist with the highest affinity for each receptor is highlighted in bold font.

cause cancer in healthy laboratory animals and NNK reproducibly induces PAC when administered to healthy rats mice or hamsters [5, 40]. In fact, an animal model for SCLC or any type of neuroendocrine lung cancer did not exist. The physiological role of PNECs, which are thought to give rise to SCLC, is to function as chemoreceptors with sensitivity for changes in pulmonary oxygen concentrations, particularly hypoxia [41, 42]. It has thus been shown that PNECs express a receptor protein that senses hypoxia and triggers the release of serotonin (5-hydroxytryptamine, (5-HT)) and MB via influx of Ca^{2+} [43–45]. In addition to modulating bronchial smooth muscle tone and respiration, 5-HT and MB act as autocrine growth factor for PNECs and SCLC. The diseased lung with impaired pulmonary ventilation thus typically demonstrates hyperplasia of PNECs [46–48]. In addition, uranium mining and other sources of exposure to radon that cause interstitial pulmonary fibrosis are documented risk factors for the development of SCLC [49, 50]. We therefore hypothesized that stimulation of the oxygen sensing receptor in PNECs by impaired pulmonary ventilation would facilitate the development of a neuroendocrine type of lung cancer in animals exposed to the nicotinic agonists nicotine or NNK as well as diethylnitrosamine (DEN) which has structural similarities with acetylcholine (fig. 1). We induced mild pulmonary interstitial fibrosis in Syrian golden hamsters by maintaining the animals in an environment of 60% oxygen. These animals developed multiple foci of hyperplastic PNECs with positive immunoreactivity for 5-HT. Hamsters that were additionally given multiple subcutaneous injections of NNK [40] or DEN [51] developed multiple neuroendocrine lung tumors at a high incidence. Similar to most human SCLCs, these tumors expressed the neuroendocrine markers 5-HT, calcitonin, MB and neuron specific enolase, they lacked activating point mutations in K-ras while over-expressing c-myc [52]. Due to their relatively small size and well-differentiated morphological appearance, these experimentally induced tumors were classified as atypical carcinoids even though they demonstrated functional and molecular features of SCLC. Hamsters with hyperoxia-induced pulmonary interstitial fibrosis and treated with multiple subcutaneous injections of nicotine developed a low but significant incidence of lung tumors with focal areas of positive immunoreactivity to the neuroendocrine markers 5-HT and neuron specific enolase [53]. Collectively, these findings support the hypothesis that the diseased lung with impaired pulmonary oxygenation and resulting hyperplasia of PNECs is more susceptible for the development of neuroendocrine lung cancers upon simultaneous exposure to the nAChR agonists nicotine, NNK or DEN. In vitro experiments corroborated this interpretation by demonstrating that SCLC or PNEC cells maintained in an environment of high CO_2 at the expense of O_2 showed induction of ERK1/2 activation [54] as well as increased proliferation response to nicotine or NNK [55, 56].

In vitro findings had shown that the influx of Ca^{2+} triggered by α_7nAChR stimulation and resulting activation of PKC, Raf-1 ERK1/2 and c-myc are important regulators of SCLC and PNEC proliferation. In accordance with these findings, the inhibitor of Ca^{2+} channels and PKC, dexniguldipine, completely blocked the development of neuroendocrine lung tumors when administered to hamsters that had been initiated for the development of such tumors by hyperoxia and DEN [57]. Dexniguldipine also completely abrogated the proliferation of SCLC and PNECs in vitro at very low concentrations in the picomolar range [58, 59]. By comparison, much higher concentrations ($100\,nM$) of verapamil, which only blocks L-type Ca^{2+}-channels, were required to yield a similar response [58]. Collectively, these in vivo and in vitro findings suggest a key role for Ca^{2+}-channels and PKC downstream of the α_7nAChR in the signaling cascade(s) involved in the development and growth regulation of neuroendocrine lung tumors such as SCLC.

Unlike G-protein coupled receptors and peptide receptors, which are downregulated by chronic exposure to agonists, the nAChRs are upregulated by chronic exposure to nicotine [15]. While this has been well documented for all nAChR types in the brain, similar effects on nAChRs expressed in lung cells have only been recently documented. It has thus been shown that treatment of pregnant monkeys from day 26–day 134 of gestation with nicotine (1 mg/kg/day by osmotic minipumps) unregulated the expression of α_7nAChRs in the lungs of their off-spring as assessed by immunohistochemitry and RT-PCR [26]. This was associated with an increase in the number of clustered PNECs. In accord with this report, we found a significant up regulation of the α_7nAChR by relative competitive RT-PCR in the lungs of hamster fetuses harvested one day prior to birth whose mothers had been given NNK ($100\,$pM) in the drinking water throughout their pregnancy [31]. These findings may have important implications for the risk assessment of individuals whose mothers smoked during pregnancy because the unregulated α_7nAChR may convey an increased susceptibility for the development of SCLC. Support for this hypothesis is lacking at this time. Epidemiological studies that have assessed cancer risk associated with maternal smoking generally looked at typical childhood cancers, such as leukemia and brain cancer. Very few recent reports have included assessments of lung cancer risk in adult individuals due to perinatal exposure to smoking with varied results. Unfortunately, a breakdown by histological lung cancer type was not provided. Animal experiments have shown that DEN or NNK when administered during the last trimester to pregnant hamsters caused the development of lung tumors in the offspring [60, 61]. However, these experiments were conducted prior to research that implicated nicotinic receptors in lung carcinogenesis, and no information is available at this time on the expression levels of nAChRs in the transplacentally induced tumors and their potential role in the carcinogenic process.

The Role of β-Adrenergic Receptors and their Signaling Cascades in Smoking-Associated Lung Carcinogenesis

Nerves of the sympathicus regulate diverse organ and cell functions by releasing the catecholamine neurotransmitters epinephrine and norepinephrine, which bind to β-adrenergic receptors (β-ARs). β-ARs are members of the super family of seven-transmembrane G-protein coupled receptors and are classified into β_1-, β_2-, and β_3-receptors [62]. Binding of agonist to β_1-ARs typically activates adenylase cyclase, cAMP, protein kinase A (PKA) and the transcription factor CREB (fig. 3). Agonist-induced stimulation of the β_2-AR may additionally activate the tyrosine kinase c-src [63], resulting in the activation of ras, Raf-1, and the mitogen-activated kinases ERK1/2 (fig. 3). In addition, activated PKA can trigger the release of arachidonic acid (AA) via phospholipase A2 while potassium channels that open in response to β_2-AR stimulation may have a similar effect [64, 65]. Direct transactivation of the epidermal growth factor receptor (EGFR) pathway by β_1- or β_2-AR stimulation may additionally occur in some cells [66, 67], resulting in the activation of ras,

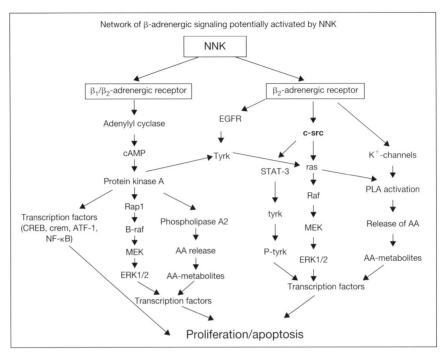

Fig. 3. Simplified scheme of signaling network potentially activated by NNK downstream of β-ARs and the EGFR, leading to modulation of cell proliferation and apoptosis.

Raf-1 and the ERK1/2 cascade in response to β-AR agonists (fig. 3). In the lungs, β$_1$- and β$_2$-receptors are widely distributed in epithelia, smooth muscle cells and endothelial cells.

It is well established that β-adrenergic signal transduction plays a key role in cardiovascular disease [62] and asthma [68]. β-AR-mediated release of AA regulates cardiovascular function by modulating smooth muscle tension, endothelial response to injury, and heart rate. The leading pharmaceuticals for the therapy and prevention of cardiovascular disease are antagonists for β-ARs (β-blockers) and aspirin, which inhibits the metabolism of AA by the enzyme cyclooxygenase-2 (COX-2). On the other hand, epinephrine and other agonists for β$_2$-ARs are widely used for the treatment of acute asthma attacks because of their broncho-dilating and mucolytic effects. These agents are also contained in numerous non-prescription drugs for the treatment of decongestion, allergies and the common cold, and they are the active ingredients of numerous dietary supplements, which promise weight loss while boosting energy.

Cardiovascular disease and PAC share the same risk factors, smoking and a high fat diet [69–71]. Over expression of the AA-metabolizing enzyme COX-2 is also commonly found in both diseases [72, 73]. In addition, human PAC frequently demonstrates activating point mutations in K-ras [74]. Disturbed β-AR signaling, which is the hallmark of cardiovascular disease, may therefore also contribute to the development of smoking-associated PAC. NNK causes PAC in all laboratory rodents and is believed to be the major cause of this cancer type in smokers. Structural analysis of classic AR agonists and NNK revealed similarities between the selective β-AR agonist, isoproterenol, and NNK (fig. 1). Radio receptor assays in CHO cells transfected with the human β_1- or β_2-AR genes showed that NNK competed successfully for β_1-AR and β_2-AR binding sites against the β-AR ligand [^{125}I]-iodocyanopindolol [75]. The affinity of NNK for the β_1-AR was about 600 times greater than that of its physiological agonist, norepinephrine while its affinity for the β_2-AR was about 2,100 times greater than that of the physiological agonist for this receptor, epinephrine. Overall, the affinity of NNK was about 22 times higher for β_1-ARs than for β_2-ASRs. By contrast, neither nicotine nor NNN bound to β-ARs. Saturation binding assays and RT-PCR revealed high levels of expression of βARs in cell membrane fractions from human PAC cell lines NCI-H322 or NCI-H441 with β_1-ARs predominating [75]. Exposure of these cells to NNK ($100\,pM$–$1\,\mu M$) caused a concentration-dependent release of AA and significant stimulation of DNA synthesis. Both of these responses were partially inhibited by antagonists for β_1- and β_2-ARs and completely blocked by the broad-spectrum β-AR antagonist, propranolol, suggesting binding of NNK to both receptors as the initiating event. In addition, aspirin and the MAP kinase inhibitor MK-886 each partially inhibited NNK-induced cell proliferation, suggesting that the AA-cascade and MAPK pathway were involved in the observed proliferative response [75]. Recent studies in an immortalized human small airway epithelial cell (SAEC) line and in the human PAC cell line NCI-H322, both of which express the Clara cell-specific CC10 protein (Schuller, unpublished), have further dissected the complex signaling events downstream of NNK-induced β_1-AR stimulation [76]. These studies showed that NNK upregulated ERK1/2 and CREB/ATF-1 phosphorylation in a PKA-dependent manner in both cell lines. This response was further increased by transient over-expression of the β_1-AR. Pre-exposure of cells to the selective β_1-AR antagonist, atenolol, attenuated the stimulatory effects of NNK, suggesting the latter upregulated ERK1/2 and CREB/ATF-1 via this receptor. In vivo labeling and immunoprecipitation assays revealed that NNK phosphorylated the EGFR at tyrosine residues, 991, 1068, and 1173. The inhibitor of EGFR-specific tyrosine kinases, AG1478, reduced NNK ability to stimulate ERK1/2 and CREB/ATF-1. Genomic analysis of the exons 18–21 of the EGFR genes showed that no mutations were present in either gene. Immunohistochemical analysis of

20 tissue samples from human non-small cell lung cancers showed simultaneous over-expression of PKA, P-CREB and P-EGFRtyrk 1173 only in 7 PACs that also expressed the Clara cell specific CC10 protein (Schuller, unpublished). Collectively, these data provide evidence, that NNK targets ERK1/2 and CREB/ATF-1 proteins via dual signaling involving the β_1-AR and EGFR pathways in PACs of Clara cell phenotype (PACC) and their putative cells of origin, SAECs and that this pathway is hyperactive in a subset of human PACs.

In support of these in vitro findings, immunohistochemical analysis of NNK-induced and Clara cell-derived PAC in hamsters recently demonstrated over expression of β_1- and β_2-ARs as well as over expression of PKA, P-CREB, P-EGFRtyrk 1173, Raf-1 and P-ERK1/2 [77]. Radio receptor assays in cell membrane fractions from NNK-induced PACs in hamsters also showed high levels of expression of β_1- and β_2-ARs, with β_1-ARs predominating [78]. These findings are in accord with the observation that a subset of human PACs simultaneously over expresses β_1- and β_2-ARs as well as COX-2 [79]. In addition, it was shown that the general β-blocker, propranolol, significantly inhibited the development of NNK-induced PAC in hamsters when given immediately prior to each NNK injection [78]. This finding provides strong support for the hypothesis that β-AR stimulation by NNK is an essential event without which PAC may not develop in this animal model. Additional evidence for the importance of β-AR signaling in the genesis of PAC came from studies which revealed a significant promoting effect of agents which stimulate β-ARs or their downstream effectors on NNK-induced PAC in hamsters [78]. The β-AR agonist, epinephrine, thus significantly increased the multiplicity of NNK induced PAC when administered after discontinuation of the tumor-initiating injections with NNK. Similarly, theopylline, which increases the intracellular concentration of the β-AR effector, cAMP, via inhibition of the enzyme phosphodiesterase, significantly promoted NNK-induced PAC development [80]. These findings on β-AR-mediated signaling by NNK are supported by a recent article which showed NNK-induced multisite Bad phosphorylation at Ser-112, Ser-136, and Ser-155 via activation of PKCiota downstream of β-AR activated c-src in association with increased survival of human lung cancer cells [81]. Additional confirmation for the discovery of NNK as a β-AR agonist has come from a recent report that showed NNK-induced stimulation of colon cancer cell growth via β-AR and cAMP-dependent signaling [82].

While the EGFR pathway has long been recognized as an important regulator of PAC growth, a potential role for CREB and ATF-1 has not been studied to date in any histological type of lung cancer. Activation of CREB has been associated with the development of endocrine tumors [83] while ATF-1 and CREB both are reportedly important stimulators of melanoma growth and metastasis [84]. The presence of this regulatory pathway in a significant subset of PAC in smokers may at least in part explain the disappointing results of clinical trials in patients with non-small cell lung cancer that targeted the EGFR pathway by the use of small molecule tyrosine kinase inhibitors [85], such as gefitinib (Iressa). These agents only yielded therapeutic responses in a small subset of individuals with PAC who were typically non-smokers whereas PACs in smokers were nonresponsive. Mutations in the tyrosine kinase domain of the EGFR have been associated with responsiveness to tyrosine kinase inhibitors [86] but neither NCI-H322 nor the immortalized SAECs harbored these mutations [76]. Successful therapeutic growth management of PACC may require the simultaneous inhibition of the β_1-AR/PKA/ CREB/ATF-1 and the EGFRtyrk/MAPK pathways.

Prevention of Smoking Associated Lung Cancer

Neither effective lung cancer therapy nor methods for the early detection of this malignancy exist. A major international effort towards controlling this disease has therefore focused on the development of lung cancer prevention strategies. Mouse models of lung cancer have been extensively used for this endeavor. While these studies have generated valuable information, it is important to note that there are some limitations to this animal model that have to be considered when extrapolating such data to human lung cancer. The entire mouse lung is a model for the human lung periphery in that a single non-stratified epithelium comprised of Clara cells and ciliated cells coats all intrapulmonary airways and even extra pulmonary stem bronchi in the mouse [87, 88]. By contrast, in the human respiratory tract, a pseudo stratified epithelium comprised of basal cells, mucous cells and ciliated cells coats the central airways (trachea, stem bronchi, lobar bronchi, segmental bronchi) while a simple non-stratified epithelium of Clara cells and ciliated cells is restricted to the bronchioles in the lung periphery [87, 88]. While the human upper airways can give rise to SCLC and SQC, human PAC arises in the lung periphery and may exhibit features of bronchiolar Clara cells or alveolar type II cells, the putative progenitor cells of these malignancies. PAC in the mouse exhibits phenotypic features of alveolar type II cells and has been shown to be derived from this cell type when experimentally induced by agents such as NNK or even when developing spontaneously [89, 90]. Accordingly, the mouse is an excellent animal model for human PAC of alveolar type II cell phenotype. On the other hand, Syrian golden hamsters reproducibly develop PAC derived from bronchiolar Clara cells and exhibiting phenotypic features of this cell type when treated with nitrosamine, including NNK [40, 91, 92]. The Syrian golden hamster is thus an excellent animal model for human PAC of Clara cell phenotype.

An α-tocopherol, β-carotene supplementation trial (ATBC) and a chemo-prevention trial with β-carotene and retinoids (β-carotene and retinoid efficacy

'CARET' trial) were conducted in populations at risk for the development of lung cancer because of previous or current exposure to smoking or asbestos [93, 94]. These trials were based on preclinical studies that had shown in vivo and in vitro inhibition of chemically induced carcinogenesis by retinoids or the pro-vitamin β-carotene [95] and on epidemiological observations that a diet rich in fruits and vegetables reduces the lung cancer risk. In addition, vitamin A deficiency was reported to cause squamous cell metaplasia in the respiratory epithelium of large airways of hamsters [96] and this change was reversed by in vitro treatment with retinoids. Both trials had to be discontinued due to an increase in lung cancer incidence (28%) and mortality (46%) and a 26% increase in cardiovascular mortality in the CARET trial and an 18% increase in lung cancer incidence and 8% increase in cardiovascular mortality in the β-carotene group of the ATBC trial. Conclusive explanations for these unfortunate outcomes have not been provided to date. It is well established that cardiovascular function is under β-adrenergic, cAMP-mediated, control, and β-blockers are widely used for the clinical management of hypertension and heart disease. In light of the recent in vitro and in vivo findings on the β-adrenergic regulation of PACC, the chemo preventive agents used in the CARET and ATBC trials may have promoted the development of PACC via stimulation of signaling component(s) of its β-adrenergic regulatory pathway. In support of this hypothesis, we found that low concentrations (1 pM–200 nM) of β-carotene or retinol increased intracellular cAMP in human SAECs and PACC cells [97]. This resulted in significant activations of PKA, CREB and ERK1/2 and increased cell proliferation. The proliferative response to β-carotene was completely abrogated by an inhibitor of adenylase cyclase, from which cAMP is formed, or by a MEK inhibitor, which blocks the activation of ERK1/2. By contrast, immortalized human bronchial epithelial cells BEAS-2B reacted with an inhibition of proliferation to β-carotene or retinol even though both agents activated PKA in these cells [97].

The pro-vitamin β-carotene is converted in the mammalian organism to vitamin A. The majority of functions of vitamin A are carried out by its metabolite, retinoic acid, a transcriptional activator operating primarily through members of the nuclear receptor family of transcription factors. Synthetic retinoids continue to be widely studied as potential preventive and/or therapeutic agents for a variety of cancers. High concentrations of vitamin A and β-carotene are also contained in numerous dietary supplements marketed as disease preventive agents. In addition to the original preclinical studies on the prevention of cancer development from epithelia of the upper airways, current research primarily concentrates on the use of retinoids for the prevention and/or therapy of head and neck cancer or leukemia. It remains undisputed that β-carotene and retinoids do have beneficial effects via inhibition of cell proliferation and/or stimulation of apoptosis in cancers of those cell lineages. In fact the MTT assays with the human large airway epithelial cells BEAS-2B confirm that. However, these data also provide compelling evidence for cell type-specific growth-stimulating effects of β-carotene on human PAC cells and their normal cells of origin (SAECs) via increase in intracellular cAMP, activation of PKA as well as phosphorylation of CREB and ERK1/2. In light of the prevalence of PAC, these novel and hitherto unknown mechanisms of action of β-carotene should be considered as key factors responsible for the unfortunate outcomes of the ATBC and CARET trials.

Pre-existing non-neoplastic pulmonary diseases; such as chronic obstructive pulmonary disease, bronchitis, and asthma increase the risk for the development of lung cancer in smokers and non-smokers [98, 99]. However, few reports have addressed the effects of members of this disease family on defined histological lung cancer types. Glucocorticoids are widely used for the long-term management of chronic obstructive pulmonary disease, chronic bronchitis and asthma because of their anti-inflammatory actions and documented synergy with $β_2$-AR agonist bronchodilators [100, 101]. They are also frequently used to minimize the side effects of cancer chemotherapy and radiation therapy [102]. A significant population of lung cancer patients is therefore exposed to glucocorticoids either before the diagnosis of lung cancer or during cancer therapy. Studies in mouse models of lung cancer have shown significant cancer preventive effects of glucocorticoids [103], leading to recent strategies to propose this family of drugs for lung cancer prevention in smokers and former smokers. In addition, glucocorticoids have been shown to inhibit cell proliferation in a number of non-pulmonary cancers and embryonic fibroblasts, an effect involving the cyclin-dependent kinase inhibitors p21 and p27 [104, 105]. Dexamethasone may therefore stimulate the β-adrenergic growth-regulating pathway in PACC via increase in cAMP because of the documented synergy of glucocorticoids with β-adrenergic agonists in the treatment of non-neoplastic pulmonary disease. In support of this hypothesis, dexamethasone (1 pM–20 μM) increased the proliferation of human SAECs and PACC cells while increasing intracellular cAMP and the activation of PKA, CREB, and ERK1/2 [106]. These findings suggest, for the first time, that dexamethasone may have tumor promoting activity in a subset of human lung cancers within the family of adenocarcinoma.

Preclinical studies in mouse and rat models of lung cancer have demonstrated that green tea, black tea and caffeine contained in these beverages inhibit the development of PAC [107, 108]. Polyphenols of the catechin family are thought to be largely responsible for the anti-tumorigenic effects of tea because of their antioxidant, anti-mutagenic, pro-apoptotic, and anti-proliferative effects [109, 110]. However, epidemiological studies on the effects of tea in human populations have been inconclusive. Some of these studies showed a reduction in lung cancer risk while others demonstrated no effect or even cancer promoting effects [111, 112]. Unfortunately, none of the epidemiological investigations have specified the effects of tea on histological lung cancer types. Thus, there may be a subpopulation of human lung cancers that responds in a similar fashion to tea as the alveolar type II cell PACs in mice or rats, while other populations of lung cancer may not respond or even get promoted. Significant amounts of the methylxanthines, caffeine and theophylline, are contained in tea. Both of these agents inhibit the enzyme phosphodiesterase that mediates the metabolism of cAMP. In turn, this causes accumulation of intracellular cAMP and activation of its downstream effectors. In vitro studies have shown that cell lines derived from human PAC of alveolar type II cell phenotype are inhibited in their growth by cAMP [113] whereas the β-adrenergic regulatory pathway discovered in human PACC cell lines [75] and in the NNK-induced hamster PACCs [78] is stimulated by cAMP. Theophylline and caffeine might therefore stimulate the growth of PACCs via activation of cAMP-dependent signaling, while inhibiting the growth of cells that are under negative growth regulation by this pathway. In support of this hypothesis, caffeine [114] and theophylline [115] each increased intracellular cAMP in human SAECs and PACC cells, resulting in activation of PKA, CREB and ERK1/2 and increased cell proliferation. On the other hand, it has been shown that theophylline inhibits the proliferation of human SCLC cells [116]. The antagonistic effects of these agents on lung cancers controlled by different signaling pathways was confirmed in a bioassay experiment that showed promotion of NNK-induced PACC by green tea or theophylline administered to hamsters after cessation of tumor induction treatments whereas the development of neuroendocrine lung tumors induced by hyperoxia and NNK and which is under growth regulation of nAChRs/Ca^{2+}PKC/Raf-1/ERK1/2 was inhibited [80]. Collectively, these findings suggest that tea as well as other beverages, dietary supplements or drugs that contain caffeine and/or theophylline may selectively promote the development of a subset of human PACs that are under positive growth control by the cAMP/PKA pathway. On the other hand, these agents may well inhibit the development of cancers that are under negative growth control by cAMP signaling.

Conclusions and Future Directions

The data summarized in this review provide compelling evidence for the involvement of nAChRs and β-ARs and their signaling pathways as important mediators of tobacco-associated tumor development in a significant subset of lung cancers. While currently available data point to SCLC and PAC, other histological lung cancer types as well as cancers in other organs may be similarly regulated. The susceptibility of these neurotransmitter receptors to the tobacco-specific agonists nicotine, NNN and NNK can be greatly modulated by preexisting disposition and environmental factors. An important difference between the two receptor families is, that chronic exposure to agonist up regulates the nAChRs whereas β-ARs are down regulated. Continued exposure of smokers to nicotine, NNN and NNK therefore increases the sensitivity of nAChR pathways whereas chronic exposure to NNK or non-tobacco agonists may decrease the sensitivity of the β-ARs. On the other hand, exposure after cessation of smoking to agents that stimulate or inhibit such receptor-mediated pathways may promote or prevent the progression of premalignant lesions and small tumors into overt cancer. Agents that prevent the metabolic activation of tobacco carcinogens are of value to reduce the carcinogenic response in individuals who actively smoke or are exposed to second-hand smoke. These agents however, should not be used for cancer prevention after cessation of smoking unless their potential modulation of signal transduction pathways involved in the regulation of cell proliferation, apoptosis and angiogenesis have been assessed. When reviewing currently available data on the interaction of nicotine, NNK and NNN with nicotinic or β-adrenergic signaling pathways, it is important to remember that the mainstream smoke of cigarettes contains between 5,000–10,000 times more nicotine and 2–3 times more NNN than NNK [117]. If all three of these nAChR agonists would bind with equal affinity to all types of nAChR, the nitrosamines would be clearly at a disadvantage to compete for these binding sites with nicotine. However, the selective high affinity of NNK for the α_7nAChR and of NNN for EB-sensitive nAChRs, which comprise the entire remaining neuronal nAChR family, counteract the prevalence of nicotine in tobacco. In healthy individuals exposed simultaneously to nicotine, NNN and NNK by tobacco smoke, nicotine and NNN will therefore occupy all of the nAChR binding sites whereas NNK binds to β-ARs. This interpretation is supported by the fact that NNK is a very potent inducer of PAC when given to healthy laboratory rodents while NNN is a weak carcinogen for this cancer type and nicotine does not cause PAC at all. The balance is shifted when the sensitivity of the α_7nAChR to agonist is selectively increased by the intracellular influx of Ca^{2+} in response to stimulation of the hypoxia receptor in PNECs in individuals with impaired pulmonary oxygenation. Under these circumstances, NNK and nicotine bind preferentially to this receptor, thus initiating a signaling cascade that selectively supports the growth of neuroendocrine lung cancers, such as SCLC. This interpretation is supported by the findings that interstitial pulmonary fibrosis induced by occupational or environmental exposure to radon is a risk factor for the development of SCLC in smokers and that hamsters with hyperoxia-induced pulmonary fibrosis develop neuroendocrine lung cancers instead of PAC when exposed to NNK or DEN.

Investigations that have used concentrations above 1 μM of nicotine, NNN, or NNK have to be interpreted with caution, as it is unlikely that such systemic concentrations are reached even in a heavy smoker. It is well established that nicotine, NNN and NNK are rapidly metabolized and excreted from the mammalian organism. Neither the parent, unmetabolized compounds nor their reactive metabolites accumulate. The argument has been made that studies on the genotoxic effects of these agents may use concentrations equivalent to the cumulative dose of a smoker over a lifespan. While this may be acceptable on the grounds that multiple hits with a small dose have similar mutational potency as few hits with a larger dose, the same argument cannot be made for studies on non-genotoxic effects of these agents. Unrealistically high concentrations of tobacco-specific receptor agonists will override their selective high affinity for certain receptors and bind non selectively to other cellular targets, making the resulting responses difficult to interpret.

Non-neoplastic diseases characterized by malfunctioning of β-AR signaling include cardiovascular disease and asthma. Long-term management of cardiovascular disease has been achieved by selective treatments with β-blockers and aspirin, while β$_2$-AR agonist inhalers or agents that stimulate its downstream effector, cAMP, and reduce inflammatory responses have been successfully employed for the management of asthma. On the other hand, a host of neurological and psychiatric disorders include malfunctioning nAChR signaling and are clinically managed by nicotinic agonists or stimulators of effectors of these receptors. If tools were available to diagnose at an early stage which signaling pathway(s) are hyperactive in individuals that have been exposed to tobacco smoke, custom-tailored strategies could be implemented to selectively inhibit those pathways. Molecular imaging that allows for the monitoring of receptors and signaling components by positron emission tomography is a very promising tool. This strategy could also be successfully implemented to prevent the recurrence of lung cancer bye selective inhibitors of its hyperactive pathway(s) after removal of the primary cancer.

References

1   Kelly A, Blair N, Pechacek TF: Women and smoking: issues and opportunities. J Womens Health Gend Based Med 2001;10:515–518.
2   Devesa SS, Bray F, Vizcaino AP, Parkin DM: International lung cancer trends by histologic type: male:female differences diminishing and adenocarcinoma rates rising. Int J Cancer 2005;117: 294–299.
3   Wynder EL, Muscat JE: The changing epidemiology of smoking and lung cancer histology. Environ Health Perspect 1995;103(suppl 8):143–148.
4   Wogan GN, Hecht SS, Felton JS, Conney AH, Loeb LA: Environmental and chemical carcinogenesis. Semin Cancer Biol 2004;14:473–486.

5 Hecht SS: Tobacco smoke carcinogens and lung cancer. J Natl Cancer Inst 1999;91:1194–1210.

6 Hecht SS: Recent studies on mechanisms of bioactivation and detoxification of 4-(methylni-trosamino)-1-(3-pyridyl)-1-butanone (NNK), a tobacco-specific lung carcinogen. Crit Rev Toxicol 1996;26:163–181.

7 Belinsky SA, Devereux TR, Anderson MW: Role of DNA methylation in the activation of proto-oncogenes and the induction of pulmonary neoplasia by nitrosamines. Mutat Res 1990;233:105–116.

8 Rodin SN, Rodin AS: Origins and selection of p53 mutations in lung carcinogenesis. Semin Cancer Biol 2005;15:103–112.

9 Hecht SS: DNA adduct formation from tobacco-specific N-nitrosamines. Mutat Res 1999;424: 127–142.

10 Pfeifer GP, Denissenko MF, Olivier M, Tretyakova N, Hecht SS, Hainaut P: Tobacco smoke car-cinogens, DNA damage and p53 mutations in smoking-associated cancers. Oncogene 2002;21: 7435–7451.

11 Gotti C, Fornasari D, Clementi F: Human neuronal nicotinic receptors. Prog Neurobiol 1997;53: 199–237.

12 Wessler I, Kirkpatrick CJ, Racke K: Non-neuronal acetylcholine, a locally acting molecule, widely distributed in biological systems: expression and function in humans. Pharmacol Ther 1998;77: 59–79.

13 Millar NS: Assembly and subunit diversity of nicotinic acetylcholine receptors. Biochem Soc Trans 2003;31(Pt 4):869–874.

14 Flores CM, Davila-Garcia MI, Ulrich YM, Kellar KJ: Differential regulation of neuronal nicotinic receptor binding sites following chronic nicotine administration. J Neurochem 1997;69: 2216–2219.

15 Vallejo YF, Buisson B, Bertrand D, Green WN: Chronic nicotine exposure upregulates nicotinic receptors by a novel mechanism. J Neurosci 2005;25:5563–5572.

16 Mousavi M, Hellstrom-Lindahl E, Guan ZZ, Bednar I, Nordberg A: Expression of nicotinic acetylcholine receptors in human and rat adrenal medulla. Life Sci 2001;70:577–590.

17 Schuller HM: Cell type specific, receptor-mediated modulation of growth kinetics in human lung cancer cell lines by nicotine and tobacco-related nitrosamines. Biochem Pharmacol 1989;38: 3439–3442.

18 Schuller HM, Hegedus TJ: Effects of endogeneous and tobacco-related amines and nitrosamines on cell growth and morphology of a cell line derived from a human neuroendocrine lung cancer. Toxicol In Vitro 1989;3:37–43.

19 Maneckjee R, Minna JD: Opioid and nicotine receptors affect growth regulation of human lung cancer cell lines. Proc Natl Acad Sci USA 1989;87:3294–3298.

20 Maneckjee R, Minna JD: Opioids induce while nicotine suppresses apoptosis in human lung cancer cells. Cell Growth Differ 1994;5:1033–1040.

21 Cattaneo MG, Codignola A, Vicentini LM, Clementi F, Sher E: Nicotine stimulates a serotonergic autocrine loop in human small-cell lung carcinoma. Cancer Res 1993;53:5566–5568.

22 Schuller HM, Orloff M: Tobacco-specific carcinogenic nitrosamines. Ligands for nicotinic acetyl-choline receptors in human lung cancer cells. Biochem Pharmacol 1998;55:1377–1384.

23 Plummer HK 3rd, Dhar M, Schuller HM: Expression of the alpha7 nicotinic acetylcholine recep-tor in human lung cells. Respir Res 2005;6:29.

24 Sheppard BJ, Williams M, Plummer HK, Schuller HM: Activation of voltage-operated Ca2+-channels in human small cell lung carcinoma by the tobacco-specific nitrosamine 4-(methylni-trosamino)-1-(3-pyridyl)-1-butanone. Int J Oncol 2000;16:513–518.

25 Schuller HM, Jull BA, Sheppard BJ, Plummer HK: Interaction of tobacco-specific toxicants with the neuronal alpha nicotinic acetylcholine receptor and its associated mitogenic signal transduc-tion pathway: potential role in lung carcinogenesis and pediatric lung disorders. Eur J Pharmacol 2000;393:265–277.

26 Sekhon HS, Jia Y, Raab R, Kuryatov A, Pankow JF, Whitsett JA, et al: Prenatal nicotine increases pulmonary alpha7 nicotinic receptor expression and alters fetal lung development in monkeys. J Clin Invest 1999;103:637–647.

27 Spindel ER: Neuronal nicotinic acetylcholine receptors: not just in brain. Am J Physiol Lung Cell Mol Physiol 2003;285:L1201–L1202.

28 West KA, Brognard J, Clark AS, Linnoila IR, Yang X, Swain SM, et al: Rapid AKT activation by nicotine and a tobacco carcinogen modulates the phenotype of normal human airway epithelial cells. J Clin Invest 2003;111:81–90.

29 Ho YS, Chen CH, Wang YJ, Pestell RG, Albanese C, Chen RJ, et al: Tobacco-specific carcinogen 4-(methylnitrosamino)-1-(3-pyridyl)-1-butanone (NNK) induces cell proliferation in normal human bronchial epithelial cells through NFkappaB activation and cyclin D1 up-regulation. Toxicol Appl Pharmacol 2005;205:133–148.

30 Jull BA, Plummer HK 3rd, Schuller HM: Nicotinic receptor-mediated activation by the tobacco-specific nitrosamine NNK of a Raf-1/MAP kinase pathway, resulting in phosphorylation of c-myc in human small cell lung carcinoma cells and pulmonary neuroendocrine cells. J Cancer Res Clin Oncol 2001;127:707–717.

31 Schuller HM, Plummer HK 3rd, Jull BA: Receptor-mediated effects of nicotine and its nitrosated derivative NNK on pulmonary neuroendocrine cells. Anat Rec 2003;270:51–58.

32 Jin Z, Gao F, Flagg T, Deng X: Tobacco-specific nitrosamine 4-(methylnitrosamino)-1-(3-pyridyl)-1-butanone promotes functional cooperation of Bcl2 and c-Myc through phosphorylation in regulating cell survival and proliferation. J Biol Chem 2004;279:40209–40219.

33 Xu L, Deng X: Tobacco-specific nitrosamine 4-(methylnitrosamino)-1-(3-pyridyl)-1-butanone induces phosphorylation of mu- and m-calpain in association with increased secretion, cell migration, and invasion. J Biol Chem 2004;279:53683–53690.

34 Wong AJ, Ruppert JM, Eggleston J, Hamilton SR, Baylin SB, Vogelstein B: Gene amplification of c-myc and N-myc in small cell carcinoma of the lung. Science 1986;233:461–464.

35 Codignola A, Tarroni P, Cattaneo MG, Vicentini LM, Clementi F, Sher E: Serotonin release and cell proliferation are under the control of alpha-bungarotoxin-sensitive nicotinic receptors in small-cell lung carcinoma cell lines. FEBS Lett 1994;342:286–290.

36 Wittau N, Grosse R, Kalkbrenner F, Gohla A, Schultz G, Gudermann T: The galanin receptor type 2 initiates multiple signaling pathways in small cell lung cancer cells by coupling to G(q), G(i) and G proteins. Oncogene 2000;19:4199–4209.

37 Seufferlein T, Rozengurt E: Galanin, neurotensin, and phorbol esters rapidly stimulate activation of mitogen-activated protein kinase in small cell lung cancer cells. Cancer Res 1996;56:5758–5764.

38 Natori T, Sata M, Washida M, Hirata Y, Nagai R, Makuuchi M: Nicotine enhances neovascularization and promotes tumor growth. Mol Cells 2003;16:143–146.

39 Zhu BQ, Heeschen C, Sievers RE, Karliner JS, Parmley WW, Glantz SA, et al: Second hand smoke stimulates tumor angiogenesis and growth. Cancer Cell 2003;4:191–196.

40 Schuller HM, Witschi HP, Nylen E, Joshi PA, Correa E, Becker KL: Pathobiology of lung tumors induced in hamsters by 4-(methylnitrosamino)-1-(3-pyridyl)-1-butanone and the modulating effect of hyperoxia. Cancer Res 1990;50:1960–1965.

41 Cutz E, Gillan JE, Bryan AC: Neuroendocrine cells in the developing human lung: morphologic and functional considerations. Pediatr Pulmonol 1985;1(3 suppl):S21–S29.

42 Cutz E: Introduction to pulmonary neuroendocrine cell system, structure-function correlations. Microsc Res Tech 1997;37:1–3.

43 Cutz E, Jackson A: Neuroepithelial bodies as airway oxygen sensors. Respir Physiol 1999;115:201–214.

44 Youngson C, Nurse C, Yeger H, Cutz E: Characterization of membrane currents in pulmonary neuroepithelial bodies: hypoxia-sensitive airway chemoreceptors. Adv Exp Med Biol 1994;360:179–182.

45 Youngson C, Nurse C, Yeger H, Cutz E: Oxygen sensing in airway chemoreceptors. Nature 1993;365:153–155.

46 Aguayo SM: Pulmonary neuroendocrine cells in tobacco-related lung disorders. Anat Rec 1993;236:122–127; discussion 127–128.

47 Alshehri M, Cutz E, Banzhoff A, Canny G: Hyperplasia of pulmonary neuroendocrine cells in a case of childhood pulmonary emphysema. Chest 1997;112:553–556.

48 Cutz E, Perrin DG, Hackman R, Czegledy-Nagy EN: Maternal smoking and pulmonary neuroendocrine cells in sudden infant death syndrome. Pediatrics 1996;98(4 Pt 1):668–672.

49 Saccomanno G: The contribution of uranium miners to lung cancer histogenesis. Recent Results Cancer Res 1982;82:43–52.

50   Land CE, Shimosato Y, Saccomano G, Tokuoka S, Aoerbach O, Tateishi R, et al: Radiation-associated lung cancer: a comparison of the histology of lung cancers in uranium miners and survivors of the atomic bombings in Hiroshima and Nagasaki. Radiat Res 1993;134:234–243.

51   Schuller HM, Becker KL, Witschi HP: An animal model for neuroendocrine lung cancer. Carcinogenesis 1988;9:293–296.

52   Miller MS, Baxter JL, Moore JD, Schuller HM: Molecular characterization of neuroendocrine lung tumors induced in hamsters by treatment with nitrosamine and hyperoxia. Int J Oncol 1994;4:5–12.

53   Schuller HM, McGavin MD, Orloff M, Riechert A, Porter B: Simultaneous exposure to nicotine and hyperoxia causes tumors in hamsters. Lab Invest 1995;73:448–456.

54   Merryman JI, Park PG, Schuller HM: Carbon dioxide, an important messenger molecule for small cell lung cancer. Chest 1997;112:779–784.

55   Schuller HM: Carbon dioxide potentiates the mitogenic effects of nicotine and its carcinogenic derivative, NNK, in normal and neoplastic neuroendocrine lung cells via stimulation of autocrine and protein kinase C-dependent mitogenic pathways. Neurotoxicology 1994;15:877–886.

56   Schuller HM, Miller MS, Park PG, Orloff MS: Promoting mechanisms of CO_2 on neuroendocrine cell proliferation mediated by nicotinic receptor stimulation. Significance for lung cancer risk in individuals with chronic lung disease. Chest 1996;109(3 suppl):20S–21S.

57   Schuller HM, Correa E, Orloff M, Reznik GK: Successful chemotherapy of experimental neuroendocrine lung tumors in hamsters with an antagonist of $Ca2+$/calmodulin. Cancer Res 1990;50:1645–1649.

58   Schuller HM, Orloff M, Reznik GK: Antiproliferative effects of the $Ca2+$/calmodulin antagonist B859–35 and the $Ca(2+)$-channel blocker verapamil on human lung cancer cell lines. Carcinogenesis 1991;12:2301–2303.

59   Schuller HM, Orloff M, Reznik GK: Inhibition of protein-kinase-C-dependent cell proliferation of human lung cancer cell lines by the dihydropyridine dexniguldipine. J Cancer Res Clin Oncol 1994;120:354–358.

60   Mohr U, Reznik-Schuller H, Reznik G, Hilfrich J: Transplacental effects of diethylnitrosamine in Syrian hamsters as related to different days of administration during pregnancy. J Natl Cancer Inst 1975;55:681–683.

61   Correa E, Joshi PA, Castonguay A, Schuller HM: The tobacco-specific nitrosamine 4-(methylnitrosamino)-1-(3-pyridyl)-1-butanone is an active transplacental carcinogen in Syrian golden hamsters. Cancer Res 1990;50:3435–3438.

62   Maki T, Kontula K, Harkonen M: The beta-adrenergic system in man: physiological and pathophysiological response. Regulation of receptor density and functioning. Scand J Clin Lab Invest Suppl 1990;201:25–43.

63   Ahn S, Maudsley S, Luttrell LM, Lefkowitz RJ, Daaka Y: Src-mediated tyrosine phosphorylation of dynamin is required for beta2-adrenergic receptor internalization and mitogen-activated protein kinase signaling. J Biol Chem 1999;274:1185–1188.

64   Muyderman H, Sinclair J, Jardemark K, Hansson E, Nilsson M: Activation of beta-adrenoceptors opens calcium-activated potassium channels in astroglial cells. Neurochem Int 2001;38:269–276.

65   Wallukat G: The beta-adrenergic receptors. Herz 2002;27:683–690.

66   Maudsley S, Pierce KL, Zamah AM, Miller WE, Ahn S, Daaka Y, et al: The beta-adrenergic receptor mediates extracellular signal-regulated kinase activation via assembly of a multi-receptor complex with the epidermal growth factor receptor. J Biol Chem 2000;275:9572–9580.

67   Pierce KL, Luttrell LM, Lefkowitz RJ: New mechanisms in heptahelical receptor signaling to mitogen activated protein kinase cascades. Oncogene 2001;20:1532–1539.

68   Rall TW: Drugs used in the treatment of asthma; in Goodman-Gilman A, Rall TW, Nies AS, Taylor P (eds): The Pharmacologiccal Basis of Therapeutics, ed 8. New York, Pergamon Press, 1990, pp 618–637.

69   Schuller HM: Mechanisms of smoking-related lung and pancreatic adenocarcinoma development. Nature Reviews Cancer 2002;2:455–463.

70   Alavanja MC, Brown CC, Swanson C, Brownson RC: Saturated fat intake and lung cancer risk among nonsmoking women in Missouri. J Natl Cancer Inst 1993;85:1906–1916.

71   Burns DM: Tobacco-related diseases. Semin Oncol Nurs 2003;19:244–249.

---

72   Hida T, Yatabe Y, Achiwa H, Muramatsu H, Kozaki K, Nakamura S, et al: Increased expression of cyclooxygenase 2 occurs frequently in human lung cancers, specifically in adenocarcinomas. Cancer Res 1998;58:3761–3764.

73   Saito T, Rodger IW, Shennib H, Hu F, Tayara L, Giaid A: Cyclooxygenase-2 (COX-2) in acute myocardial infarction: cellular expression and use of selective COX-2 inhibitor. Can J Physiol Pharmacol 2003;81:114–119.

74   Mitsudomi T, Viallet J, Mulshine JL, Linnoila RI, Minna JD, Gazdar AF: Mutations of ras genes distinguish a subset of non-small-cell lung cancer cell lines from small-cell lung cancer cell lines. Oncogene 1991;6:1353–1362.

75   Schuller HM, Tithof PK, Williams M, Plummer H 3rd: The tobacco-specific carcinogen 4-(methylnitrosamino)-1-(3-pyridyl)-1-butanone is a beta-adrenergic agonist and stimulates DNA synthesis in lung adenocarcinoma via beta-adrenergic receptor-mediated release of arachidonic acid. Cancer Res 1999;59:4510–4515.

76   Laag E, Majidi M, Cekanova M, Masi T, Takahashi T, Schuller HM: NNK activates ERK1/2 and CREB/ATF-1 via beta-1-AR and EGFR signaling in human lung adenocarcinoma and small airway epithelial cells. Int J Cancer 2006;(in press).

77   Schuller HM, Cekanova M: NNK-induced hamster lung adenocarcinomas over-express beta2-adrenergic and EGFR signaling pathways. Lung Cancer 2005;49:35–45.

78   Schuller HM, Porter B, Riechert A: Beta-adrenergic modulation of NNK-induced lung carcinogenesis in hamsters. J Cancer Res Clin Oncol 2000;126:624–630.

79   Schuller HM, Plummer HK 3rd, Bochsler PN, Dudric P, Bell JL, Harris RE: Co-expression of beta-adrenergic receptors and cyclooxygenase-2 in pulmonary adenocarcinoma. Int J Oncol 2001;19:445–449.

80   Schuller HM, Porter B, Riechert A, Walker K, Schmoyer R: Neuroendocrine lung carcinogenesis in hamsters is inhibited by green tea or theophylline while the development of adenocarcinomas is promoted: implications for chemoprevention in smokers. Lung Cancer 2004;45:11–18.

81   Jin Z, Xin M, Deng X: Survival function of protein kinase C{iota} as a novel nitrosamine 4-(methylnitrosamino)-1-(3-pyridyl)-1-butanone-activated bad kinase. J Biol Chem 2005;280:16045–16052.

82   Wu WK, Wong HP, Luo SW, Chan K, Huang FY, Hui MK, et al: 4-(Methylnitrosamino)-1-(3-pyridyl)-1-butanone from cigarette smoke stimulates colon cancer growth via beta-adrenoceptors. Cancer Res 2005;65:5272–5277.

83   Rosenberg D, Groussin L, Jullian E, Perlemoine K, Bertagna X, Bertherat J: Role of the PKA-regulated transcription factor CREB in development and tumorigenesis of endocrine tissues. Ann N Y Acad Sci 2002;968:65–74.

84   Jean D, Harbison M, McConkey DJ, Ronai Z, Bar-Eli M: CREB and its associated proteins act as survival factors for human melanoma cells. J Biol Chem 1998;273:24884–24890.

85   Janne PA, Gurubhagavatula S, Yeap BY, Lucca J, Ostler P, Skarin AT, et al: Outcomes of patients with advanced non-small cell lung cancer treated with gefitinib (ZD1839, 'Iressa') on an expanded access study. Lung Cancer 2004;44:221–230.

86   Paez JG, Janne PA, Lee JC, Tracy S, Greulich H, Gabriel S, et al: EGFR mutations in lung cancer: correlation with clinical response to gefitinib therapy. Science 2004;304:1497–1500.

87   Reznik-Schuller H, Reznik G: Experimental pulmonary carcinogenesis. Int Rev Exp Pathol 1979;20:211–281.

88   Plopper CG, Hill LH, Mariassy AT: Ultrastructure of the nonciliated bronchiolar epithelial (Clara) cell of mammalian lung. III. A study of man with comparison of 15 mammalian species. Exp Lung Res 1980;1:171–180.

89   Rehm S, Takahashi M, Ward JM, Singh G, Katyal SL, Henneman JR: Immunohistochemical demonstration of Clara cell antigen in lung tumors of bronchiolar origin induced by N-nitrosodiethylamine in Syrian golden hamsters. Am J Pathol 1989;134:79–87.

90   Nikitin AY, Alcaraz A, Anver MR, Bronson RT, Cardiff RD, Dixon D, et al: Classification of proliferative pulmonary lesions of the mouse: recommendations of the mouse models of human cancers consortium. Cancer Res 2004;64:2307–2316.

91   Reznik-Schuller H: Ultrastructural alterations of nonciliated cells after nitrosamine treatment and their significance for pulmonary carcinogenesis. Am J Pathol 1976;85:549–554.

92   Reznik-Schuller H: Sequential morphologic alterations in the bronchial epithelium of Syrian golden hamsters during N-nitrosomorpholine-induced pulmonary tumorigenesis. Am J Pathol 1977;89:59–66.

93   The Alpha-Tocopherol, Beta Carotene Cancer Prevention Study Group: The effect of vitamin E and beta carotene on the incidence of lung cancer and other cancers in male smokers. N Engl J Med 1994;330:1029–1035.

94   Omenn GS, Goodman GE, Thornquist MD, Balmes J, Cullen MR, Glass A, et al: Effects of a combination of beta carotene and vitamin A on lung cancer and cardiovascular disease. N Engl J Med 1996;334:1150–1155.

95   Sporn MB, Dunlop NM, Newton DL, Smith JM: Prevention of chemical carcinogenesis by vitamin A and its synthetic analogs (retinoids). Fed Proc 1976;35:1332–1338.

96   Harris CC, Sporn MB, Kaufman DG, Smith JM, Jackson FE, Saffiotti U: Histogenesis of squamous metaplasia in the hamster tracheal epithelium caused by vitamin A deficiency or benzo[a]pyrene-Ferric oxide. J Natl Cancer Inst 1972;48:743–761.

97   Al-Wadei HAN, Takahashi T, Schuller H: Growth stimulation of human pulmonary adenocarcinoma cells and small airway epithelial cells by beta-carotene via activation of cAMP, PKA, CREB and ERK1/2. Int J Cancer 2006;118:1370–1380.

98   Osann KE: Lung cancer in women: the importance of smoking, family history of cancer, and medical history of respiratory disease. Cancer Res 1991;51:4893–4897.

99   Mayne ST, Buenconsejo J, Janerich DT: Previous lung disease and risk of lung cancer among men and women nonsmokers. Am J Epidemiol 1999;149:13–20.

100 Rennard SI: Treatment of stable chronic obstructive pulmonary disease. Lancet 2004;364:791–802.

101 Kuyucu S, Unal S, Kuyucu N, Yilgor E: Additive effects of dexamethasone in nebulized salbutamol or L-epinephrine treated infants with acute bronchiolitis. Pediatr Int 2004;46:539–544.

102 Herr I, Ucur E, Herzer K, Okouoyo S, Ridder R, Krammer PH, et al: Glucocorticoid cotreatment induces apoptosis resistance toward cancer therapy in carcinomas. Cancer Res 2003;63: 3112–3120.

103 Yao R, Wang Y, Lemon WJ, Lubet RA, You M: Budesonide exerts its chemopreventive efficacy during mouse lung tumorigenesis by modulating gene expressions. Oncogene 2004;23:7746–7752.

104 Rogatsky I, Hittelman AB, Pearce D, Garabedian MJ: Distinct glucocorticoid receptor transcriptional regulatory surfaces mediate the cytotoxic and cytostatic effects of glucocorticoids. Mol Cell Biol 1999;19:5036–5049.

105 Wang Z, Garabedian MJ: Modulation of glucocorticoid receptor transcriptional activation, phosphorylation, and growth inhibition by p27Kip1. J Biol Chem 2003;278:50897–50901.

106 Al-Wadei HAN, Takahashi T, Schuller HM: PKA-dependent growth stimulation by dexamethasone of cells derived from human pulmonary adenocarcinoma and small airway epithelium. Eur J Cancer Prev 2005.

107 Chung FL: The prevention of lung cancer induced by a tobacco-specific carcinogen in rodents by green and black Tea. Proc Soc Exp Biol Med 1999;220:244–248.

108 Chung FL, Wang M, Rivenson A, Iatropoulos MJ, Reinhardt JC, Pittman B, et al: Inhibition of lung carcinogenesis by black tea in Fischer rats treated with a tobacco-specific carcinogen: caffeine as an important constituent. Cancer Res 1998;58:4096–4101.

109 Lee KW, Lee HJ, Lee CY: Antioxidant activity of black tea vs. green tea. J Nutr 2002;132:785; discussion 786.

110 Dhawan A, Anderson D, de Pascual-Teresa S, Santos-Buelga C, Clifford MN, Ioannides C: Evaluation of the antigenotoxic potential of monomeric and dimeric flavanols, and black tea polyphenols against heterocyclic amine-induced DNA damage in human lymphocytes using the Comet assay. Mutat Res 2002;515:39–56.

111 Bertram B, Bartsch H: Cancer prevention with green tea: reality and wishful thinking. Wien Med Wochenschr 2002;152:153–158.

112 Bushman JL: Green tea and cancer in humans: a review of the literature. Nutr Cancer 1998;31: 151–159.

113 Adissu HA, Schuller HM: Antagonistic growth regulation of cell lines derived from human lung adenocarcinomas of Clara cell and aveolar type II cell lineage: implications for chemoprevention. Int J Oncol 2004;24:1467–1472.

114 Al-Wadei HAN, Takahashi T, Schuller HM: Caffeine stimulates the proliferation of human lung adenocarcinoma cells and small airway epithelial cells via activation of PKA, CREB and ERK1/2. Oncol Rep 2006;15:431–435.

115 Al-Wadei HAN, Takahasi T, Schuller HM: Theophylline stimulates cAMP-mediated signaling associated with growth regulation in human cells from pulmonary adenocarcinoma and small airway epithelia. Int J Oncol 2005;27:155–160.

116 Shafer SH, Phelps SH, Williams CL: Reduced DNA synthesis and cell viability in small cell lung carcinoma by treatment with cyclic AMP phosphodiesterase inhibitors. Biochem Pharmacol 1998;56:1229–1236.

117 Hecht SS, Hoffmann D: Tobacco-specific nitrosamines, an important group of carcinogens in tobacco and tobacco smoke. Carcinogenesis 1988;9:875–884.

Hildegard M. Schuller
Experimental Oncology Laboaratory, Department of Pathobiology
College of Veterinary Medicine, University of Tennessee, 2407 River Drive
Knoxville, TN 37996 (USA)
Tel. +1 865 974 8217, Fax +1 8640974 5616, E-Mail hmsch@utk.edu

Cho CH, Purohit V (eds): Alcohol, Tobacco and Cancer.
Basel, Karger, 2006, pp 229–236

....

Tobacco and Cancer in the Digestive Tract

Helge L. Waldum, Tom C. Martinsen

NTNU, Department of Cancer Research and Molecular Medicine and
Department of Medicine, St. Olavs Hospital, Trondheim University Hospital,
Trondheim, Norway

Abstract

Tobacco smoking has well-known negative health effects in the lungs (cancer and
obstructive lung disease) as well as on the cardiovascular system. In the digestive tract
tobacco smoking increases the risk of oral cancers, and does also slightly increase the risk of
cancers at other locations, particularly the pancreas and the oesophagus. Use of smokeless
tobacco has a slight carcinogenic effect in the mouth, but otherwise the data incriminating
smokeless tobacco in the carcinogenesis at other locations of the digestive tract are scares
and unconvincing.

Tobacco is used mainly for the central nervous effects of nicotine. Nicotine
induces well-feeling, but at the same time addiction [1]. Smoking is the most
popular route of tobacco administration, presumably due to the rapid uptake
from the huge lung surface resulting in a peak of nicotine concentration in arterial blood. Thereby the central nervous effects of nicotine are most pronounced.
Tobacco may also be administered as snuff or chew (smokeless tobacco) to be
absorbed from the nasal and/or oral cavities. Tobacco contains a large quantity
of different substances, some of which experimentally have been shown to be
carcinogenic when administered in high doses [2]. At combustion, many
polynuclear aromatic hydrocarbons are produced, which themselves are carcinogens [2]. Tobacco smoke also contains carbon monoxide which has a high
affinity for oxygen binding sites. The factors mainly responsible for the negative health effect of tobacco intake/smoking are, however, not known. While it
is no doubt that tobacco smoking predisposes or causes lung cancer [3] and

atherosclerosis [4], the association between tobacco and diseases/cancers in the digestive tract is less obvious. In the present review, the role of tobacco in carcinogenesis at different anatomical sites of the digestive tract will be covered.

Oral Cavity/Pharynx Cancer

Carcinomas of the oral cavity include cancers of the tongue, tonsils, salivary glands as well as the lips, the palate and the cheek. Tobacco consumption is a major risk factor for the development of cancers in the oral cavity and the pharynx [5, 6]. Heavy alcohol and tobacco intake are the most important causal factors for cancers at these locations [7]. So-called smokeless tobacco intake also disposes for oral cancer [8], although there has been some uncertainty regarding Swedish snuff [9]. Nevertheless, Swedish snuff has also been shown slightly to increase the likelihood of oral cancers [10] including lip cancer [10]. It is easily understood and accepted that continuous exposure of the mucus surfaces of the mouth by smokeless tobacco may dispose for oral cancers. On the other hand, few seem to know that tobacco smoking also increases the risk of oral carcinomas [11]. Oral cancers show great variability in frequency between different countries and even between regions in the same country [5]. Besides, tobacco smoking and alcohol consumption diet may play a role in the aetiology of cancers at this location [5], which is not unexpected taking into consideration that the mouth is the first place where food is exposed to a mucus membrane. The role of smokeless tobacco in oral cancers has more recently been challenged [12]. Rodu and Jansson [12] showed that the data for the carcinogenic effect of moist snuff was insufficient, and they also described cancer-inhibiting properties of substances found in tobacco.

Oesophageal Cancer

Most carcinomas of the oesophagus are either squamous cell carcinomas or adenocarcinomas [13]. The latters are associated with Barrett's oesophagus related to gastroesopohageal reflux diseases [14]. Tobacco smoking and moderate to heavy alcohol consumption are the major predisposing factors for squamous cell carcinoma of the oesophagus [15, 16]. Tobacco smoking has been reported to increase the risk of squamous oesophageal cancer about three times, whereas moderate to heavy alcohol consumption does so by a factor about seven in men [17]. In a large case control study on the role of tobacco smoking on the risk of adenocarcinomas of the oesophagus, Wu et al. [18] also found that tobacco smoking is a risk factor also for this type of tumour. There also seems to be a correlation between the magnitude of smoking as well as the time of smoking, with the risk of development of oesophageal carcinoma [17, 18]. Snuff seems not to be associated with increased risk of oesoopageal adenocarcinomas, and in the same study even tobacco smoking was only weakly correlated to that tumour [15].

To conclude that part, tobacco smoking unequivocally increases the risk of squamous cell carcinoma of the oesophagus whereas the association to adenocarcinomas is weaker and snuff seems not to increase the risk of oesophageal cancer.

Gastric Cancer and Tobacco

Gastric carcinomas occur much more frequently in Japan than in the western world. This difference in prevalence is not based upon genetic differences since Japanese moving to the United States soon show the same frequency in gastric cancer as other American citizens [19]. There is, however, also a form of familial gastric cancer due to mutation of the E-cadherin gene [20]. The dominating cause of gastric cancer, however, seems to be infection with *Helicobacter pylori* [21]. *Helicobacter pylori* infection is associated, but of course not caused by smoking [22].

The gastric carcinomas may be divided anatomically in two parts according to the mucosa where they originate (distal or antral carcinomas and proximal or oxyntic carcinomas). Previously, a third type of gastric muosa, cardiac mucosa, was claimed to be a separate normal variant, but this has been challenged in recent years [23]. In the latter study it was claimed that the so-called cardiac mucosa represents metaplasia from oxyntic mucosa in a border zone to the squamous cell epithelium in the oesophagus. It therefore seems natural to divide gastric carcinomas topographically only into proximal and distal carcinomas. Based upon histological examination Laurén [24] divided gastric carcinomas into two types, intestinal and diffuse. During a long time period, we have given evidence for the concept that the diffuse gastric carcinomas originate from neuroendocrine or more specifically, from enterochromaffin like cells [25–29].

The role of tobacco smoking in gastric carcinogenesis has been disputed, but there seems to be a slight increase in the risk for gastric carcinoma in tobacco smokers [30], although other studies have not confirmed that [31]. Tobacco smoking seems to dispose only to differentiated (intestinal) type of cancer with a distal location [32]. In a rat long-term study, we could not find any effect of long-term nicotine inhalation on the gastrointestinal tract, including the stomach [33, 34]. Similarly, carbon monoxide inhalations for 18 months did not either induce macroscopical or microscopical changes in the stomach (Sørhaug S,

Nilsen OG, Steinshamn S, Waldum HL, to be published). Accordingly, two of the most and best characterised constituents of tobacco smoke, nicotine and carbon monoxide, do not induce changes in the stomach in long-term rat studies. Anyhow, tobacco smoking moderately increases the risk of gastric cancer.

Small Intestinal Malignancies

Malignant tumours in the small intestine are rare compared to the stomach and colon. The reason for this difference in frequencies of malignant lesions within the gastrointestinal tract, is not known, but could be due to microbiological differences as the small intestine is less contaminated than the colon. Furthermore, *Helicobacter pylori* in the stomach is an important carcinogen [21]. Both carcinomas and carcinoid tumours may occur in the small intestine, but to our knowledge tobacco smoking does not influence the likelihood for any of these malignancies. In this connection it should be added that due to the rarity of these small intestinal malignancies, large epidemiological studies are not so easy to perform.

Colonic Malignancies

Tobacco and Chronic Inflammatory Bowel Diseases

While tobacco affects the health negatively in many organs and tissues, it is an established fact that tobacco smoking has a positive effect on ulcerative colitis [35], reducing the recurrence or delaying the debut of this disease. Since ulcerative colitis predisposes to colonic cancer [36], this effect of tobacco smoking may be expected to reduce colonic cancer secondary to ulcerative colitis. Moreover, there are indications that the beneficial effect of tobacco smoking on the course of ulcerative colitis is due to nicotine [37], although nicotine treatment has not gained any established place in the treatment of ulcerative colitis. On the other hand, tobacco smoking may worsen Crohn's disease [35], a disease which is less incriminated in colonic carcinogenesis than ulcerative colitis [38].

Colon Cancers

Most colon cancers develop from polyps. There are studies indicating an increase in the occurrence of colorectal carcinomas in smokers [39, 40]. There are also studies suggesting that tobacco smoking increases the risk of development of hyperplastic [40] as well as adenomatous [41] colonic polyps. Taking together, tobacco smoking seems to increase the risk of colonic carcinoma slightly, perhaps doubling the incidence.

Pancreas Cancers

Malignancies of the pancreas occurred with increasing frequency during the last century [42], but during the last decades the incidence levelled off [43]. The aetiology of carcinomas of this organ is as for most malignant diseases, not known. However, there is epidemiological evidence for a role of tobacco smoking also in this disease [44], which has such a bad prognosis. The mechanism by which tobacco smoking predisposes to carcinomas of the pancreas are not known, but in Syrian hamsters N-nitroso compounds, which are found in tobacco, disposes for cancer development in the pancreas [45]. Not only tobacco smoking, but also smokeless tobacco may increase the risk of pancreatic cancer [46, 47]. One of these studies [46] was, however, criticised [48, 49] casting doubt upon the association between smokeless tobacco and pancreatic cancer. Like for carcinoids in the small intestine, we do not know of any connection between use of tobacco and endocrine pancreatic malignancies.

Liver Cancers

Primary liver cancers may be divided into hepatocellular carcinomas and cholangio-carcinomas. Chronic infections with the hepatotropic viruses, hepatitis B and hepatitis C virus, increase the risk for hepatocellular carcinoma directly and also indirectly by inducing inflammation and liver cirrhosis. Sclerosing cholangitis, a disease closely associated to ulcerative colitis, disposes to cholangiocarcinoma. Like ulcerative colitis, sclerosing cholangitis is also associated to non-smoking [50, 51]. However, neither oral [52] nor transdermal [53] nicotine improves sclerosing cholangitis when established. Concerning cholangiocarcinoma, the effect of smoking on the risk to develop this disease is unknown, but by the beneficial effect on the underlying condition sclerosing cholangitis, it would be expected to reduce the risk. Overall, however, the association between tobacco use and liver cancers has previously been uncertain. Newer studies have, however, established that there is also a doubling of the risk of liver cancers in smokers, compared with non-smokers [54]. Interestingly, alcohol and tobacco used together has a more than additive effect on the risk for hepatocellular carcinoma [55]. Similarly, carriers of hepatitis B virus may be particularly susceptible for the liver carcinogenic effect of tobacco smoking [56].

Conclusion

In conclusion, use of tobacco increases the risk of cancers at most sites of the digestive tract. Although, the increased risk due to tobacco is small compared with for instance lung cancer, the fact that these cancers are prevalent, makes even a moderate increase in risk quantitatively important.

References

1   Surgeon General: The Health Consequences of Smoking: Nicotine Addiction. Washington (DC), U.S. Gov Print, 1988.
2   Hecht SS: Tobacco smoke carcinogens and lung cancer. J Natl Cancer Inst 1999;91:1194–1210.
3   Blot WJ, Fraumeni JF Jr: Cancers of the lung and pleura; in Schottenfeld D, Fraumneni J Jr (eds): Cancer Epidemiology and Prevention. New York (NY), Oxford University Press, 1966, pp 637–665.
4   Centers for Disease Control and Prevention: Cigarette smoking-attributable mortality and years of potential life lost – United States (1990). MMWR Morb Mortal Wkly Rep 1993;42:645–649.
5   La Vecchia C, Tavani A, Franceschi S, Levi F, Corrao G, Negri E: Epidemiology and prevention of oral cancer. Oral Oncol 1997;33:302–312.
6   Blot WJ, McLaughlin JK, Winn DM: Smoking and drinking in relation to oral and pharyngeal cancer. Cancer Res 1988;48:3282–3287.
7   Moreno-López LA, Esparza-Gómez GC, González-Navarro A, Cerero-Lapiedra R, Gonxãlez-Hernãndez MJ, Domingues-Rojas V: Risk of oral cancer associated with tobacco smoking, alcohol consumption and oral hygiene: a case-control study in Madrid, Spain. Oral Oncol 2000;36: 170–174.
8   Gupta PC: A study of dose-response relationship between tobacco habits and oral leukoplakia. Br J Cancer 1984;50:527–531.
9   Axéll TE: Oral mucosal changes related to smokeless tobacco usage: research findings in Scandinavia. Eur J Cancer B Oral Oncol 1993;29B:299–302.
10 Warnakulasuriya S, Sutherland G, Scully C: Tobacco, oral cancer, and treatment of dependence. Oral Oncol 2005;41:244–260.
11 Warnakulasuriya KAAS, Harris CK, Scarrott DM, Watt R, Gelbier S, Peters TJ, Johnson NW: An alarming lack of public awareness towards oral cancer. Br Dent J 1999;187:319–322.
12 Rodu B, Jansson C: Smokeless tobacco and oral cancer: a review of the risks and determinants. Crit Rev Oral Biol Med 2004;15:256–263.
13 Daly JM, Fry WA, Little AG, et al: Esophageal cancer: results of an American College of Surgeons patient care evaluation study. J Am Coll Surg 2000;19:562–572.
14 Lagergren J, Bergström R, Lindgren A, Nyrén O: Symptomatic gastroesophageal reflux as a risk factor for esophageal adenocarcinomas. N Engl Med 1999;340:825–831.
15 Lagergren J, Bergstrom R, Lindgren A, Nyrén O: The role of tobacco, snuff and alcohol, use in the aetiology of cancer on the oesophagus and gastric cardia. Int J Cancer 2000;85:340–346.
16 Engel LS, Chow W-H, Vaughan TL, Gammon MD, Risch HA, Stanford JL, Schoenberg JB, Mayne ST, Dubrow R, Rotterdam H, West AG, Blaser M, Blot WJ, Gail MH, Fraumeni JP: Population attributable risks of esophageal and gastric cancers. J Natl Cancer Inst 2003;95: 1404–1413.
17 Brown LM, Hoover R, Silverman D, Baris D, Hayes R, Swanson GM, Schoenberg J, Greenberg R, Liff J, Schwartz A, Dosemeci M, Pottern L, Fraumeni JF Jr: Excess incidence of squamous cell esophageal cancer among US black men: role of social class and other risk factors. Am J Epidemiol 2001;153:114–122.
18 Wu AH, Wan P, Bernstein L: A multiethnic population-based study of smoking, alcohol on body size and risk of adenocarcinomas of the stomach and esophagus (United States). Cancer Causes Control 2001;12:721–732.
19 Kamineni A, Williams MA, Schwarz SM, et al: The incidence of gastric carcinoma in Asian migrants to the United States and their descendants. Cancer Causes Control 1999;10:77–83.
20 Guilford P, Hopkins J, Harraway J, McLeod N, Harawira P, Taite H, et al: E-cadherin germline mutations in familial gastric cancer. Nature 1998;392:402–405.

21 The Eurogast Study Group: An international association between Helicobacter pylori infection and gastric cancer. Lancet 1993;341:1359–1362.

22 Parasher G, Fastwood GL: Smoking and peptic ulcer in the Helicobacter era. Eur J Gastroenterol Hepatol 2000;12:843–853.

23 Chandrasoma PT, Der R, Ma Y, Dalton P, Taira M: Histology of the gastroesophageal junction. An autopsy study. Am J Surg Pathol 2000;24:402–409.

24 Laurén P: The two histological main types of gastric carcinoma: diffuse and so-called intestinal-type carcinoma. Acta Pathol Microbiol Scand 1965;64:31–49.

25 Waldum HL, Haugen OA, Isaksen C, Mecsei R, Sandvik AK: Are diffuse gastric carcinomas neuroendicrine tumours (ECL-omas)? Eur J Gastroenterol Hepatol 1991;3:245–249.

26 Waldum HL, Aase S, Kvetnoi I, Brenna E, Sandvik AK, Syversen U, Johnsen G, Vatten L, Polak JM: Neuroendocrine differentiation in human gastric carcinoma. Cancer 1998;83:435–444.

27 Qvigstad G, Falkmer S, Westre B, Waldum HL: Clinical and histopoathological tumour progression in ECL cell carcinoids. APMIS 1999;107:1985–1092.

28 Qvigstad G, Sandvik AK, Brenna E, Aase S, Waldum HL: Detection of chromogranin A in human gastric adenocarcinomas using a sensitive immunohistochemical technique. Histochem J 2000; 32:551–556.

29 Qvigstad G, Qvigstad T, Westre B, Sandvik AK, Brenna E, Waldum HL: Neuroendokrine differentiation in gastric adenocarcinomas associated with severe hypergastrinemia and/or pernicious anemia. APMIS 2002;110:132–139.

30 Doll R, Peto R, Wheatley K, Gray R, Sutherland I: Mortality in relation to smoking: 40 years' observations on male. BMJ 1994;309:901–911.

31 Nomura AMY, Stemmermann GN, Chyou P-H: Gastric cancer among the Japanese in Hawaii. Jpn J Cancer Res 1995;86:916–923.

32 Sasazuki S, Sasaki S, Tsugane S: Cigarette smoking, alcohol consumption and subsequent gastric cancer risk by subsite and histologic type. Int J Cancer 2002;101:560–566.

33 Brenna E, Zahlsen K, Mårvik R, Nilsen T, Nilsen OG, Waldum HL: Effect of nicotine on the enterochromaffin-like cells of the oxyntic mucosa of the rat. Life Sci 1993;53:21–29.

34 Waldum HL, Nilsen OG, Nilsen T, Rørvik H, Syversen U, Sandvik AK, Haugen OA, Torp SH, Brenna E: Long-term effects of inhaled nicotine. Life Sci 1996;58:1339–1346.

35 Vessey M, Jewell D, Smith A, Yeates D, McPherson K: Chronic inflammatory bowel disease, cigarette smoking, and use of oral contraceptives: findings in a large cohort study of women of childbearing age. BMJ 1986;292:1101–1103.

36 Greenstein AJ, Sacher DB, Smith H, Pucillo A, Papatestas AE, Kreel I, et al: Cancer in universal and left-sided ulcerative colitis: factors determining risk. Gastroenterology 1979;89:290–294.

37 Sandorn WJ, Tremaine WJ, Offord KP, Lawson GM, Petersen BT, Batts KP, Croghan IT, Dale LC, Schroeder DR, Hurt RD: Transdermal nicotine for mildly to moderately active ulcerative colitis. A randomized, double-blind, placebo-controlled trial. Ann Intern Med 1997;126:364–371.

38 Goldman H: Significance and detection of dysplasia in chronic colitis. Am Cancer Soc 1996;78: 2261–2263.

39 Rimm GE, Stampfer MJ, Colditz GA, Ascherio A, Kearney J, Willett WC: A prospective study of cigarette smoking and risk of colorectal adenoma and colorectal cancer in U.S. men. J Natl Cancer Inst 1994;86:162–164.

40 Lieberman DA, Prindiville S, Weiss DG, Willett W: Risk factors for advanced colonic neoplasia and hyperplastic polyps in asymptomatic individuals. JAMA 2003;290:2959–2967.

41 Kikendall JW, Bowen PE, Burgess MB, Magnetti C, Woodward J, Langenberg P: Cigarettes and alcohol as independent risk factors for colonic adenomas. Gastroenterology 1989;97:660–664.

42 Krain LS: The rising incidence of carcinoma of the pancreas: real or apparent? J Surg Oncol 1970;2:115–124.

43 Gold EB, Goldin SB: Epidemiology of and risk factors for pancreatic cancer. Surg Oncol Clin N Am 1998;7:67–91.

44 Boyle P, Maisonneuve P, Bueno de Mesquita B, Ghadirian P, Howe GR, Zatonski W, Baghurst P, Moerman CJ, Simard A, Miller AB, Przewoniak K, McMichael AJ, Hsieh C-C, Walker AM: Cigarette smoking and pancreas cancer: a case-control study of the search programme of the IARC. Int J Cancer 1961;67:63–71.

45 Pour PM, Salmasi SZ, Runge RG: Selective induction of pancreatic ductular tumors by single doses of N-nitrosobis(2-oxopropyl)amine in Syrian golden hamsters. Cancer Lett 1978;4: 3317–3323.

46 Boffetta P, Aagnes B, Weiderpass E, Andersen AA: Smokeless tobacco use and risk of cancer of the pancreas and other organs. Int J cancer 2005;114:992–995.

47 Alguacil J, Silverman DT: Smokeless and other noncigarette tobacco use and pancreatic cancer: a case-control study based on direct interviews. Cancer Epidemiol Biomark Prev 2004;13:55–58.

48 Ramström L: Smokeless tobacco use and risk of cancer of the pancreas and other organs. Letter to the Editor. Int J Cancer 2006;118:1584.

49 Rodu B, Cole P: A deficient study of smokeless tobacco use and cancer. Letter to the Editor. Int J Cancer 2005;118:1585.

50 Loftus EV, Sandborn WJ, Tremaine WJ, Mahoney DW, Zinsmeister AR, Offord KP, et al: Primary sclerosing cholangitis is associated with non-smoking: a case-control study. Gastroenterology 1996;110:1496–1502.

51 Erpecum KJ, Smits SJHM, van de Meeberg PC, Linn FHH, Wolfhagen FHJ, van Berge-Henegouwen GP, et al: Risk of primary sclerosing cholangitis is associated with non-smoking behaviour. Gastroenterology 1996;110:1503–1506.

52 Angulo P, Bharucha AE, Jorgensen RA, DeSotel CK, Sandborn WJ, Larusso NF, Lindor KD: Oral nicotine in treatment of primary sclerosing cholangitis. A pilot study. Dig Dis Sci 1999;44:602–607.

53 Jørgensen G, Waldum HL: Lack of effect of transdermal nicotine on 3 cases of primary sclerosing cholangitis. Dig Dis Sci 1999;44:2484.

54 Sasco AJ, Secretan MB, Straif K: Tobacco smoking and cancer: a brief review of recent epidemiological evidence. Lung Cancer 2004;45:53–59.

55 Kuper H, Tzonou A, Kaklamani E, Hsieh C-C, Lagiou P, Adami H-O, Trichopoulos D, Stuver SO: Tobacco smoking, alcohol consumption and their interaction in the causation of hepatocellular carcinoma. Int J Cancer 2000;85:498–502.

56 Yu M-W, Pai C-I, Yang S-Y, Hsiao T-J, Chang H-C, Lin S-M, Liaw Y-F, Chen P-J, Chen C-J: Role of N-acetyltransferase polymorphisms in hepatitis B related hepatocellular carcinoma: impact of smoking on risk. Gut 2000;47:703–709.

Helge L. Waldum
Department of Cancer Research and Molecular Medicine, St. Olavs Hospital
Trondheim University Hospital
NO–7006 Trondheim (Norway)
Tel. +47 73 86 8541, Fax +47 73 86 7546, E-Mail helge.waldum@ntnu.no

Cho CH, Purohit V (eds): Alcohol, Tobacco and Cancer.
Basel, Karger, 2006, pp 237–252

Modification of Experimental Carcinogenesis by Cigarette Smoke and its Constituents

Akiyoshi Nishikawa

Division of Pathology, National Institute of Health Sciences,
Kamiyoga, Setagaya-ku, Tokyo, Japan

Abstract

Epidemiologically, it has been suggested that inhalation of cigarette smoke (CS) is closely associated with increased risk of human cancers in various organs such as the lung, oropharynx, stomach, pancreas, liver and colon. Nevertheless, except for a recent study showing that lifetime exposure to high doses of CS strongly increases lung cancer incidence in B6C3F$_1$ mice, there have been only studies reporting modifying effects of CS exposure on chemically induced or spontaneous carcinogenesis. We have showed that CS exposure induces hepatic CYP isozymes, especially CYP1A2, in both rats and hamsters, and in fact S9 fraction from their livers exposed to CS specifically increases the mutagenicity in Ames assay of various heterocyclic amines (HCAs), in good agreement with the fact that HCAs are principally activated by CYP1A2 to proximate carcinogens. Actually, CS exposure enhanced hepatocarcinogenesis in rats induced by 2-amino-3,8-dimethylimidazo[4,5-*f*]quinoxaline, a major HCA. In clear contrast to the effects of CS exposure per se, it is well known that some CS constituents such as HCAs, nitrosamines, and polyaromatic hydrocarbons readily induce cancers in several organs of rodents. In addition to carcinogenicity, it has been shown that various CS constituents including catechol and nicotine can modify experimental carcinogenesis. In this review article, the modifying effects of CS and its constituents on experimental carcinogenesis are overviewed and the significance of these effects on human risk is discussed.

Epidemiological studies have suggested that cigarette smoke (CS) closely associated with increased risk of cancers in various organs including the lung, oropharynx, esophagus, stomach, colon, liver, pancreas, kidney, urinary bladder, and breast [1, 2]. CS is a complex aerosol of minute liquid droplets suspended within a mixture of gases and semi-volatile compounds [3]. The International

Agency for Research on Cancer has classified 81 chemical constituents reported in mainstream CS as carcinogens (11 Group 1, 14 Group 2A and 56 Group 2B) [3]. In addition to the experimental data showing that inhaled CS per se increases the incidence of spontaneous lung tumors in A/J mice [4–8] and promotes upper respiratory tract tumorigenesis in hamsters initiated with diethylnitrosamine (DEN) [9], very recently, it was demonstrated that lifetime whole-body exposure of $B6C3F_1$ mice to high doses of CS strongly increases lung cancer incidence compared with sham exposed animals [10].

Meanwhile, it has also been shown that inhaled CS modifies experimental carcinogenesis [9, 11–21]. The mechanisms underlying the modification effects are complicated in association with the fact that CS is a complex aerosol [3]. Most environmental carcinogens are metabolically activated and detoxified by phases I and II enzymes [22]. Because CS can affect drug metabolizing enzymes of both phases I and II enzymes [23], it is possible that the modifying effects of CS on such drug metabolizing enzymes may be associated with metabolic activation and/or detoxification of carcinogens which target various organs. In fact, we have shown that CS exposure induces hepatic CYP1A enzyme, especially CYP1A2, in both rats and hamsters [24], and consequently S9 fraction from their livers exposed to CS specifically increases the *Salmonella* mutagenicity of various heterocyclic amines (HCAs) [24, 25] contained in CS as well as cooked food [26] in line with the fact that HCAs are activated by CYP1A2 to proximate carcinogens [27]. However, there are limited data of experiments showing associations between CS and drug metabolism and carcinogenesis in the whole body of animals including rats, mice, hamsters, guinea pigs, and ferrets [11, 23–25, 28–39].

In this review article, the modifying effects of CS and its constituents on experimental carcinogenesis are overviewed and the significance of these effects on human risk is discussed.

Carcinogenicity of Inhaled CS and its Constituents

CS plays a major role in the epidemiology of lung cancer, and CS constituents have extensively been investigated in carcinogenicity studies using experimental animals. In fact, it is well documented that definite animal carcinogens are included in CS although the amounts are very low. For example, it has been shown that tobacco-specific nitrosamines such as 4-(methylnitrosamino)-1-(3-pyridyl)-1-butanone (NNK) and 4-(methylnitrosamino)-1-(3-pyridyl)-1-butanol (NNAL), and polyaromatic hydrocarbons such as benzo[*a*]pyrene (B(*a*)P) can induce respiratory tract tumors in rodents [40]. However, it was much more difficult to reproduce the tumorigenicity of CS mixture in pre-clinical models [41].

Table 1. Modifying effects of cigarette smoke (CS) on experimental carcinogenesis

Species/organ	Initiators	Timing of treatment	Summarized effects	References
Rat liver	MeIQx	Initiation	Enhancement	[11]
Rat lung	B(*a*)P	Initiation	Enhancement	[12]
Rat upper respiratory and alimentary tracts	4-NQO	Throughout	No effect	[41]
Rat skin	Irradiation	Post-initiation	Enhancement	[13]
Mouse skin	DMBA	Post-initiation	Enhancement or reduction	[14]
Mouse skin	DMBA	Post-initiation	Enhancement	[16]
Mouse skin	DMBA	Post-initiation	Enhancement	[18]
Mouse lung	NNK	Post-initiation	No effect	[43]
Mouse colon	DSS	Simultaneous	Enhancement	[19]
Hamster pancreas and respiratory tract	BOP	Initiation Post-initiation	Reduction Reduction	[20]
Hamster respiratory tract	DEN	Post-initiation	Enhancement	[9]
Hamster buccal pouch	B(*a*)P	Simultaneous	Enhancement	[21]

B(*a*)P = benzo[*a*]pyrene;   BOP = *N*-nitrosobis(2-oxopropyl)amine;   DEN = diethylnitrosamine. DMBA = 7,12-dimethylbenz[*a*]-anthracene; 4-NQO = 4-nitroquinoline*N*-oxide; DSS = dextran sulfate sodium; MeIQx = 2-amino-3,8-dimethylimidazo[4,5-*f*]quinoxaline; NNK = 4-(methylnitrosamino)-1-(3-pyridyl)-1-butanone.

Likewise, although CS has been epidemiologically associated with lung cancer in humans for many years, lung carcinogenicity of CS has been mostly detected in A/J mice that develop lung adenomas spontaneously [4–8]. Thus animal models of CS-induced lung cancer have been actually lacking. However, it was very recently demonstrated that life time whole body exposures of female B6C3F$_1$ mice to mainstream CS at 250 mg total particulate matter/m^3 for 6 h per day, 5 days a week induces marked increases in the incidence of focal alveolar hyperplasias, pulmonary adenomas, papillomas, and adenocarcinomas [10]. CS-exposed mice had a 10-fold increase in the incidence of hyperplastic lesions, and a 4.6-fold (adenomas and papillomas), 7.25-fold (adenocarcinomas) and 5-fold (metastatic pulmonary adenocarcinomas) increase in primary lung neoplasms compared with sham-exposed mice [10].

Modification of Carcinogenesis by CS

Effects of CS inhalation or condensate on various experimental carcinogenesis are summarized in table 1.

Rat Liver

The modifying effects of CS on liver carcinogenesis were examined in male F344 rats fed 2-amino-3,8-dimethylimidazo[4,5-*f*]quinoxaline (MeIQx) at a dose of 300 ppm for 4 or 16 weeks, simultaneously with or without CS exposure or followed by CS exposure for 12 weeks [11]. CS was transnasally administered 7 days a week by using a Hamburg type II smoking machine under the following conditions: exposure period, 6 min/day; doses, 30 cigarettes/time; inhalation volume, 35 ml; inhalation flow, 17.5 ml/s; dilution of CS, 1/7. At the 16th week of the experiment, the mean number and area of glutathione *S*-transferase (GST) placental form positive-liver cell foci (GST-P$^+$ foci) were significantly higher in rats given MeIQx plus CS for 16 weeks than in those receiving MeIQx alone for the same period [11].

In order to assess the effect of CS on metabolic enzymes, male rats were exposed for 2 weeks to smoke produced in the Hamburg type II smoking machine [24]. The livers were then used for Ames liquid incubation and western immunoblot assays. Mutagenic activities of seven HCAs in *Salmonella typhimurium* TA98 in the presence of rat liver S9 were elevated up to 3.7 times above controls. Western immunoblot analyses of liver microsomes using anti-rat cytochrome P-450 (CYP) anti-bodies revealed that CS exposure increased the levels of rat CYP1A2 and CYP1A1, without significant change in the levels of CYP2E1 and 3A. The observed selective induction of HCA activation and CYP isozymes due to CS supported the idea that CS may contribute to enhancing effects on initiation by carcinogens that are metabolically activated by hepatic CYP1A1/1A2 [24].

Furthermore, in order to elucidate the mechanism on the enhancement by CS of MeIQx-induced rat hepatocarcinogenesis, hepatic levels of CYP enzymes, mutagenic activation of various carcinogens and UDP-glucuronyltransferase (UDPGT) activities were assayed in male F344 rats [23]. Immunoblot analyses for microsomal CYP proteins revealed induction of CYP1A1 and constitutive CYP1A2, but not CYP2B1/2, 2E1 or 3A2, by CS exposure for 1, 12 or 16 weeks. CS also elevated the mutagenic activities of MeIQx and five other HCAs. Combined treatment with CS and MeIQx showed a summation effect on induction of UDPGT1A6 activity. In conjunction with the finding of *N*-hydroxy-MeIQx being a poor substrate for rat liver UDPGT, these results clearly indicated that enhancement by CS of MeIQx-induced hepatocarcinogenesis in F344 rats can be attributed to an increase in metabolic activation of MeIQx by hepatic CYP1A2 during the initiation phase [23].

Rat Lung

Inhalation of CS during initiation and post-initiation phases of (B(*a*)P)-induced carcinogenesis resulted in higher tumor burden as compared to the same observed in the animals exposed to CS during the post-initiation phase only [12]. Stimulation in tumor burden by CS was increased further by vitamin A deficiency [12].

Rat Lip, Oral and Nasal Cavities, Esophagus, and Forestomach

A canal in the lower lip to function as a reservoir for snuff was surgically created in male Sprague-Dawley rats [42]. The animals received painting on the hard palate with 4-nitroquinoline N-oxide (4-NQO) dissolved in propylene glycol, 3 times a week for 4 week, and/or snuff twice a day, 5 days a week for up to 108 week. Although squamous cell tumors of the lip, oral and nasal cavities, esophagus, and forestomach were seen evenly in all the groups, hyperplastic lesions of the lip, palate, and forestomach were significantly more frequent in the rats treated with snuff than in those without snuff. The results showed that snuff and 4-NQO by themselves have the potential to induce malignant tumors, but initiation with 4-NQO followed by snuff does not significantly enhance tumor formation [42].

Rat Skin

Skin tumors from male albino Charles River CD rats (outbred Sprague-Dawley descended) were classified into 2 groups: carcinomas and other non-carcinoma tumors [13]. A statistically significant increase in tumor yield occurred after CS condensate treatment that began 2 months after irradiation, and carcinoma yield was also increased albeit without statistical significance [13]. When CS condensate treatment began immediately after irradiation, the yield of non-carcinoma tumors was significantly reduced, whereas the carcinoma yield increased without statistical significance [13].

Mouse Skin

Short-term in vivo analyses conducted within the context of a mouse skin, tumor-promotion protocol (i.e. comparative measures of epidermal thickness, proliferative index, myeloperoxidase activity, leukocyte invasion, mutation of Ha-*ras*, and formation of modified DNA bases) showed that reference CS condensate induced statistically significant and dose-dependent increases (relative to vehicle control) for nearly all indices examined, while prototype CS condensate possessed a significantly reduced potential for inducing changes such as epidermal hyperplasia and/or inflammation [14].

CS condensates possessing significantly different tumorigenic potentials according to a standardized 30-week mouse skin tumor-promotion protocol could likewise be discriminated utilizing short-term indices of sustained hyperplasia and/or inflammation [14]. Employing a truncated initiation–promotion protocol to further evaluate CS condensate-induced hyperplasia, condensate application induced treatment-related increases for epidermal thickness, proliferative index as assessed by 5-bromo-2′-deoxyuridine labeling, and ornithine decarboxylase expression. Interestingly, observed increases for interfollicular 5-bromo-2′-deoxyuridine labeling and ornithine decarboxylase expression were partially reversed but still elevated upon cessation of promotion, while increases within the perifollicular epidermis remained elevated at a level similar to that observed during CS condensate application [15].

Workers engaged in processing tobacco for the manufacture of bidis, the most popular smoking devices in India, have been exposed to tobacco dust, volatile components and flakes via nasopharyngeal and cutaneous routes [16]. In order to evaluate the risk of occupational CS, the complete carcinogenic action of an aqueous extract of bidi tobacco, its ability to initiate and promote skin papillomas and to convert these to carcinomas, was tested in hairless S/RV Cri-ba mice using the skin tumorigenesis protocol. Epidermal cell kinetics and tissue alterations were recorded after a single or multiple applications of the extract to 7,12-dimethylbenz[a]-anthracene (DMBA)-initiated mouse skin. While the extract did not exhibit complete carcinogenic, initiating or progressor activity, it effectively promoted skin papilloma formation in DMBA-initiated mice [16]. An increase in papilloma yield per mouse above the control was noted only after 30 weeks of promotion, and at week 40 of promotion with 5 and 50 mg of the extract it was significantly higher than that in the control mice [16]. Mild epidermal hyperplasia, increase in mitotic activity and dermal thickness induced by a single application of the extract persisted upon multiple treatment and correlated well with its tumor-promoting activity [16].

The weakly acidic fraction of CS particulate matter was fractioned by silica get chromatography [17]. Various primary subfractions for potential tumor-promoting activity were assayed by measuring the incorporation of tritiated thymidine into mouse epidermal DNA as induced by these subfractions. Based on these results and on chemical composition, the primary subfractions were then combined into four major subfractions and tested on initiated mouse skin for tumor-promoting activity by long-term application. Two of these subfractions were inactive, whereas the other two showed tumor-promoting activity. The two active portions were then further chromatographed and tested by the short-term bioassay. Some major components of the resulting active fractions included alkyl-2-cyclopenten-2-ol-1-ones, catechols, hydroquinone, fatty acids, and 3-hydroxypyridines. Among these components, catechol, hydroquinone, 3-hydroxypyridine, 6-methyl-3-hydroxypyridine, linolenic acid, and linoleic acid were inactive as tumor promoters in the experimental animal [17].

Earlier studies showed that aqueous extracts of tobacco exhibit tumor-promoting activity, the activity requiring the simultaneous presence of two agents, one of which was methanol soluble and the other, methanol insoluble [18]. The 80% methanol insoluble fraction was further fractionated using dialysis through controlled pore membranes. Each resulting sub-fraction was then combined with the methanol soluble fraction and tested as a promoting stimulus in mice treated with DMBA. The subfraction with a presumptive molecular weight greater than 13,000 produced a significantly higher tumor incidence and tumor yield together with a significantly shorter latent period than the other subfractions [18]. The subfraction contained about 12% of the total 80% methanol insoluble material, and all of the other subfractions exhibited significant but less pronounced co-promoting activity [18].

Mouse Lung

A six-month bioassay in A/J mice was conducted to test the hypothesis that chronically inhaled mainstream CS would either induce lung cancer or promote lung carcinogenicity induced by NNK [43]. Female A/J mice were exposed to filtered air or CS, injected with NNK, or exposed to both CS and NNK. Mice were injected once with NNK; 3 days later, they were exposed to CS for 6 h/day, 5 days/week, for 26 weeks at a mean 248 mg total particulate matter/m^3 concentration. Tumor incidences, as determined from gross observation of lung nodules, were similar between the CS-exposed and filtered air groups, and the NNK and CS + NNK groups [43]. Furthermore, tumor multiplicity in tumor-bearing animals was not significantly different among any of the three groups in which tumors were observed [43]. Thus, CS exposure neither induced lung tumors nor promoted NNK-induced tumors [43].

Mouse Colon

Because both chronic ulcerative colitis and smoking are associated with colorectal cancer in humans, the effects of CS exposure on inflammation-associated tumorigenesis in the mouse colon were investigated in male Balb/c mice [19]. Mice were given water or 3% dextran sulfate sodium (DSS) in drinking water for 7 days to induce colitis, with or without 1 h daily exposure to 2% or 4% CS, and then allowed to drink water for 14 days. The cycle of these treatments was repeated. CS exposure dose-dependently increased colon adenoma formation in mice with inflamed mucosa 1 month after the three cycles of treatments. CS exposure plus colitis was strongly associated with a high incidence of dysplasia and adenocarcinoma formation compared with induction of colitis alone [19]. CS significantly attenuated the apoptotic effect by dextran sulfate sodium, but increased vascular endothelial growth factor (VEGF) in the colon of mice with colitis [19]. Thus CS promoted inflammation-associated adenoma/adenocarcinoma formation in the mouse colon in a dose-dependent manner in association with inhibited cellular apoptosis and increased angiogenesis [19].

As several studies have suggested that there is a link between 5-lipoxygenase (5-LOX) and carcinogenesis in humans and animals, exposure to unfiltered mainstream CS enhanced the 5-LOX protein expression in the inflammation-associated colonic adenomas [44]. It was accompanied with an up-regulation of the key angiogenic factors for tumorigenesis, matrix metalloproteinase-2 and VEGF. 5-LOX-inhibitors decreased the incidence of colonic adenoma formation and reduced angiogenesis, matrix metalloproteinase-2 activity and VEGF protein expression in the colons of these animals [44]. These results strongly suggested that CS induces 5-LOX expression that plays an important role in activation of matrix metalloproteinase-2 and VEGF to induce angiogenic process and promotion of inflammation-associated adenoma formation in mice [44].

In line with the fact that the arachidonic acid-metabolizing enzymes COX-2 and 5-LOX are overexpressed during the process of colon adenoma formation promoted by CS, pretreating colon cancer cells with CS extract promoted colon cancer growth in the nude mouse xenograft model [45]. Inhibition of COX-2 or 5-LOX reduced the tumor size, and in the group treated with COX-2-inhibitor, the PGE_2 level decreased while the LTB_4 level increased [45]. In contrast, in the 5-LOX-inhibitor treated group, the LTB_4 level was reduced, and the PGE_2 level was unchanged, however, combined treatment with both COX-2 and 5-LOX-inhibitors further inhibited the tumor growth promoted by CSE over treatment with either COX-2-inhibitor or 5-LOX-inhibitor alone [45]. This was accompanied by the down-regulation of PGE_2 and LTB_4, indicating that inhibition of COX-2 leads to a shunt of arachidonic acid metabolism towards the leukotriene pathway during colon tumorigenesis promoted by CS extract [45]. Because suppression of 5-LOX did not induce such a shunt and produced a better response, it was suggested that 5-LOX-inhibitor is more effective than COX-2-inhibitor, and blocker of both COX-2 and 5-LOX may present a superior anti-cancer profile in cigarette smokers [45].

Hamster Pancreas and Respiratory Tract

The influences of CS on *N*-nitrosobis(2-oxopropyl)amine (BOP)-induced pancreatic duct and respiratory tract tumorigenesis were investigated in male Syrian hamsters subcutaneously injected with BOP at a dose of 10 mg/kg once a week for 3 weeks as an initiation treatment together with or without CS exposure for 4 weeks or followed by CS exposure for 26 weeks [20]. CS was transnasally administered 7 days a week by using the Hamburg type II smoking machine under the following conditions: exposure period, 9 min/day twice; doses, 30 cigarettes/time; inhalation volume, 35 ml; inhalation flow, 17.5 ml/s; dilution of CS, 1/7. At the termination of 30 weeks after the first BOP injection, the incidence of pancreatic adenocarcinomas was significantly decreased in animals given BOP plus CS as compared to those receiving BOP alone [20]. In contrast, the incidence of laryngeal and tracheal proliferative lesions (hyperplasias and papillomas) was significantly increased in hamsters with BOP followed by CS than those receiving BOP alone [20].

Male Syrian golden hamsters were exposed for 1 or 2 weeks to CS produced by commercial non-filter cigarettes for 5 consecutive [25]. Post-mitochondrial fractions (S9) prepared from the liver, lungs, and pancreas were used in the Ames liquid incubation assay, in order to assess the effect of CS on the metabolic activation of four groups of procarcinogens. The mutagenic activities of five HCAs on strain TA98 in the presence of liver S9 mix were induced up to 3.7 times above sham smoke control, while no significant alteration of mutagenicity was observed with BOP on TA100. A similar stimulation of metabolic activation was also observed for 3-amino-1,4-dimethyl-5H-pyrido[4,3-b] indole with S9 from the lungs but not from the pancreas [25]. CS exposure of hamsters might selectively induce hepatic CYP1A2 that cannot activate BOP [25]. The findings further supported the idea that CS markedly stimulates the metabolic activation of food-derived carcinogens, which may contribute to the overall carcinogenic effects of CS [25].

Hamster Respiratory Tract

The post-initiation effects of CS on the development of respiratory tumors were also investigated in male Syrian hamsters receiving a single *sc* injection of 100 mg/kg body weight of DEN and then exposed to non-filter CS, filter-tip CS or sham smoke, in the Hamburg II type smoking machine from week 1 to 12 [9]. CS was transnasally administered as described above [20]. In the DEN-treated animals, epithelial hyperplasias and/or papillomas were induced, the incidences and multiplicities being significantly increased by the CS treatments [9].

Hamster Buccal Pouch

Tumor incidence in hamster buccal pouches following topical exposures to B(*a*)P (either daily for 10 days or thrice weekly for 6 months) alone or in combination with extract of tobacco, betel nut or betel leaf was examined [21]. Given alone, the B(*a*)P treatment yielded, 6 months after the last treatment, tumors in both B(*a*)P exposures. When B(*a*)P and betel quid ingredients were painted concomitantly for 10 days, there was suppression of tumor production [21]. But when B(*a*)P-plus-tobacco or B(*a*)P-plus-betel nut treatments were given for 6 months, there was a considerable increase in tumor incidence [21]. Betel leaf extract, in both short-term and long-term studies, expressed its inhibitory influence on B(*a*)P-induced tumorigenesis [21].

Modification of Carcinogenesis by CS Constituents

Rat Stomach

Catechol (1,2-dihydroxybenzene) is present in the environment, being a major industrial chemical as well as a major phenolic component of CS. Continuous oral treatment of rats with 0.8% catechol for 51 weeks after a single intragastric dose of 150 mg/kg of *N*-methyl-*N'*-nitro-*N*-nitrosoguanidine strongly enhanced both forestomach and glandular stomach carcinogenesis [46]. In addition, and more importantly, catechol alone induced adenocarcinomas and adenomatous hyperplasias in the pyloric region of the glandular stomach [46]. These results clearly indicated that this environmental contaminant merits classification as an enhancer of forestomach and glandular stomach carcinogenesis with complete carcinogenic potential for the glandular stomach [46].

Rat Esophagus

In line with the fact that alcohol consumption and CS are synergistic etiologic factors for squamous cell carcinoma of the esophagus in Western countries, catechol was previously found to be a co-carcinogen with methyl-*N*-amylnitrosamine (MNAN) for esophageal tumors in rats, when given in the diet [47]. Male MRC-Wistar rats were injected 3 times intraperitoneally with 25 mg MNAN/kg or simultaneously given 10% ethanol and/or 0.2% catechol in the drinking water for life. Incidences of esophageal papillomas and carcinomas showed a tendency for increase in the MNAN + ethanol + catechol group, and the multiplicities were significantly greater in the MNAN + ethanol + catechol group than in the MNAN alone group [47]. These findings indicated that, in the esophagus, catechol alone was not co-carcinogenic with MNAN, but the combined treatment with ethanol and catechol was co-carcinogenic with MNAN [47].

Mouse Skin

The weakly acidic fraction of CSC was fractionated by preparative high-pressure liquid chromatography into major subfractions I–IV, and major subfractions II and III were fractionated further into subfractions A–J [48]. Subfractions A–J were tested for co-carcinogenicity on the skin of non-inbred Ha:ICR Swiss albino mice by application with 0.003% B(*a*)P. Subfractions A–C and F–J showed significant co-carcinogenic activity; subfractions A, F, and H were the most active [48]. Catechol was a major component of subfraction A and was also detected in subfractions B–D and F [48]. Major components of the other subfractions included hydroquinone (B), coniferyl alcohol (C and H), hydroxyphenyl alcohols (D), alkyl-2-hydroxy-2-cyclopenten-1-ones (C, D, and F), hydroxyacetophenones (F), phenolic cyano compounds (F), and fatty acids (F). The results demonstrate the importance of catechol as a co-carcinogen in the weakly acidic fraction of CSC and indicate the presence of other co-carcinogens [48].

Catechol is a major phenolic compound present in the co-carcinogenic fraction of cigarette tar [49], being a potent co-carcinogen with B(*a*)P in mouse skin. The co-carcinogenic and co-initiating activities of catechol with 7,8-dihydroxy-7,8-dihydrobenzo[*a*]pyrenes (B(*a*)P-7,8-diols) were examined in mouse skin. Similar to enhancement of B(*a*)P carcinogenesis, repeated concurrent applications of catechol with or without B(*a*)P-7,8-diol to mouse skin strongly enhanced B(*a*)P-7,8-diol tumor multiplicity and tumor incidence, and decreased latency [49]. Co-application of catechol with B(*a*)P-7,8-diol in a two-stage initiation–promotion protocol increased the tumor initiating activity of racemic B(*a*)P-7,8-diol, similar to that of B(*a*)P, but had no statistically significant effect on the tumor initiating activity of the (+)- or (−)-enantiomers in mouse skin [49]. Thus catechol was as potent a co-carcinogen with (+/−)-B(*a*)P-7,8-diol as it is with B(*a*)P, but only a weak co-initiator when applied with (+/−)-B(*a*)P-7,8-diol or B(*a*)P [49].

A series of 21 CS components and related compounds were applied to mouse skin 3 times weekly with B(*a*)P [50]. The following compounds enhanced remarkably the carcinogenicity of B(*a*)P: catechol, pyrogallol, decane, undecane, pyrene, benzo[*e*]pyrene, and fluoranthene [50]. The following compounds inhibited B(*a*)P carcinogenicity completely: esculin, quercetin, squalene, and oleic acid. Phenol, eugenol, resorcinol, hydroquinone, hexadecane, and limonene partially inhibited B(*a*)P carcinogenicity [50]. Six of the 21 compounds were also tested as tumor promoters in two-stage carcinogenesis. No direct correlation existed between tumor-promoting activity and co-carcinogenic activity, and the co-carcinogens pyrogallol and catechol did not show tumor-promoting activity [50]. Decane, tetradecane, anthralin, and phorbol myristate acetate showed both types of activity [50].

Mouse Stomach

It has been revealed that CS promotes gastric cancer growth through the induction of cyclooxygenase-2 (COX-2), and nicotine, one of the active ingredients in CS, has detrimental effects in the stomach [51]. To date, there is no direct evidence to validate the effect of nicotine on gastric tumor growth and its carcinogenic mechanisms. Athymic nude mice, with gastric cancer cells orthotopically implanted into the gastric wall, treated with nicotine in their drinking water for 3 months developed larger tumor areas than mice in the control group [51]. Nicotine further increased proliferating cellular nuclear antigen staining and microvessel density with concomitant activation of ERK phosphorylation, COX-2 and VEGF expression in the tumors [51]. Intraperitoneal administration of a selective COX-2-inhibitor prevented the nicotine-induced tumor growth and neovascularization dose-dependently [51]. These findings revealed a direct promoting action of nicotine on the growth of gastric tumor and neovascularization through sequential activation of the ERK/COX-2/VEGF signaling pathway, which can be targeted for chemoprevention of gastric cancer, particularly in cigarette smokers [51].

Hamster Pancreas

The modifying effects of NNK, a tobacco-specific nitrosamine, during the post-initiation phase of carcinogenesis were examined in female Syrian hamsters receiving a single *sc* injection of BOP at a dose of 10 mg/kg and then administered 3 ppm or 1 ppm NNK in their drinking water for the following 87 weeks [52]. At week 88 of the experiment, however, there was no statistically significant influence on tumorigenesis in the pancreas, lung, liver or kidney, suggesting that NNK does not enhance BOP-induced hamster tumorigenesis when given in the promotion phase [52].

Similarly, the modifying effects of NNAL, another tobacco-specific nitrosamine, were investigated in female Syrian hamsters given a single *sc* injection of BOP at a dose of 10 mg/kg and then administered 2 or 5 ppm NNAL in their drinking water for 52 weeks [53]. At week 53 of the experiment, the total incidence of combined adenocarcinomas and atypical hyperplasias of the exocrine pancreas was significantly higher in the BOP/NNAL 5 ppm group than in the BOP alone group while the NNAL treatment did not exert any influence on endocrine pancreas, lung, liver or kidney tumorigenesis, suggesting that NNAL enhances BOP-induced exocrine pancreatic tumorigenesis in hamsters when given in the post-initiation phase [53]. Thus, with regard to the promoting capability of pancreatic carcinogenesis, NNAL may be more powerful than NNK.

Hamster Respiratory Tract

Syrian golden hamsters were given intratracheal instillations of glass fibers with or without B(*a*)P suspended in saline, once a fortnight for 52 weeks; the experiment was terminated at week 85 [54]. No tumors of the respiratory tract were observed in hamsters treated with glass fibers alone, and there was no indication that glass fibers enhanced the development of respiratory tract tumors induced by B(*a*)P [54]. In another study, Syrian golden hamsters were exposed to fresh air or to a mixture of 4 major vapor phase components of CS, viz. isoprene, methyl chloride, methyl nitrite and acetaldehyde for up to 23 months. Some of the animals were also given repeated intratracheal instillations of B(*a*)P or norharman in saline. Laryngeal tumors were found in male and hamsters exposed only to the vapor mixture, whereas no laryngeal tumors occurred in controls [54]. The tumor response of the larynx most probably has to be ascribed entirely to the action of acetaldehyde. Simultaneous treatment with norharman or B(a)P did not affect the tumor response of the larynx [54]. Chronic inhalation exposure of rats to acetaldehyde at levels of 750 ppm or more resulted in a high incidence of nasal carcinomas, both squamous cell carcinomas of the respiratory epithelium and adenocarcinomas of the olfactory epithelium [54].

Conclusions

Articles concerning the carcinogenic effects of CS and its constituents, and their modifying effects on experimental carcinogenesis were reviewed. A very recent study demonstrated that lifetime whole-body exposure of B6C3F$_1$ mice to high doses of CS robustly increases lung cancer incidence compared with sham exposed animals [10], being the first to demonstrate a strong effect of inhaled CS on lung cancer in an animal model [55]. Except for this study, it seems much more difficult to reproduce the tumorigenicity of CS per se in preclinical models [41], otherwise significant lung carcinogenicity of CS has been mostly detected in A/J mice that develop lung adenomas spontaneously [4–8]. On the other hand, excluding a few studies reporting little or no effects, inhaled CS enhanced liver, respiratory tract and colon carcinogenesis in rats, mice or hamsters induced chemically [9, 11, 12, 19, 20]. Painted CSC also promoted skin tumorigenesis in mice or rats induced chemically or by irradiation [13–18]. It is now recognized that approximately 80 CS constituents are carcinogenic to human [3]. In fact, some CS constituents such as HCAs, nitrosamines and polyaromatic hydrocarbons experimentally induce cancers in several organs of rodents [3]. In addition to the promotional or co-carcinogenic properties of such carcinogens like NNAL [53], some other constituents including catechol and nicotine proved to induce tumors in the stomach, esophagus or skin of rats and mice [46–51].

One of the mechanisms underlying the modifying effects of CS and its constituents on experimental carcinogenesis is suggested to be altered metabolism of carcinogens including or excluding CS. Thus it can be concluded that such modifying effects of CS and its constituents may be critical in understanding overall carcinogenesis of CS, especially in organs other than the respiratory tract.

Acknowledgment

This work was supported in part by SRF grants for Biomedical Research.

References

1   Doll R: Cancers weakly related to smoking. Br Med Bull 1996;52:35–49.
2   Thun MJ, Henley SJ, Calle EE: Tobacco use and cancer: an epidemiologic perspective for geneticists. Oncogene 2002;21:7307–7325.
3   Smith CJ, Perfetti TA, Garg R, Hansch C: IARC carcinogens reported in cigarette mainstream smoke and their calculated log P values. Food Chem Toxicol 2003;41:807–817.
4   Hecht SS, Isaacs S, Trushin N: Lung tumor induction in A/J mice by the tobacco smoke carcinogens 4-(methylnitrosamino)-1-(3-pyridyl)-1-butanone and benzo[a]pyrene: a potentially useful model for evaluation of chemopreventive agents. Carcinogenesis 1994;15:2721–2725.
5   Morse MA, Amin SG, Hecht SS, Chung FL: Effects of aromatic isothiocyanates on tumorigenicity, O^6-methylguanine formation, and metabolism of the tobacco-specific nitrosamine 4-(methylnitrosamino)-1-(3-pyridyl)-1-butanone in A/J mouse lung. Cancer Res 1989;49:2894–2897.
6   Witschi H, Espiritu I, Peake JL, Wu K, Maronpot RR, Pinkerton KE: The carcinogenicity of environmental tobacco smoke. Carcinogenesis 1997;18:575–586.
7   Witschi H: A/J mouse as a model for lung tumorigenesis caused by tobacco smoke: strengths and weaknesses. Exp Lung Res 2005;31:3–18.
8   Witschi H: Tobacco smoke as a mouse lung carcinogen. Exp Lung Res 1998;24:385–394.
9   Takahashi M, Imaida K, Mitsumori K, Okamiya H, Shinoda K, Yoshimura H, Furukawa F, Hayashi Y: Promoting effects of cigarette smoke on the respiratory tract carcinogenesis of Syrian golden hamsters treated with diethylnitrosamine. Carcinogenesis 1992;13:69–72.
10 Hutt JA, Vuillemenot BR, Barr EB, Grimes MJ, Hahn FF, Hobbs CH, March TH, Gigliotti AP, Seilkop SK, Finch GL, Mauderly JL, Belinsky SA: Life-span inhalation exposure to mainstream cigarette smoke induces lung cancer in B6C3F$_1$ mice through genetic and epigenetic pathways. Carcinogenesis 2005;26:1999–2009.
11 Nishikawa A, Furukawa F, Miyauchi M, Son HY, Okazaki K, Koide A, Mori Y, Hirose M: Enhancement by cigarette smoke exposure of 2-amino-3,8-dimethylimidazo[4,5-f]quinoxaline-induced rat hepatocarcinogenesis in close association with elevation of hepatic CYP1A2. Jpn J Cancer Res 2002;93:24–31.
12 Gupta MP, Khanduja KL, Koul IB, Sharma RR: Effect of cigarette smoke inhalation on benzo[a]pyrene-induced lung carcinogenesis in vitamin A deficiency in the rat. Cancer Lett 1990; 55:83–88.
13 McGregor JF: Enhancement of skin tumorigenesis by cigarette smoke condensate following beta-irradiation in rats. J Natl Cancer Inst 1982;68:605–611.
14 Curtin GM, Hanausek M, Walaszek Z, Mosberg AT, Slaga TJ: Short-term in vitro and in vivo analyses for assessing the tumor-promoting potentials of cigarette smoke condensates. Toxicol Sci 2004;81:14–25.
15 Curtin GM, Hanausek M, Walaszek Z, Zoltaszek R, Swauger JE, Mosberg AT, Slaga TJ: Short-term biomarkers of cigarette smoke condensate tumor promoting potential in mouse skin. Toxicol Sci 2006;89:66–74.
16 Bagwe AN, Ramchandani AG, Bhisey RA: Skin-tumour-promoting activity of processed bidi tobacco in hairless S/RV Cri-ba mice. J Cancer Res Clin Oncol 1994;120:485–489.
17 Hecht SS, Thorne RL, Maronpot RR, Hoffmann D: A study of tobacco carcinogenesis. XIII. Tumor-promoting subfractions of the weakly acidic fraction. J Natl Cancer Inst 1975;55: 1329–1336.
18 Bock FG, Clausen DF: Further fractionation and co-promoting activity of the large molecular weight components of aqueous tobacco extracts. Carcinogenesis 1980;1:317–321.
19 Liu ES, Ye YN, Shin VY, Yuen ST, Leung SY, Wong BC, Cho CH: Cigarette smoke exposure increases ulcerative colitis-associated colonic adenoma formation in mice. Carcinogenesis 2003;24:1407–1413.
20 Nishikawa A, Furukawa F, Imazawa T, Yoshimura H, Ikezaki S, Hayashi Y, Takahashi M: Effects of cigarette smoke on N-nitrosobis(2-oxopropyl)amine-induced pancreatic and respiratory tumorigenesis in hamsters. Jpn J Cancer Res 1994;85:1000–1004.
21 Rao AR: Modifying influences of betel quid ingredients on B(a)P-induced carcinogenesis in the buccal pouch of hamster. Int J Cancer 1984;33:581–586.

22   Sheweita SA: Drug-metabolizing enzymes: mechanisms and functions. Curr Drug Metab 2000;1:107–132.

23   Mori Y, Koide A, Kobayashi Y, Furukawa F, Hirose M, Nishikawa A: Effects of cigarette smoke and a heterocyclic amine, MeIQx on cytochrome P-450, mutagenic activation of various carcinogens and glucuronidation in rat liver. Mutagenesis 2003;18:87–93.

24   Koide A, Fuwa K, Furukawa F, Hirose M, Nishikawa A, Mori Y: Effect of cigarette smoke on the mutagenic activation of environmental carcinogens by rodent liver. Mutat Res 1999;428: 165–176.

25   Mori Y, Iimura K, Furukawa F, Nishikawa A, Takahashi M, Konishi Y: Effect of cigarette smoke on the mutagenic activation of various carcinogens in hamster. Mutat Res 1995;346:1–8.

26   Wakabayashi K, Nagao M, Esumi H, Sugimura T: Food-derived mutagens and carcinogens. Cancer Res 1992;52(suppl 7):2092S–2098S.

27   Kato R, Kamataki T, Yamazoe Y. N-hydroxylation of carcinogenic and mutagenic aromatic amines. Environ Health Perspect 1983;49:21–25.

28   Eke BC, Vural N, Iscan M: Combined effects of ethanol and cigarette smoke on hepatic and pulmonary xenobiotic metabolizing enzymes in rats. Chem Biol Interact 1996;2:155–167.

29   Eke BC, Vural N, Iscan M: Age dependent differential effects of cigarette smoke on hepatic and pulmonary xenobiotic metabolizing enzymes in rats. Arch Toxicol 1997;71:696–702.

30   Eke BC, Iscan M: Effects of cigarette smoke with different tar contents on hepatic and pulmonary xenobiotic metabolizing enzymes in rats. Hum Exp Toxicol 2002;21:17–23.

31   Nair UJ, Ammigan N, Kulkarni JR, Bhide SV: Species difference in intestinal drug metabolising enzymes in mouse, rat and hamster and their inducibility by masheri, a pyrolysed tobacco product. Indian J Exp Biol 1991;29:256–258.

32   Wardlaw SA, Nikula KJ, Kracko DA, Finch GL, Thornton-Manning JR, Dahl AR: Effect of cigarette smoke on CYP1A1, CYP1A2 and CYP2B1/2 of nasal mucosae in F344 rats. Carcinogenesis 1998;19:655–662.

33   Bilimoria MH, Ecobichon DJ: Protective antioxidant mechanisms in rat and guinea pig tissues challenged by acute exposure to cigarette smoke. Toxicology 1992;72:131–144.

34   Graziano MJ, Gairola C, Dorough HW: Effects of cigarette smoke and dietary vitamin E levels on selected lung and hepatic biotransformation enzymes in mice. Drug Nutr Interact 1985;3: 213–222.

35   Lee CZ, Royce FH, Denison MS, Pinkerton KE: Effect of in utero and postnatal exposure to environmental tobacco smoke on the developmental expression of pulmonary cytochrome P450 monooxygenases. J Biochem Mol Toxicol 2000;14:121–130.

36   Sindhu RK, Rasmussen RE, Yamamoto R, Fujita I, Kikkawa Y: Depression of hepatic cytochrome P450 monooxygenases after chronic environmental tobacco smoke exposure of young ferrets. Toxicol Lett 1995;76:227–238.

37   Liu C, Russell RM, Wang XD: Exposing ferrets to cigarette smoke and a pharmacological dose of beta-carotene supplementation enhance in vitro retinoic acid catabolism in lungs via induction of cytochrome P450 enzymes. J Nutr 2003;133:173–179.

38   Villard PH, Herber R, Seree EM, Attolini L, Magdalou J, Lacarelle B: Effect of cigarette smoke on UDP-glucuronosyltransferase activity and cytochrome P450 content in liver, lung and kidney microsomes in mice. Pharmacol Toxicol 1998;82:74–79.

39   Nishikawa A, Mori Y, Lee IS, Tanaka T, Hirose M: Cigarette smoking, metabolic activation and carcinogenesis. Curr Drug Metab 2004;5:363–373.

40   Hecht SS: DNA adduct formation from tobacco-specific N-nitrosamines. Mutat Res 1999;424: 127–142.

41   De Flora S, Izzotti A, D'Agostini F, Bennicelli C, You M, Lubet RA, Balansky RM: Induction and modulation of lung tumors: genomic and transcriptional alterations in cigarette smoke-exposed mice. Exp Lung Res 2005;31:19–35.

42   Johansson SL, Hirsch JM, Larsson PA, Saidi J, Osterdahl BG: Snuff-induced carcinogenesis: effect of snuff in rats initiated with 4-nitroquinoline N-oxide. Cancer Res 1989;49:3063–3069.

43   Finch GL, Nikula KJ, Belinsky SA, Barr EB, Stoner GD, Lechner JF: Failure of cigarette smoke to induce or promote lung cancer in the A/J mouse. Cancer Lett 1996;99:161–167.

44   Ye YN, Liu ES, Shin VY, Wu WK, Cho CH: Contributory role of 5-lipoxygenase and its association with angiogenesis in the promotion of inflammation-associated colonic tumorigenesis by cigarette smoking. Toxicology 2004;203:179–188.

45   Ye YN, Wu WK, Shin VY, Bruce IC, Wong BC, Cho CH: Dual inhibition of 5-LOX and COX-2 suppresses colon cancer formation promoted by cigarette smoke. Carcinogenesis 2005;26:827–834.

46   Hirose M, Kurata Y, Tsuda H, Fukushima S, Ito N: Catechol strongly enhances rat stomach carcinogenesis: a possible new environmental stomach carcinogen. Jpn J Cancer Res 1987;78:1144–1149.

47   Mirvish SS, Weisenburger DD, Hinrichs SH, Nickols J, Hinman C: Effect of catechol and ethanol with and without methylamylnitrosamine on esophageal carcinogenesis in the rat. Carcinogenesis 1994;15:883–887.

48   Hecht SS, Carmella S, Mori H, Hoffmann D: A study of tobacco carcinogenesis. XX. Role of catechol as a major cocarcinogen in the weakly acidic fraction of smoke condensate. J Natl Cancer Inst 1981;66:163–169.

49   Melikian AA, Jordan KG, Braley J, Rigotty J, Meschter CL, Hecht SS, Hoffmann D: Effects of catechol on the induction of tumors in mouse skin by 7,8-dihydroxy-7,8-dihydrobenzo[a]pyrenes. Carcinogenesis 1989;10:1897–1900.

50   Van Duuren BL, Goldschmidt BM: Cocarcinogenic and tumor-promoting agents in tobacco carcinogenesis. J Natl Cancer Inst 1976;56:1237–1242.

51   Shin VY, Wu WK, Ye YN, So WH, Koo MW, Liu ES, Luo JC, Cho CH: Nicotine promotes gastric tumor growth and neovascularization by activating extracellular signal-regulated kinase and cyclooxygenase-2. Carcinogenesis 2004;25:2487–2495.

52   Furukawa F, Nishikawa A, Yoshimura H, Mitsui M, Imazawa T, Ikezaki S, Takahashi M: Effects of 4-(methylnitrosamino)-1-(3-pyridyl)-1-butanone (NNK) on N-nitrosobis(2-oxopropyl)amine (BOP)-initiated carcinogenesis in hamsters. Cancer Lett 1994;86:75–82.

53   Furukawa F, Nishikawa A, Enami T, Mitsui M, Imazawa T, Tanakamaru Z, Kim HC, Lee IS, Kasahara K, Takahashi M: Promotional effects of 4-(methylnitrosamino)-1-(3-pyridyl)-1-butanol (NNAL) on N-nitrosobis(2-oxopropyl)amine (BOP)-initiated carcinogenesis in hamsters. Food Chem Toxicol 1997;35:387–392.

54   Feron VJ, Kuper CF, Spit BJ, Reuzel PG, Woutersen RA: Glass fibers and vapor phase components of cigarette smoke as cofactors in experimental respiratory tract carcinogenesis. Carcinog Compr Surv 1985;8:93–118.

55   Hecht SS: Carcinogenicity studies of inhaled cigarette smoke in laboratory animals: old and new. Carcinogenesis 2005;26:1488–1492.

Akiyoshi Nishikawa
Division of Pathology, National Institute of Health Sciences
1–18–1 Kamiyoga, Setagaya-ku
Tokyo 158–8501 (Japan)
Tel. +81 3 3700 9819, Fax +81 3 3700 1425
E-Mail nishikaw@nihs.go.jp

Cho CH, Purohit V (eds): Alcohol, Tobacco and Cancer.
Basel, Karger, 2006, pp 253–267

....

Nicotine and Cancer

William Ka Kei Wu[a]*, Helen Pui Shan Wong*[a]*, Le Yu*[a]*, Chi Hin Cho*[a,b]

[a]Department of Pharmacology; [b]Research Center of Infection and Immunology,
Faculty of Medicine, The University of Hong Kong, Hong Kong, SAR, China

Abstract

Cigarette smoking is a risk factor for various types of cancer. Nicotine, a major alkaloid in tobacco, is responsible for different aspects in the pathogenesis of smoking-related malignancies. Nicotine only not perpetuates smoking behavior in smokers, which results in further intake of tobacco-derived carcinogens, but also directly increases cellular mutagenic events. Current evidence also supports that nicotine can be metabolized into highly carcinogenic nitrosamines. In addition, nicotine by itself stimulates cancer cell proliferation through multiple mitogenic signaling pathways. Nicotinic stimulation also provides pro-survival signals to cancer cells such that they are more resistant to apoptosis induced by chemotherapeutic agents or ionizing radiation. Moreover, there is evolving evidence suggesting that nicotine can stimulate tumor-associated angiogenesis, a biological process essential for tumor growth and metastasis. It has also been reported that nicotine enhances cancer cell invasiveness and weakens host cancer-killing immunity. Taken together, nicotine seems to play an important role in the initiation, promotion, and progression of smoking-related cancers. A controlled surveillance study on cancer risk of current nicotine users especially those on nicotine replacement therapy is therefore justified.

Smoking, Nicotine, and Cancer: An Overview

Cigarette smoking is an established risk factor for many types of cancers. In 1986, International Agency for Research on Cancer (IARC) published their first monograph on cigarette smoking, revealing the causational relationship between cigarette smoking and cancers of the lung, oral cavity, pharynx, larynx, oesophagus (squamous-cell carcinoma), pancreas, urinary bladder, and renal pelvis [1]. More recent evidence further unravels that cancers of the nasal cavities and nasal sinuses, oesophagus (adenocarcinoma), stomach, liver, kidney (renal-cell carcinoma), uterine cervix, and myeloid leukemia can be caused by cigarette smoking [2].

Cigarette smoke, however, contains thousands of chemicals, making it difficult to delineate the toxicologic mechanism relevant to the pathogenesis of smoking-related malignancies. Great effort has therefore been put forward to identify and characterize individual carcinogenic compound contained in cigarette smoke. To date, over 60 tobacco-derived compounds have been evaluated by International Agency for Research on Cancer and found to be carcinogenic in either laboratory animals or humans [3]. Most of these carcinogens (e.g. polycyclic aromatic hydrocarbons, aromatic amines, and N-nitrosamines) are strong mutagens, which may result in formation of DNA adducts and thereby leads to activation of proto-oncogenes or inactivation of tumor suppressor genes. Nicotine is a major alkaloid in tobacco and is responsible for the additive use of cigarettes. Nicotine is not considered to be a full carcinogen because by itself nicotine fails to induce tumor formation in experimental animals. Chemical carcinogenesis, however, should be conceived as an interconnection of cellular events involving tumor initiation, promotion, and subsequent progression. There is evolving evidence suggesting that nicotine and its metabolites may play critical roles in these processes by acting as mutagenic, mitogenic, anti-apoptotic, pro-angiogenic, and immunosuppressive agents. In this chapter, we would review some of the current evidence regarding the pharmacologic contribution of nicotine to the process of cancer development.

Basic Pharmacology of Nicotine

Nicotine is a tertiary amine consisting of a pyridine and a pyrrolidine ring. It is distilled from burning tobacco, and small droplets of tar containing nicotine are inhaled and deposited in the small airways and alveoli. Each cigarette contains 15–30 mg nicotine, which can be rapidly absorbed through mucous membrane, skin, alveoli, and the gastrointestinal tract. Levels of nicotine in the plasma decline rapidly owing to distribution to peripheral tissues and elimination. Therefore, when smokers smoke multiple cigarettes during the day, oscillations between peaks and troughs plasma nicotine levels occur. The half-life of nicotine is 30–60 min and it is primarily metabolized in the liver, but to a lesser extent in the lung and the brain. About 70–80% of nicotine is metabolized by cytochrome P450 enzymes and flavin monooxygenases to cotinine via C-oxidation, and another 4% to nicotine-N-oxide. A considerable portion is also excreted unchanged in acidic urine. Cotinine has a much longer half-life than nicotine and therefore is used as a marker of nicotine intake. In smokers, venous nicotine levels range from 5 to 15 ng/ml and arterial nicotine levels can peak as high as 80 ng/ml [4, 5]. Remarkably and extremely high levels of nicotine in saliva and gastric juice have been reported in smokers, reaching more than 1,300 and 800 ng/ml, respectively [6].

Nicotine has been used as an adjuvant therapeutic in the treatment of different neuropsychological disorders, such as Alzheimer's disease, Parkinson's disease, anxiety, and Tourette's syndrome and as a replacement therapy to aid smoking cessation [7–11]. Nicotine is also used to enhance prostate function and to alleviate the symptoms of ulcerative colitis [12]. Activation of classical and nonclassical nicotinic receptors has been known to mediate the pharmacologic effect of nicotine [13, 14]. The classical receptors, which operate as ligand-gated ion channels, encompass membrane proteins encoded by nicotinic acetylcholine receptors (nAChRs) gene superfamily, which includes ten α (α1–α10), four β (β1–β4), one γ, one δ, and one ε subunits. Functional nAChRs in the neuromuscular junction are composed of five subunits ($\alpha_2\beta\gamma\delta$ or $\alpha_2\beta\varepsilon\delta$) arranged around a central pore region whereas neuronal nAChRs only consist of α and β subunits. Variations in combination of subunits govern the ion-gating and ligand-binding properties of nAChRs [15]. When nicotine binds to the receptor, there are changes in the allosteric configuration of the receptor subunits, leading to nonselective flow of cations across the membrane. Notably, nAChRs can be desensitized by continuous or repeated exposure to agonists (e.g. nicotine), leading to a decrease or loss of biological response. The mechanism by which activation of nAChRs is coupled to the downstream signaling cascade is presently unknown but long-term modulation of signaling pathways by nAChR-mediated Ca^{2+} influx has been proposed. In this connection, the Ca^{2+}/Na^+ permeability ratios of neuronal nAChRs are higher than those of nAChRs in the neuromuscular junction, which may be associated with the role of pre-synaptic nAChRs in Ca^{2+}-dependent release of neurotransmitters [16]. Unlike nAChRs, the nonclassical nicotine receptor pathway is signaling through a noncholinergic metabotropic receptor, which is G-protein-coupled and positively linked to phospholipase C. The molecular identity of this receptor, however, has not yet been revealed [14].

Nicotine Dependence and Cancer: A Behavioral Perspective

Although tobacco-specific carcinogens contained in cigarette smoke play an important role in tumor initiation, without the additive property of nicotine, cigarette will simply be a commercial product with negligible health-damaging effect. Nicotine and tobacco-specific carcinogens are deadly amalgamation in the carcinogenesis of smoking-related malignancies. Behaviorally, the addictive influence of nicotine perpetuates people to consume cigarette. The maintenance of smoking behavior results in further intake of nicotine and carcinogenic substances, which initiates a vicious cycle of health destruction. It is estimated that 30% of cancer-related deaths in developed countries can be attributed to this fatal partners [17]. The neural and psychological basis of nicotine dependence is complex and beyond the scope of this chapter, but the involvement of meso-corticolimbic dopaminergic release and inactivation of nAChRs in the central nervous system have been suggested [18].

Nicotine as a Mutagen

There are several lines of evidence supporting that nicotine is mutagenic. In one of the studies, nicotine induces sister-chromatid exchanges and increases chromosome aberration frequency in Chinese hamster ovary cells at concentrations achievable in the saliva of tobacco chewers in a time- and dose-dependent manner [19, 20]. In addition, genotoxicity test utilizing cytokinesis-block micronucleus method demonstrates that nicotine causes a statistically significant increase of micronucleus frequency, which is accompanied by an increased level of oxidative stress. The mutagenicity of nicotine can be abrogated by pre-incubation with *N*-acetyl-cysteine, a substrate for glutathione synthesis, and catalase, the oxygen free radical scavenger [21]. Rodent bioassays also show that nicotine, which previously demonstrates to have no association between its exposure and induction of Leydig cell hyperplasia or adenomas in humans, is able to induce Leydig cell tumor in rats [22]. Recently, more sensitive tests have been employed to study the genotoxicity of nicotine. For instance, as revealed by accelerator mass spectrometry, nicotine dose-dependently forms adducts with liver DNA, lung DNA, histone H1/H3, Hb, and albumin in mice [23, 24]. The formation of DNA adduct is a crucial step in the process of chemical carcinogenesis. Moreover, cotinine and nicotine-*N*-oxide, two of the major metabolites of nicotine, demonstrate direct mutagenicity in the Mutatox test [25]. Another study also shows that cotinine is mutagenic in the bacterial luminescence genotoxicity test where the S9 metabolic activation system is absent and nicotine can potentiate the mutagenicity of cotinine in the absence or presence of S9 [26]. All these findings converge to suggest that nicotine and its metabolites could have a direct genotoxic effect.

Nicotine as a Pre-Carcinogen

Cigarette smoke contains many nicotine-derived tobacco specific nitrosamines, such as 4-(methylnitrosamino)-1-(3-pyridyl)-1-butanone (NNK) and *N*′-nitrosonornicotine (NNN), which are strong carcinogens that play critical roles in the initiation and promotion of smoking-related malignancies in humans. Although these compounds are mainly formed in the processing and curing of tobacco plants, there is evolving evidence suggesting that these carcinogenic compounds can be metabolized from nicotine endogenously. One of the studies shows that nicotine can be metabolized by 2'-hydroxylation to aminoketone, which is a precursor of NNK [27]. Urine of animals treated with nicotine is also reported to contain NNN, N'-nitrosoanabasine and N'-nitrosoanatabine, all of which are nitrosamines derived from nicotine [28]. However, another study shows that the use of nicotine patch do not alter the levels of 4-(methylnitrosamino)-1-(3-pyridyl)-1-butanol (NNAL) and NNAL-glucosiduronic acid, which are the immediate products metabolized from NNK, in the urine and blood of human subjects, indicating that NNK may not be formed endogenously from nicotine [29]. The evidence available so far remains equivocal and inconclusive and more investigations on the disposition of nicotine are warranted.

Nicotine as a Mitogen

Retrospective studies reveal that patients with smoking-related malignancies who continue to smoke after their diagnosis have lower response rates and shorter median survival compared with those who cease smoking [30–32]. This clinical observation can be partly explained by the finding that cigarette smoking can directly promote cancer growth, in which nicotine may be a key player. In association with this observation, nicotine has been shown to promote the proliferation of cultured lung, colon, gastric, pancreatic, and cervical cancer cells [33–37]. Moreover, nicotine can enhance the cell growth of mesothelioma, osteoblast, oral epidermoid carcinoma, vascular smooth muscle, and endothelium [38–42].

In small-cell lung carcinoma cells, the stimulatory action of nicotine is mediated by nAChRs, in which different subtypes are found to be expressed. Stimulation of these receptors with nicotine can induce Ca^{2+}-dependent release of serotonin, which in turn stimulates cell proliferation via serotonergic receptors [33]. Nicotine can therefore affect the proliferation of small-cell lung carcinoma cells by stimulating a serotonergic autocrine loop. A later study demonstrates that the stimulatory effect of nicotine on serotonin release and cell proliferation is mainly mediated by α-bungarotoxin-sensitive α7 subtype of nAChRs [43]. In addition, nicotine can stimulate small-cell lung carcinoma cell proliferation through rapid induction of Akt phosphorylation with concomitant upregulation of cyclin D1 expression, an effect mediated by both α3/α4 and α7 subtypes of nAChRs. Multiple downstream substrates of Akt, including glycogen synthase kinase-3, forkhead transcription factor, tuberin, mammalian target of rapamycin and ribosomal protein S6 kinase 1, are also phosphorylated upon nicotinic stimulation [44]. Cross-talk between nAChRs and other membrane receptors relevant to the pathogenesis of lung cancer has also been reported. For instance, μ, δ and κ opioid receptors and $\alpha7$ nAChR are expressed in diverse histologic types of lung cancer cell lines. Stimulation of opioid receptors inhibits lung cancer growth, in which the effect can be abolished by nicotine. It is also found that lung cancer cells express various opioid peptides, such as β-endorphin, enkephalin, and dynorphin, suggesting the participation of opioids in a negative autocrine loop. These data indicate that opioids can act as part of a 'tumor suppressor' system and that nicotine can function to circumvent this system [45]. Beside mediating the pharmacologic actions of exogenous chemicals (e.g. nicotine), nAChRs are also involved in the modulation of cellular response of lung cancer to endogenously secreted molecules. In this regard, it has been shown that small-cell lung carcinoma cells synthesize and secrete acetylcholine, whose release is of fundamental importance to the maintenance of basal cell growth [46].

The promoting effect of nicotine on the growth of gastrointestinal tumors is also widely studied. In this respect, nicotine has been shown to enhance *N*-methyl-*N'*-nitro-*N*-nitrosoguanidine-induced gastric carcinogenesis in rats [47]. Our laboratory also shows that nicotine can increase proliferation of cultured gastric and colon cancer cells in vitro. Moreover, nicotine in drinking water enhances the growth of gastric and colon cancers in nude mice xenograft model [34, 35]. We are also able to demonstrate that the stimulation of gastric and colon cancer cell proliferation by nicotine is mediated by activation of mitogen activated protein kinases (MAPKs) and the subsequent induction of arachidonic acid cascade. In gastric cancer cells, nicotine dose-dependently increases extracellular signal-regulated protein kinase (ERK) phosphorylation, cyclooxygenase-2 (COX-2) expression, and prostaglandin E_2 (PGE_2) release. The induction of COX-2 and PGE_2 by nicotine can be attenuated by pre-treatment with mitogen-activated protein kinase inhibitor, suggesting that the activation of arachidonic acid cascade is downstream to that of ERK (fig. 1). PGE_2, to this end, is known to stimulate tumor cell migration, proliferation and tumor-associated neovascularization while inhibiting programmed cell death [48]. In colon cancer cells, nicotine transactivates epidermal growth factor receptor and phosphorylates intracellular tyrosine kinase c-Src, in which activation of both kinases leads to upregulation of 5-lipooxgenase (fig. 2). Several studies have suggested that there is a link between 5-lipooxgenase and carcinogenesis in humans and animals.

With respect to smoking-associated cervical cancer, physiologically attainable concentrations of nicotine is shown to enhance the proliferation of human cervical cells in vitro [37]. Several mechanisms by which nicotine exerts

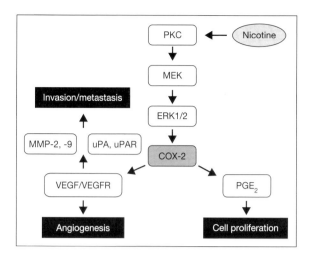

Fig. 1. Proposed mechanisms by which nicotine promotes gastric cancer growth.

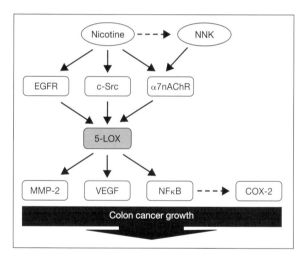

Fig. 2. Illustration of the proposed mechanisms by which nicotine and its active metabolite NNK (4-(methylnitrosamino)-1-(3-pyridyl)-1-butanone) induce colon cancer growth.

the mitogenic effect on cervical cancer cells have been proposed. First of all, nicotine can inhibit the degradation of platelet-derived growth factor AA at the lysosomal level and increase the nuclear accumulation of the growth factor, resulting in activation of RNA synthesis and cell proliferation [49]. Second, nicotine can upregulate EGF and epidermal growth factor receptor expression in cervical cancer cells [50]. More recent evidence also demonstrates that anti-proliferative cytokine TGF-β can be downregulated by nicotine treatment [51].

There is a growing body of evidence suggesting that nicotine may indirectly promote cancer growth through induction of epinephrine secretion from adrenal glands and the subsequent stimulation of adrenoceptors on tumor cells. To this end, β-adrenergic receptor activation has recently been implicated in the carcino-genesis of various kinds of cancers. Stimulation of β-adrenergic receptors is known to stimulate pulmonary, pancreatic, and colon carcinoma cell growth [52–54]. In addition, recent clinical studies also revealed that the use of β-blockers is negatively associated with cancer risk, suggesting that an increased level of circulating epinephrine may play a crucial role in the development of cancer [55–57]. It is therefore possible that the elevated level of epinephrine induced by nicotine may be responsible for one of the cancer-promoting mechanisms of cig-arette smoking. Alternatively, nicotine may be metabolized endogenously to NNK, whose structure resembles to those of β-adrenergic agonists. Indeed, β-adrenergic receptor is known to mediate the mitogenic signals induced by NNK in pulmonary, pancreatic ductal, and colon cancer cells [52–54].

Nicotine as an Anti-Apoptotic Agent

There is culminating evidence indicating that apoptotic cell death may be a crucial mechanism to prevent the development of neoplastic diseases. Pharmacological agents that suppress apoptosis may, therefore, act as tumor promoters. In addition, agents that inhibit the process of apoptotic cell death may also increase the resistance of cancer cells to the cytotoxic effects of chemotherapeutic drugs, and thereby reduce the efficacy of cancer therapies. The effects of nicotine on apoptosis have therefore been extensively investi-gated. Nicotine, in this respect, has been shown to inhibit tumor necrosis factor-ultraviolet light-, chemotherapeutic drugs-, and calcium ionophore-induced apoptosis in normal and transformed cells derived from a variety of species and tissues. Apoptosis can also be inhibited by cotinine [13].

In small-cell lung carcinoma, nicotine has been shown to reduce the apop-totic effect of anti-cancer agents by disrupting the critical balance between pro- and anti-apoptotic members of Bcl2 family. In this connection, nicotine activates the MAPK, specifically ERK1/2, and PKC signaling pathways in lung cancer cells, leading to upregulation of the anti-apoptotic Bcl2 protein and inhibition of apoptosis. The anti-apoptotic signal of nicotine is independent of c-jun NH2-terminal protein kinase and *p38* MAP kinases, which have also been shown to be involved in apoptosis [58]. Nicotine also induces PKCα-, ERK1/2-, and

PLC-dependent Bcl2 phosphorylation exclusively at the Ser-70 site in small-cell lung cancer, which is associated with prolonged survival of these cells [59]. Moreover, nicotine can potently induce ERK1/2-, PKC-, protein kinase A (PKA)-, and phosphoinositide-3 kinase (PI3K)/Akt-dependent phosphorylation of the pro-apoptotic Bad protein, in which phosphorylation negatively regulates its activity, at Ser-112, Ser-136, and Ser-155 [60]. Similarly, nicotine induces Bax phosphorylation at Ser-184, which is another pro-apoptotic member of Bcl2 family, through PI3K/Akt pathway, resulting in reduced stress-induced translocation of Bax from cytosol to mitochondria, impaired Bax insertion into mitochondrial membranes, and shortened half-life of Bax protein [61]. Surprisingly, nicotine-induced activation of ERK1/2, Akt, PKA, Bad phosphorylation, and cell survival can be blocked by β-adrenergic receptor blocker propranolol, indicating that Bad phosphorylation induced by nicotine may occur through the transactivation of β-adrenergic receptors [60]. Furthermore, recent finding suggests that the protection from apoptosis caused by serum deprivation and/or chemotherapy conferred by nicotine is NFκB-dependent [44].

The effect of nicotine on the phenotype of normal human airway epithelial cells has also been studied, where nicotine can attenuate apoptosis caused by etoposide, ultraviolet irradiation, or hydrogen peroxide. Nicotine also induces a transformed phenotype manifested as loss of contact inhibition and loss of dependence on exogenous growth factors or adherence to extracellular matrix in these nonimmortalized human airway epithelial cells, suggesting that nicotine may evoke intracellular signals that enhance cellular transformation [62].

Contradictory evidence regarding the role of nicotine in apoptotic signaling also pervades the literature. For instances, nicotine dose-dependently induces cytotoxicity in human glioma and glioblastoma cell lines [63]. Nicotine also induces internucleosomal DNA cleavage in various leukemic cell lines but not in normal peripheral blood lymphocytes and polymorphonuclear cells [64]. The pro-apoptotic effect of nicotine on these cell lines is dependent on rapid increase in the intracellular Ca^{2+} concentration. Furthermore, nicotine can increase oxidative stress, activate NF-κB, and induce apoptosis in colon cancer cells [65].

Nicotine as a Pro-Angiogenic Agent

Angiogenesis is of paramount importance to the development of cancer in a sense that the growth of new blood vessels not only supplies the tumor with nutrients and oxygen, but also facilitates the spread of cancer cells to secondary sites through hematological route. The growth of all solid tumors is dependent on angiogenesis, in which they are unable to grow larger than a few millimeters in diameter unless they are able to recruit their own vascular bed [66]. The discovery that nicotine can induce angiogenesis and contribute to tumor growth through increased neovascularization has shed new light on the pathogenesis of smoking-related neoplastic diseases. Nicotine not only increases endothelial-cell growth and tube formation in vitro, but also accelerates fibrovascular growth in vivo. In addition, nicotine enhances lung cancer growth in association with an increase in lesion vascularity in a mouse model, where the effect is mediated by nAChRs and partially dependent on the elaboration of nitric oxide, prostacyclin, and vascular endothelial growth factor (VEGF) [42]. It is later found that nicotine-induced neovascularization is mainly mediated by α7nAChRs on endothelial cells, whose blockade results in reduction of the angiogenic response to neoplasia. The angiogenic effect of α7nAChR activation is transduced through MAPK and PI3K/Akt pathways and the subsequent NK-κB activation [67]. In parallel to these findings, our laboratory found that cigarette smoke can increase ulcerative colitis-associated colonic adenoma formation in mice, which is associated with increased angiogenesis [68]. We later verify that nicotine can upregulate VEGF, matrix metalloproteinase (MMP)-2, and MMP-9 in colon and gastric cancer xenograft in athymic mice, which is mediated by upregulation of arachidonic acid cascade [34, 69].

Nicotine as a Motogen

The effect of nicotine on cancer cell migration, invasion, and metastasis is poorly understood. In this regard, our laboratory is the first to demonstrate that nicotine can enhance the invasiveness of cancer cells, in which nicotine increases the invasiveness of cultured gastric cancer cells by 4-fold as determined by Matrigel membrane invasion assay. The increased invasiveness is accompanied by increases in the activity of MMP-2 and MMP-9 and protein expressions of urokinase-type plasminogen activator and its receptor, an effect abolished by COX-2 and VEGF receptor blockers, suggesting that the pro-invasive action of nicotine is mediated by upregulation of arachidonic acid cascade and the subsequent enhanced release of VEGF [69].

Nicotine as an Immunosuppressive Agent

Chronic exposure to cigarette smoke is known to attenuate the antibody-forming cell response and antigen-mediated T cell proliferation in animals [70]. While no significant difference is observed in the distribution of lymphocyte subsets, long-term treatment with nicotine inhibits the antibody-forming cell response, impairs the antigen-mediated signaling in T cells, and induces T cell anergy in animals as manifested by the induction of G_0/G_1 phase arrest in T cells, which are unresponsive to stimulation with concanavalin A or anti-CD3 [71]. In association with this finding, nicotine is shown to deplete inositol-1,4,5-trisphosphate-sensitive Ca^{2+} stores in T cells and decrease the ability to raise intracellular Ca^{2+} levels in response to T cell receptor ligation [72]. The immunosuppressive effect of nicotine is further substantiated by the observation that gestational exposure to the chemical can cause long-term suppression of the proliferative response of the offspring B and T cells [73].

Beside the direct inhibitory action on B and T cells, nicotine can also alter the function of antigen-presenting cells. Recent evidence suggests that nicotine has a direct immunosuppressive effect on human dendritic cells, which are critical for initiation of cell-mediated immunity against neoplastic diseases. In that study, nicotine reduces the endocytic and phagocytic activities of monocyte-derived dendritic cells, which is accompanied by decreased levels of pro-inflammatory cytokines, particularly interleukin (IL)-12, and reduced ability to stimulate antigen-presenting cell-dependent T-cell responses. Dendritic cells treated with nicotine also exhibit a diminished ability to induce differentiation and expansion of interferon-γ-producing effector cells [74]. In addition, nicotine inhibits lipopolysaccharide-induced IL-1, IL-8, and PGE_2 expression at the transcriptional level in cultured macrophages, presumably through inhibition of NF-κB [75]. These findings suggest that nicotine can impair the immune system and may thereby increase the susceptibility to the development of cancer.

The Road Ahead

The experimental evidence culminating so far demonstrates that nicotine is an agent actively involved in the pathogenesis of smoking-related malignancies (fig. 3). Nicotine not only possesses intrinsic mutagenicity, but also enhances tumor cell proliferation, increases resistance to apoptosis, and promotes the growth of tumor-associated blood vessels. Nicotine may also facilitate the spread of cancer by increasing the invasiveness of cancer and weakening the immune system. Elucidation of the molecular and cellular mechanisms underlying the promoting action of nicotine on cancer development is still far from completion. The use of high-throughput gene expression profiling methods and knockout animals should provide a better understanding on the involvement of specific genes in the growth-related signaling pathways that are modulated by nicotine. Moreover, although there is still no epidemiologic evidence to date indicating that the use of nicotine can increase cancer incidence, all the pre-clinical evidence converge to suggest that nicotine may pose a safety concern to current

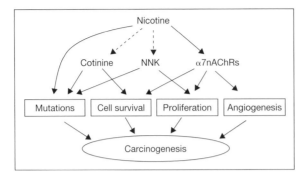

Fig. 3. Carcinogenic action of nicotine. Dotted lines represent metabolic conversion. nAChRs: nicotinic acetylcholine receptors and NNK: 4-(methylnitrosamino)-1-(3-pyridyl)-1-butanone.

nicotine users. A close surveillance on the cancer risk of these subjects is therefore justified.

References

1   International Agency for Research on Cancer (IARC): Tobacco smoking; in IARC Monographs on the Evaluation of the Carcinogenic Risk of Chemicals to Humans. Lyon, France, IARC, 1986, vol 38.
2   International Agency for Research on Cancer (IARC): Tobacco smoking and tobacco smoke; in IARC Monographs on the Evaluation of the Carcinogenic Risk of Chemicals to Humans. Lyon, France, IARC, 2002, vol 83.
3   Hecht SS: Tobacco smoke carcinogens and lung cancer. J Natl Cancer Inst 1999;91:1194–1210.
4   Benowitz NL: Drug therapy. Pharmacologic aspects of cigarette smoking and nicotine addiction. N Engl J Med 1988;319:1318–1330.
5   Henningfield JE: Nicotine medications for smoking cessation. N Engl J Med 1995;333:1196–1203.
6   Lindell G, Farnebo LO, Chen D, Nexo E, Rask Madsen J, Bukhave K, Graffner H: Acute effects of smoking during modified sham feeding in duodenal ulcer patients. An analysis of nicotine, acid secretion, gastrin, catecholamines, epidermal growth factor, prostaglandin E2, and bile acids. Scand J Gastroenterol 1993;28:487–494.
7   Jones RW: Have cholinergic therapies reached their clinical boundary in Alzheimer's disease? Int J Geriatr Psychiatry 2003;18(suppl 1):S7–S13.
8   Quik M, Kulak JM: Nicotine and nicotinic receptors; relevance to Parkinson's disease. Neurotoxicology 2002;23:581–594.
9   Picciotto MR, Brunzell DH, Caldarone BJ: Effect of nicotine and nicotinic receptors on anxiety and depression. Neuroreport 2002;13:1097–1106.
10   Sanberg PR, Silver AA, Shytle RD, Philipp MK, Cahill DW, Fogelson HM, McConville BJ: Nicotine for the treatment of Tourette's syndrome. Pharmacol Ther 1997;74:21–25.
11   Marlow SP, Stoller JK: Smoking cessation. Respir Care 2003;48:1238–1254; discussion 1254–1256.
12   Birrenbach T, Bocker U: Inflammatory bowel disease and smoking: a review of epidemiology, pathophysiology, and therapeutic implications. Inflamm Bowel Dis 2004;10:848–859.
13   Wright SC, Zhong J, Zheng H, Larrick JW: Nicotine inhibition of apoptosis suggests a role in tumor promotion. FASEB J 1993;7:1045–1051.

14   Garnier M, Lamacz M, Tonon MC, Vaudry H: Functional characterization of a nonclassical nicotine receptor associated with inositolphospholipid breakdown and mobilization of intracellular calcium pools. Proc Natl Acad Sci USA 1994;91:11743–11747.

15   Sargent PB: The diversity of neuronal nicotinic acetylcholine receptors. Annu Rev Neurosci 1993;16:403–443.

16   Rathouz MM, Vijayaraghavan S, Berg DK: Elevation of intracellular calcium levels in neurons by nicotinic acetylcholine receptors. Mol Neurobiol 1996;12:117–131.

17   Hecht SS: Tobacco carcinogens, their biomarkers and tobacco-induced cancer. Nat Rev Cancer 2003;3:733–744.

18   Clarke PB: Nicotine dependence – mechanisms and therapeutic strategies. Biochem Soc Symp 1993;59:83–95.

19   Riebe M, Westphal K: Studies on the induction of sisterchromatid exchanges in Chinese hamster ovary cells by various tobacco alkaloids. Mutat Res 1983;124:281–286.

20   Trivedi AH, Dave BJ, Adhvaryu SG: Assessment of genotoxicity of nicotine employing in vitro mammalian test system. Cancer Lett 1990;54:89–94.

21   Argentin G, Cicchetti R: Genotoxic and antiapoptotic effect of nicotine on human gingival fibroblasts. Toxicol Sci 2004;79:75–81.

22   Cook JC, Klinefelter GR, Hardisty JF, Sharpe RM, Foster PM: Rodent Leydig cell tumorigenesis: a review of the physiology, pathology, mechanisms, and relevance to humans. Crit Rev Toxicol 1999;29:169–261.

23   Li XS, Wang HF, Shi JY, et al: Genotoxicity study on nicotine and nicotine-derived nitrosamine by accelerator mass spectrometry. Radiocarbon 1996;38:347–353.

24   Wu XH, Wang HF, Liu YF, Lu XY, Wang JJ, Li K: Histone adduction with nicotine: a bio-AMS study. Radiocarbon 1997;39:293–297.

25   Yim SH, Hee SS: Bacterial mutagenicity of some tobacco aromatic nitrogen bases and their mixtures. Mutat Res 2001;492:13–27.

26   Yim SH, Hee SS: Genotoxicity of nicotine and cotinine in the bacterial luminescence test. Mutat Res 1995;335:275–283.

27   Hecht SS, Hochalter JB, Villalta PW, Murphy SE: 2′-Hydroxylation of nicotine by cytochrome P450 2A6 and human liver microsomes: formation of a lung carcinogen precursor. Proc Natl Acad Sci USA 2000;97:12493–12497.

28   Carmella SG, Borukhova A, Desai D, Hecht SS: Evidence for endogenous formation of tobacco-specific nitrosamines in rats treated with tobacco alkaloids and sodium nitrite. Carcinogenesis 1997;18:587–592.

29   Hecht SS, Carmella SG, Chen M, Dor Koch JF, Miller AT, Murphy SE, Jensen JA, Zimmerman CL, Hatsukami DK: Quantitation of urinary metabolites of a tobacco-specific lung carcinogen after smoking cessation. Cancer Res 1999;59:590–596.

30   Johnston-Early A, Cohen MH, Minna JD, Paxton LM, Fossieck BE Jr, Ihde DC, Bunn PA Jr, Matthews MJ, Makuch R: Smoking abstinence and small cell lung cancer survival. An association. JAMA 1980;244:2175–2179.

31   Browman GP, Wong G, Hodson I, Sathya J, Russell R, McAlpine L, Skingley P, Levine MN: Influence of cigarette smoking on the efficacy of radiation therapy in head and neck cancer. N Engl J Med 1993;328:159–163.

32   Videtic GM, Stitt LW, Dar AR, Kocha WI, Tomiak AT, Truong PT, Vincent MD, Yu EW: Continued cigarette smoking by patients receiving concurrent chemoradiotherapy for limited-stage small-cell lung cancer is associated with decreased survival. J Clin Oncol 2003;21:1544–1549.

33   Cattaneo MG, Codignola A, Vicentini LM, Clementi F, Sher E: Nicotine stimulates a serotonergic autocrine loop in human small-cell lung carcinoma. Cancer Res 1993;53:5566–5568.

34   Ye YN, Liu ES, Shin VY, Wu WK, Luo JC, Cho CH: Nicotine promoted colon cancer growth via epidermal growth factor receptor, c-Src, and 5-lipoxygenase-mediated signal pathway. J Pharmacol Exp Ther 2004;308:66–72.

35   Shin VY, Wu WK, Ye YN, So WH, Koo MW, Liu ES, Luo JC, Cho CH: Nicotine promotes gastric tumor growth and neovascularization by activating extracellular signal-regulated kinase and cyclooxygenase-2. Carcinogenesis 2004;25:2487–2495.

36   Bose C, Zhang H, Udupa KB, Chowdhury P: Activation of p-ERK1/2 by nicotine in pancreatic
     tumor cell line AR42J: effects on proliferation and secretion. Am J Physiol Gastrointest Liver
     Physiol 2005;289:G926–G934.

37   Waggoner SE, Wang X: Effect of nicotine on proliferation of normal, malignant, and human
     papillomavirus-transformed human cervical cells. Gynecol Oncol 1994;55:91–95.

38   Trombino S, Cesario A, Margaritora S, Granone P, Motta G, Falugi C, Russo P: Alpha7-nicotinic
     acetylcholine receptors affect growth regulation of human mesothelioma cells: role of mitogen-
     activated protein kinase pathway. Cancer Res 2004;64:135–145.

39   Fang MA, Frost PJ, Iida-Klein A, Hahn TJ: Effects of nicotine on cellular function in UMR
     106–01 osteoblast-like cells. Bone 1991;12:283–286.

40   Cheng YA, Shiue LF, Yu HS, Hsieh TY, Tsai CC: Interleukin-8 secretion by cultured oral epider-
     moid carcinoma cells induced with nicotine and/or arecoline treatments. Kaohsiung J Med Sci
     2000;16:126–133.

41   Villablanca AC: Nicotine stimulates DNA synthesis and proliferation in vascular endothelial cells
     in vitro. J Appl Physiol 1998;84:2089–2098.

42   Heeschen C, Jang JJ, Weis M, Pathak A, Kaji S, Hu RS, Tsao PS, Johnson FL, Cooke JP: Nicotine
     stimulates angiogenesis and promotes tumor growth and atherosclerosis. Nat Med 2001;7:833–839.

43   Codignola A, Tarroni P, Cattaneo MG, Vicentini LM, Clementi F, Sher E: Serotonin release and
     cell proliferation are under the control of alpha-bungarotoxin-sensitive nicotinic receptors in
     small-cell lung carcinoma cell lines. FEBS Lett 1994;342:286–290.

44   Tsurutani J, Castillo SS, Brognard J, Granville CA, Zhang C, Gills JJ, Sayyah J, Dennis PA:
     Tobacco components stimulate Akt-dependent proliferation and NFkappaB-dependent survival in
     lung cancer cells. Carcinogenesis 2005;26:1182–1195.

45   Maneckjee R, Minna JD: Opioid and nicotine receptors affect growth regulation of human lung
     cancer cell lines. Proc Natl Acad Sci USA 1990;87:3294–3298.

46   Song P, Sekhon HS, Jia Y, Keller JA, Blusztajn JK, Mark GP, Spindel ER: Acetylcholine is syn-
     thesized by and acts as an autocrine growth factor for small cell lung carcinoma. Cancer Res
     2003;63:214–221.

47   Gurkalo VK, Volfson NI: Nicotine influence upon the development of experimental stomach
     tumors. Arch Geschwulstforsch 1982;52:259–265.

48   Backlund MG, Mann JR, Dubois RN: Mechanisms for the prevention of gastrointestinal cancer:
     the role of prostaglandin E2. Oncology 2005;69(suppl 1):28–32.

49   Rakowicz-Szulczynska EM, McIntosh DG, Perry M, Smith ML: PDGF AA as mediator in nico-
     tine-dependent carcinogenesis. Carcinogenesis 1996;17:1813–1818.

50   Mathur RS, Mathur SP, Young RC: Up-regulation of epidermal growth factor-receptors (EGF-R)
     by nicotine in cervical cancer cell lines: this effect may be mediated by EGF. Am J Reprod
     Immunol 2000;44:114–120.

51   Lane D, Gray EA, Mathur RS, Mathur SP: Up-regulation of vascular endothelial growth factor-C
     by nicotine in cervical cancer cell lines. Am J Reprod Immunol 2005;53:153–158.

52   Schuller HM, Cole B: Regulation of cell proliferation by beta-adrenergic receptors in a human
     lung adenocarcinoma cell line. Carcinogenesis 1989;10:1753–1755.

53   Weddle DL, Tithoff P, Williams M, Schuller HM: Beta-adrenergic growth regulation of human
     cancer cell lines derived from pancreatic ductal carcinomas. Carcinogenesis 2001;22:473–479.

54   Wu WK, Wong HP, Luo SW, Chan K, Huang FY, Hui MK, Lam EK, Shin VY, Ye YN, Yang YH,
     Cho CH: 4-(Methylnitrosamino)-1-(3-pyridyl)-1-butanone from cigarette smoke stimulates colon
     cancer growth via beta-adrenoceptors. Cancer Res 2005;65:5272–5277.

55   Jick H, Jick S, Derby LE, Vasilakis C, Myers MW, Meier CR: Calcium-channel blockers and risk
     of cancer. Lancet 1997;349:525–528.

56   Pahor M, Guralnik JM, Salive ME, Corti MC, Carbonin P, Havlik RJ: Do calcium channel block-
     ers increase the risk of cancer? Am J Hypertens 1996;9:695–699.

57   Algazi M, Plu-Bureau G, Flahault A, Dondon MG, Le MG: Could treatments with beta-blockers
     be associated with a reduction in cancer risk? Rev Epidemiol Sante Publique 2004;52:53–65.

58   Heusch WL, Maneckjee R: Signalling pathways involved in nicotine regulation of apoptosis of
     human lung cancer cells. Carcinogenesis 1998;19:551–556.

59   Mai H, May WS, Gao F, Jin Z, Deng X: A functional role for nicotine in Bcl2 phosphorylation and suppression of apoptosis. J Biol Chem 2003;278:1886–1891.

60   Jin Z, Gao F, Flagg T, Deng X: Nicotine induces multi-site phosphorylation of Bad in association with suppression of apoptosis. J Biol Chem 2004;279:23837–23844.

61   Xin M, Deng X: Nicotine inactivation of the proapoptotic function of Bax through phosphorylation. J Biol Chem 2005;280:10781–1079.

62   West KA, Brognard J, Clark AS, Linnoila IR, Yang X, Swain SM, Harris C, Belinsky S, Dennis PA: Rapid Akt activation by nicotine and a tobacco carcinogen modulates the phenotype of normal human airway epithelial cells. J Clin Invest 2003;111:81–90.

63   Yamamura M, Amano Y, Sakagami H, Yamanaka Y, Nishimoto Y, Yoshida H, Yamaguchi M, Ohata H, Momose K, Takeda M: Calcium mobilization during nicotine-induced cell death in human glioma and glioblastoma cell lines. Anticancer Res 1998;18:2499–2502.

64   Yoshida H, Sakagami H, Yamanaka Y, Amano Y, Yamaguchi M, Yamamura M, Fukuchi K, Gomi K, Ohata H, Momose K, Takeda M: Induction of DNA fragmentation by nicotine in human myelogenous leukemic cell lines. Anticancer Res 1998;18:2507–2511.

65   Crowley-Weber CL, Dvorakova K, Crowley C, Bernstein H, Bernstein C, Garewal H, Payne CM: Nicotine increases oxidative stress, activates NF-kappaB and GRP78, induces apoptosis and sensitizes cells to genotoxic/xenobiotic stresses by a multiple stress inducer, deoxycholate: relevance to colon carcinogenesis. Chem Biol Interact 2003;145:53–66.

66   Folkman J: What is the evidence that tumors are angiogenesis dependent? J Natl Cancer Inst 1990;82:4–6.

67   Heeschen C, Weis M, Aicher A, Dimmeler S, Cooke JP: A novel angiogenic pathway mediated by non-neuronal nicotinic acetylcholine receptors. J Clin Invest 2002;110:527–536.

68   Liu ES, Ye YN, Shin VY, Yuen ST, Leung SY, Wong BC, Cho CH: Cigarette smoke exposure increases ulcerative colitis-associated colonic adenoma formation in mice. Carcinogenesis 2003;24:1407–1413.

69   Shin VY, Wu WK, Chu KM, Wong HP, Lam EK, Tai EK, Koo MW, Cho CH: Nicotine induces cyclooxygenase-2 and vascular endothelial growth factor receptor-2 in association with tumor-associated invasion and angiogenesis in gastric cancer. Mol Cancer Res 2005;3:607–615.

70   Savage SM, Donaldson LA, Cherian S, Chilukuri R, White VA, Sopori ML: Effects of cigarette smoke on the immune response. II. Chronic exposure to cigarette smoke inhibits surface immunoglobulin-mediated responses in B cells. Toxicol Appl Pharmacol 1991;111:523–529.

71   Geng Y, Savage SM, Johnson LJ, Seagrave J, Sopori ML: Effects of nicotine on the immune response. I. Chronic exposure to nicotine impairs antigen receptor-mediated signal transduction in lymphocytes. Toxicol Appl Pharmacol 1995;135:268–278.

72   Kalra R, Singh SP, Savage SM, Finch GL, Sopori ML: Effects of cigarette smoke on immune response: chronic exposure to cigarette smoke impairs antigen-mediated signaling in T cells and depletes IP3-sensitive Ca(2+) stores. J Pharmacol Exp Ther 2000;293:166–171.

73   Basta PV, Basham KB, Ross WP, Brust ME, Navarro HA: Gestational nicotine exposure alone or in combination with ethanol down-modulates offspring immune function. Int J Immunopharmacol 2000;22:159–169.

74   Nouri-Shirazi M, Guinet E: Evidence for the immunosuppressive role of nicotine on human dendritic cell functions. Immunology 2003;109:365–373.

75   Sugano N, Shimada K, Ito K, Murai S: Nicotine inhibits the production of inflammatory mediators in U937 cells through modulation of nuclear factor-kappaB activation. Biochem Biophys Res Commun 1998;252:25–28.

Prof. Chi Hin Cho
L2-55, Department of Pharmacology, Laboratory Block
Faculty of Medicine Building
21 Sassoon Road, Pokfulam
The University of Hong Kong
Hong Kong (China)
Tel +86 852 2819 9252, Fax +86 852 2817 0859
E-Mail chcho@hkusua.hku.hk

Cho CH, Purohit V (eds): Alcohol, Tobacco and Cancer.
Basel, Karger, 2006, pp 268–287

....

Phytochemicals in the Prevention of Lung Cancer Induced by Tobacco Carcinogens

Fung-Lung Chung, Lixin Mi

Lombardi Comprehensive Cancer Center, Georgetown University Medical Center,
Washington, D.C., USA

Abstract

Over the past decades a number of structurally diverse small compounds of plant origin
have been identified to possess activity to inhibit lung tumorigenesis caused by tobacco carcinogens. The two families of compounds most extensively investigated are cruciferous
isothiocyanates and tea polyphenols. This chapter presents a salient account of these studies
on tumor bioassays and their mechanisms on the roles of these compounds in the prevention
of lung cancer with considerations of epidemiological evidence.

Lung cancer is the most prevalent lethal cancer in western countries, and the
primary cause of lung cancer is cigarette smoking. Because the prevalence of cigarette smoking has risen in the last few decades in many developing countries, lung
cancer mortality has also rapidly increased worldwide. Notably, lung cancer in
China has become the number one killer among all cancers in recent years, sur-
passing liver and stomach [1]; it is estimated that lung cancer deaths will reach
1.2 million worldwide in 2005 [2]. Moreover, radiotherapy and chemotherapy of
lung cancer are highly toxic and ineffective; the lung cancer mortality rate currently stands at 80% or higher [3]. Therefore, lung cancer represents the most
urgent public health issue for many countries to face in the new century. The best
way to reduce the lung cancer risk is to avoid exposure to the carcinogens in
tobacco. To educate its citizens about the harm of smoking and to curtail its use is
becoming the most important health policy for many developing countries.
However, this approach has had limited successes, as evident by the continuing
significant smoking populations in the US and West European countries, despite

years of smoking cessation programs [4]. Recently, even the tobacco industries have attempted to respond through efforts to develop less harmful cigarettes [5]. While this goal is to be applauded, considerable uncertainties, and even doubts, have been raised about these products. Questions about the safety of these cigarettes and about how the smokers' behavior will be influenced still need to be addressed. To reduce the risk of lung cancer by discovering and developing drugs that can slow down or block the formation of early lung lesions as well as progression from early lesions to malignancies is therefore not only an imminent goal, but also a viable and practical alternative to prevent the epidemic of lung cancer.

Research conducted over the years has generated detailed information regarding mechanisms of smoking-induced lung carcinogenesis. The knowledge obtained from the studies in cell culture, animals, and humans on the mechanisms of lung cancer development after exposure to tobacco carcinogens has been applied to prevention research. Figure 1 outlines the multiple steps involved in lung carcinogenesis induced by cigarette smoking. The identification of carcinogens in cigarette smoke (CS), and understanding their metabolic activation and detoxification, as well as what DNA damage and the molecular and cellular effects they can cause have fueled the advance in lung cancer prevention studies in recent years. The knowledge accumulated over the past decades has offered an unprecedented opportunity for us to study and design chemopreventive strategy for lung cancer caused by smoking.

While the urgency and promises to develop chemopreventive agents for lung cancer are recognized, it is surprising that there is a paucity of candidate agents for clinical trials. The failed β-carotene trial has had a negative impact on the clinical prevention of lung cancer [6]. Many lessons have been learned from this trial, and an important issue concerns agent selection. A comprehensive and in-depth evaluation of agents based on evidence from epidemiological studies, and more importantly, the understanding of mechanisms from cell culture and animal studies needs to be implemented.

Carcinogens in Cigarette Smoke

CS is a complex mixture containing more than 4,000 toxic and carcinogenic compounds. A number of these compounds have been suspected to be carcinogenic and responsible for its harmful effects [7]. The compounds in CS that are known to cause DNA damage can be categorized into several major classes: nitrosamines, catechols, PAHs, aldehydes, and inorganics (metals). The concentration of these substances varies widely in CS. Their roles in carcinogenesis have been suggested by their activities to induce DNA damage in tissues and cells,

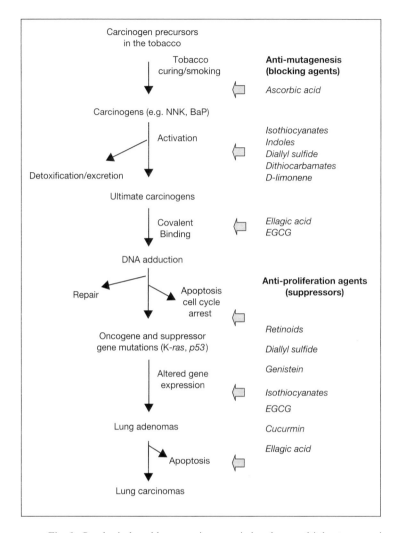

Fig. 1. Smoke-induced lung carcinogenesis involves multiple steps serving as possible targets of phytochemicals.

mutations, transforming cells in cultures, and tumors in animals. A number of structurally diverse DNA adducts caused by these agents have been identified from the carcinogens in tobacco: alkylated bases, oxidized bases, and cyclic adducts [8]. Some of these DNA lesions are mutagenic; therefore, if they are not repaired, they are likely to be involved in the development of lung cancer. More global damage in DNA caused by CS also occurs, including strand cross-linking, strand breakage, chromosomal breakage, and sister-chromatin exchange. Among

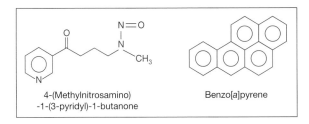

Fig. 2. Structures of the two most widely-studied lung carcinogens found in cigarette smoke: NNK and BaP.

the tobacco carcinogens, 4-(methylnitrosamino)-1-(3-pyridyl)-1-butanone (NNK) and benzo[*a*]pyrene (BaP) (fig. 2) are the most investigated, and arguably most relevant ones because substantial evidence supports their roles in lung cancer. Both NNK and BaP yield promutagenic DNA adducts, like O^6- or N^2-alkylated guanine adducts. The importance of BaP as a lung carcinogen in smokers is further illustrated by its binding at the codon 249 to human *p53* gene, a widely occurring mutation in lung cancer of smokers [9]. DNA damage in smokers' lungs occurs by continuous exposure to these carcinogens and, ultimately, it overwhelms the cellular defense mechanisms, i.e. DNA repair, cell growth arrest, and apoptosis. The subsequent mutations at critical sites can then activate oncogenes and deactivate suppressor genes, cause loss of genes, and genomic instability, eventually leading to uncontrolled cell proliferation and cancer.

Animal Models for Lung Carcinogenesis Prevention

F344 rats and A/J mice are the two most widely used animal models in the study of lung carcinogenesis and its prevention. While a number of carcinogens can cause lung cancer in these models, NNK and BaP are the most frequently used as a single agent or in combination. A single dose of NNK can induce lung adenomas, the benign tumors, in A/J mice within 14 weeks [10]. BaP administered intraperitoneally also induces lung adenomas, however, forestomach tumors can also form with multiple dosing [11]. The lung tumors in A/J mice can be easily quantifiable by counting the nodules on the surface of the lung; expressed as number of animals with tumors per group (incidence) and number of tumors per mouse (multiplicity). F344 rats are also susceptible to lung cancer induction by NNK and BaP [12]. However, unlike A/J mice, a considerably longer period of time is needed for tumors to develop in this model. Lung adenocarcinomas and, less frequently, squamous cell carcinomas are formed within two years after the treatment begins with multiple doses of carcinogens. Because of the similar histopathology of the lung cancer in F344 rats with that in smokers and the broader spectrum of tumors, the F344 rat is thought to be a model more closely related to human lung cancer in smokers. A/J mice also developed lung adenocarcinomas about 40 weeks after treatment with carcinogens [13].

Other animal models, e.g. the Syrian golden hamster have been used to study lung cancer [14, 15]. A unique feature of this model is that NNK induces neuroendocrine tumors in animals with hyperoxic injury, whereas it induces adenocarcinomas in healthy hamsters. Another model developed in recent years is the A/J mice with germline $p53$ mutation and/or Ink4A/Arf heterozygous-deficient mice [16]. These mice are remarkably sensitive to lung tumorigenesis induced by tobacco carcinogens. Exposure to total CS is generally ineffective in inducing lung cancer in laboratory animals. Therefore, it was somewhat unexpected when studies first showed that chronic CS exposure could cause lung adenomas in A/J mice [17]. However, CS is a considerably weaker carcinogen than the purified carcinogens, such as NNK. All these models mentioned have been applied to study chemopreventive agents. Recent studies also reported that a wide range of tumors from hyperplasia to adenocarcinomas in the lung was induced in F344 rats and B6C3F$_1$ mice with total body exposure to CS [18, 19]. The application of the CS models in chemoprevention research has not yet been fully explored, but it certainly represents a potential important model for future studies of lung cancer and its prevention.

Phytochemicals as Chemopreventive Agents for Lung Cancer

The discoveries that dietary compounds can prevent lung cancer primarily are prompted by epidemiological observations. The epidemiological studies have established a link between high intake of fruits and vegetables and lowered risk of lung cancer, suggesting potential benefits of these foods in warding off cancer in people [20]. Based on these findings, an increased consumption of vegetables and fruits is recommended to the public at large. Facing the challenges of changing dietary habits in people, it is realized that a realistic, perhaps more promising, goal is to identify the active compounds in these foods and develop them into clinically useful agents for high risk population such as smokers and ex-smokers.

In past decades, many small bioactive molecules in plants or phytochemicals with potential to reduce the risk of cancer have been identified through animal bioassays and mechanism studies at molecular and cellular levels. These studies not only reinforce the evidence from epidemiological observations but also help us to understand the mechanistic basis of further developing more efficacious agents through structure–activity relationship studies.

Over the years, a number of phytochemicals have been identified as possessing the activity to block lung tumorigenesis; some of them are outlined in figure 1. Some critical steps involved in lung cancer development have been identified as targets. The most effective first line of defense is to block carcinogen formation from its precursors; e.g. ascorbic acid inhibits nitrosation of nicotine to form the tobacco-specific carcinogens, such as NNK. Since most carcinogens require metabolic activation, compounds can also exert their protective effects by modulating phase I enzymes (i.e. cytochrome p450 enzymes) or phase II enzymes (i.e. glutathione transferases or glucuronyl transferases) involved in the activation or deactivation of tobacco carcinogens, respectively. The scavenging of the 'ultimate carcinogens' or highly reactive electrophiles can block or reduce the damage to DNA, thus altering the tumor outcome. The agents with all these activities are called 'blocking' or 'anti-mutagenesis' agents. In principal, compounds can also up- or downregulate repair after DNA adducts are formed, although few compounds have so far been identified with this activity. On the other hand, compounds may induce apoptosis or cell cycle arrest caused by DNA damage to eliminate the 'initiated cells.' Alternatively, compounds may display anti-proliferative activity by slowing down early tumor growth and/or by inducing apoptosis through cell signaling pathways. These are called 'suppressors' or 'anti-proliferation agents.' One or more of these mechanisms may be involved in inhibition of tumorigenesis by phytochemicals.

To date, two major classes of phytochemicals have been studied quite extensively and have been shown to inhibit lung tumorigenesis: tea flavonoids and cruciferous vegetable isothiocyanates (ITCs). In addition, a number of structurally diverse phytochemicals have been shown as chemopreventive agents for lung cancer. Some of these compounds can also inhibit tumorigenesis at sites other than lung, such as mammary gland, colon, prostate, and esophagus. This chapter represents a salient account of these compounds with emphasis on their efficacies in tumor bioassays and mechanisms in cultured cells and it is also an attempt to explain how the knowledge gained from these laboratory studies is used to aid epidemiological and clinical studies.

Cruciferous Isothiocyanates

Studies in Cell Culture and Animals

The inhibitory activity of ITCs against lung tumorigenesis in animals was first described for benzyl ITC (BITC) and phenethyl ITC (PEITC) in dimethyl benz[a]anthracene (DMBD)-induced lung tumorigenesis in mice by Wattenberg [21]. The identification of nicotine-derived nitrosamines as potent lung carcinogens prompted a search for dietary inhibitors against lung tumorigenesis caused by tobacco-specific nitrosamines. Studies were conducted to screen phytochemicals to inhibit NNK metabolic activation in vitro and in vivo [22–24]. These studies showed that the pretreatment with PEITC reduced the levels of DNA adducts, i.e. O^6-methylguanine, in the target tissue caused by NNK. In agreement with these activities, they are effective inhibitors against lung tumor development in A/J mice and F344 rats when orally administered either as a single dose or multiple daily doses before or during NNK or BaP treatment. In rat liver microsomes, PEITC was shown to act as a competitive and non-competitive inhibitor of the metabolism of NNK. These results suggest that the naturally-occurring ITC compounds, BITC in garden cress and PEITC in watercress, could block cytochrome p450s involved in the activation of carcinogens. Based on these studies, a structure–activity relationship study was conducted to examine the effect of alkyl chain length in the aromatic ITC compounds on lung cancer inhibition, and this study showed that a number of synthetic homologs of BITC and PEITC with longer chain up to six carbons are even more efficacious than the naturally-occurring ITCs [25].

Not all ITC compounds can inhibit lung carcinogenesis in A/J mice. Sulforaphane (SFN) and allyl ITC are inducers in phase II enzyme in cultured human cells, suggesting their potential to inhibit tumorigenesis, yet the pretreatment with these ITCs was ineffective for lung tumorigenesis induced by NNK in A/J mice [26]. Similarly, pretreatment of PEITC and BITC did not inhibit lung tumorigenesis in A/J mice treated with NNK. These results suggest that these ITC compounds are not capable of modulating phase I enzyme activities involved in NNK activation, and the specific phase II enzyme assayed in cultured human cell is not important for NNK metabolism. Unlike in lung carcinogenesis, SFN was reported to inhibit mammary gland, skin, and stomach cancer development [27–29]. Interestingly, although SFN was not active when administered before carcinogen exposure, PEITC, SFN and their N-acetylcysteine (NAC) conjugates administered after carcinogen exposure (post-initiation) were shown to inhibit NNK- or BaP-induced lung adenoma formation in A/J mice [30]. PEITC–NAC and SFN–NAC are the major urinary metabolites formed by the mercapturic acid pathway by conjugating ITCs with glutathione, a reaction known to involve glutathione transferases (GSTs). Evidence suggests that the post-initiation effect of these compounds is likely to be attributed to its anti-proliferative activity by inducing apoptosis or cell cycle arrest. Moreover, a recent report indicates that PEITC, SFN and their NAC derivatives can slow down the progression of lung adenomas to adenocarcinomas in A/J mice treated with a combination of NNK and BaP after benign lung tumors have already formed [31]. Table 1 summarizes the studies showing inhibition of lung tumorigenesis by ITC compounds. These studies demonstrate that structurally diverse ITC are capable of inhibiting lung tumorigenesis in various animal models [21, 22, 24, 25, 31–41].

Table 1. Phytochemical isothiocyanates shown to inhibit lung carcinogenesis in rodents

Compounds	Carcinogens	Species	References
Benzyl isothiocyanate	DMBA	Mouse	[21]
	BaP	Mouse	[32, 33]
Phenethyl isothiocyanate	DMBA	Mouse	[21]
	NNK	Rat	[22, 24, 34]
		Mouse	[25, 31, 35–40]
	BaP	Mouse	[31]
Phenylpropyl isothiocyanate	NNK	Mouse	[25, 35, 36]
		Hamster	[41]
Phenylbutyl isothiocyanate	NNK	Mouse	[25, 35, 36]
Hexyl isothiocyanate	NNK	Mouse	[25]
Sulforaphane	NNK	Mouse	[31]
	BaP	Mouse	[31]

The bioactive phytochemicals can act as switches for signaling pathways in cells, including mitogen-activated protein kinases (MAPKs) [extracellular-signal-regulated kinase, JUN N-terminal kinase (JNK) and *p38*], protein kinase C, and phosphatidylinositol 3-kinase (PI3K) [42–44]. These upstream kinases in turn affect a distinct set of transcriptional factors, including nuclear factor κB and activator protein 1 (AP1). Other possible targets to control cell cycle include cyclin-dependent kinases and their inhibitors. Mechanisms of ITCs involving signaling pathways have emerged in the recent years, mostly from cell culture studies. These studies showed that ITCs display activities to induce apoptosis and cell cycle arrest. Certain molecular pathways leading to apoptosis by ITCs have been identified. ITCs regulate phase II enzyme through intracellular signaling pathways [45]. For example, SFN has been shown to disrupt complexes of Kelch-like ECH-associated protein 1 with the NEH2 domain of Nrf2 by reacting directly with certain cysteines of Kelch-like ECH-associated protein 1 stoichiometrically [46]. Thus, the released Nrf2 translocates to the nucleus and activates detoxifying genes by binding to anti-oxidant-responsive elements [47]. Also, it is demonstrated that MAPK can regulate the antioxidant-responsive element in a Nrf2-dependent way [47, 48]. Also PI3K is shown to be involved in Nrf2/anti-oxidant-responsive element activation [49, 50]. The potency of inducing apoptosis by ITCs has been related to JNK activation and the suppression of phosphatase activity against JNK [51, 52]. Activation of caspases, executors of apoptosis, is closely related to JNK activation [53], but it seems that the activation of various caspases is selective in different cell lines for PEITC-induced apoptosis [54, 55]. The involvement of *p53*, a tumor suppression protein, is also cell-type dependent in PEITC-induced apoptosis [56, 57]. A recent study on non-small cell lung carcinoma A549 [58] showed that PEITC and BITC induced apoptosis is associated with increased expression of *p53* and *p21*.[WAF1/CIP1] Besides *p53*, AP1 and upstream JNK are also involved in the signal transduction leading to A549 apoptosis induced by PEITC–NAC [59]. Only limited data on the mechanisms of apoptosis by ITCs are from animal studies; these studies have shown that PEITC, SFN and the NAC conjugates can induce apoptosis in lung of A/J mice treated with NNK or BaP, and the tumor inhibition was prominently associated with induction of AP1 [30, 31].

The current data indicate that the mechanisms by which ITCs inhibit tumorigenesis are multifactorial. The versatility of naturally-occurring ITCs is consistent with the observations from animal tumor bioassays in that these compounds are chemopreventive during different stages of lung cancer development, including initiation, post-initiation, and progression. Taken together, these studies suggest the potential of certain ITCs as chemopreventive agents for lung cancer in current and ex-smokers even with early lesions.

Studies in Humans

The promising results of PEITCs, SFN and other-related ITCs in cell culture and tumor bioassays raise questions about the actual role the dietary ITCs play in the protection against lung cancer in smokers. Clinical trials are needed to prove this, but they are costly and time-consuming. Recent epidemiological studies have shown that dietary ITC intake is associated with lowering the risk of certain cancer, including breast, colon, and lung [60–65]. A urinary marker of dietary ITC uptake was developed and used in some of these studies. In a Shanghai cohort study the urinary marker was measured in 232 lung cancer patients (mostly smokers) and 710 controls [60]. The study found that dietary ITC intake is inversely related to the risk of lung cancer and that individuals with GSTM1 and/or T1 nulls are better protected than those with GSTM1 and/or T1 positive (table 2). This is the first study providing evidence to support a relationship between dietary ITCs and a reduced risk of lung cancer in humans. More importantly, it demonstrates that individual genetic disposition may be an important factor in determining protection against lung cancer by dietary ITCs.

Tea Flavonoids

Studies in Cells Culture and Animals

Lung cancer prevention by flavonoids is mostly conducted in animal models with green tea extract and its catechins or with black tea extract and its

Table 2. Urinary total isothiocyanate (ITC) level in relation to risk of lung cancer in a cohort of men in Shanghai, China

ITC level	Genotype	Cases	Controls	RR (95% CI) Age-adjusted	RR (95% CI) Smoking-adjusted
Undetectable	–	52	111	1.00	1.00
Detectable		180	599	0.64 (0.44–0.93)	0.66 (0.44–0.99)
Undetectable	GSTM1 (+)	21	62	1.00	1.00
Detectable		89	221	1.20 (0.69–2.09)	1.27 (0.69–2.33)
Undetectable	GSTM1 (–)	31	49	1.00	1.00
Detectable		91	378	0.38 (0.23–0.62)	0.35 (0.20–0.63)
Undetectable	GSTT1 (+)	18	54	1.00	1.00
Detectable		80	230	1.05 (0.58–1.90)	0.99 (0.52–1.89)
Undetectable	GSTT1 (–)	34	57	1.00	1.00
Detectable		100	369	0.45 (0.28–0.73)	0.51 (0.30–0.87)
Undetectable	GSTM1 (+) and	28	82	1.00	1.00
Detectable	GSTT1 (+)	119	353	0.99 (0.61–1.60)	1.04 (0.62–1.74)
Undetectable	GSTM1 (–) and	24	29	1.00	1.00
Detectable	GSTT1 (–)	61	246	0.30 (0.16–0.55)	0.27 (0.13–0.56)

RR = Relative risk.

theaflavins. Numerous animal bioassay studies have reported that tea and its components (polyphenols and caffeine) can inhibit lung tumorigenesis [66–81]. A summary of these studies is shown in table 3. These studies consistently demonstrate that with different treatment protocols in various species, tea, mostly green tea, and its catechins, can inhibit lung cancer development during the promotion and progression of lung tumorigenesis. A majority of the studies were conducted in A/J mice treated with NNK. Green and black tea infusion, the total tea extract or their active polyphenols or caffeine as pure compounds were administered in drinking water throughout the tumor bioassays. Lung adenomas and adenocarcinomas in mice were shown to be inhibited by these treatments. To our best knowledge, only one tumor bioassay study was conducted in F344 rats with NNK. In this two-year lifetime study, black tea administered as drinking water was shown to inhibit lung carcinogenesis induced by NNK, however, the activity was attributed by and large to caffeine in the tea. Together with the data from A/J mice showing that caffeine decreased lung tumor multiplicity induced by NNK, these results suggest caffeine is one of the active ingredients in tea responsible for its chemopreventive activity for lung cancer. One of the possible mechanisms of tea polyphenols is related to their anti-oxidant activity. This notion was supported by showing in vivo that levels of 8-hydroxydeoxyguanosine (8-oxodG), a marker of oxidative damage, in lung DNA of A/J mice treated with NNK were reduced by

Table 3. Tea and its components shown to inhibit lung carcinogenesis in rodents

Compounds	Carcinogens	Species	References
Black tea	NNK	Mouse	[66, 73, 77, 78]
	NNK	Rat	[68]
	Spontaneous	Mouse	[82]
Green tea	N-nitrosodiethylamine	Mouse	[67]
	NNK	Hamster	[75]
	NNK	Mouse	[67, 73, 74, 76, 77]
	N-methyl-N'-nitro-N-nitrosoguanidine	Mouse	[79]
	Spontaneous	Mouse	[81]
Black tea extracts (BTE)	N-nitrosodiethylamine	Mouse	[66]
Decaffeinated green and black tea	N-nitrosodiethylamine	Mouse	[69]
	NNK	Mouse	[77]
Polyphenolic fraction from green tea	BaP	Mouse	[70]
	N-nitrosodiethylamine	Mouse	[70]
Water extract of green tea (WEGT)	BaP	Mouse	[70–72]
	N-nitrosodiethylamine	Mouse	[70–72]
	NNK	Mouse	[72]
	Crocidolite plus BaP	Rat	[80]
EGCG	NNK	Mouse	[76]
Caffeine	NNK	Mouse	[76]
	NNK	Rat	[68]
Methylxanthine theophylline	NNK	Hamster	[75]

green tea [76]. Other mechanisms of tea have also been investigated, mostly in cell culture, but only a few of these cell culture studies were actually conducted in lung cells, although the mechanisms seen in other cells may also operate in lung cancer cells. In addition to its anti-oxidant activity, tea polyphenols display anti-proliferative activities, by the induction of apoptosis and cell cycle arrest. (–)-Epigallocatechin-3-gallate (EGCG), the most abundant and active catechin in green tea, has been shown to have chemopreventive effects toward various cancers including cervical, colon, stomach, and prostate cancers [82–85]. Some of the mechanisms include the inhibition of telomerase, MAPK, AP1, nuclear factor-B, binding of epidermal growth factor to its receptor, angiogenesis, and activation of apoptosis [86–93]. Recent studies indicate that the catechins are metabolized by a variety of biotransformations, including methylation, glucuronidation, sulfation and ring-fission [94–97]. Studies in human lung cancer cell lines reveal that EGCG can trigger H_2O_2 production, thus induce cell growth inhibition and apoptosis [98], because in these cell lines the presence of catalase either partially or completely blocks apoptosis by EGCG.

Table 4. Epidemiologic studies on tea drinking and lung cancer

Type of studies	Association	Origin of study
Ecological	+ (increased risk)	USA & Europe
Cohort	+	Britain
Case-control	+	Hong Kong
Ecological	– (decreased risk)	Japan
Case-control	–	Japan
Case-control	–	China
Prospective	No effect	Japan
Case-control	–	Uruguay

Studies in Humans

Unlike the results of animal and cell culture studies, epidemiological studies on tea and lung cancer prevention have shown mixed results (Table 4). Several ecological, case-control, and cohort studies were conducted in different parts of the world. Two Japanese studies showed that tea drinking was associated with reduced risk of lung cancer, whereas studies from Britain, US, and Hong Kong observed an increased risk [99, 100]. More recently, a case-control study showed daily consumption of tea is linked to a 72% reduction in risk of lung cancer in OGG1 polymorphic population [101]. Because OGG1 is a gene involved in repair of oxidative DNA damage, this finding suggests that tea protects against lung cancer in individual under more oxidative stress due to lack of the repair activity. Another prospective study conducted in Hiroshima and Nagasaki involving 38,540 people found that tea drinking had no effect on all cancer risks, including lung cancer [102]. However, a case-control study in Uruguay reported heavy drinkers of black tea appeared to have a lowered risk of lung cancer [103]. The conflicting results, possibly due to a combination of confounding factors and the moderate effect of tea, emphasize the importance of conducting clinical intervention trials. One recent such trial using urinary 8-oxodG as a marker of oxidative damage in smokers showed that a protective effect of drinking four cups of green tea per day reduced 8-oxodG levels in the urine of smokers, suggesting drinking green tea can protect against smoking-induced DNA damage [104]. More direct cellular and molecular cancer markers than 8-oxodG have been suggested for lung cancers, such as proliferation antigen, LOH (3p, 8p, 9p), K-*ras*, *p53*, DNA methylation, and apoptosis [105]. To verify the validity of these markers in future clinical intervention trials is a priority. Clearly, well-designed and more detailed mechanism-based intervention trials are needed in order to confirm the protective role of tea in human lung cancer.

Other Phytochemicals as Inhibitors of Lung Tumorigenesis

Besides cruciferous ITCs and tea polyphenols, a number of phytochemicals have been shown to inhibit lung tumorigenesis in animal models. D-limonene is a major bioactive compound found in citrus rind extract. Wattenberg and Coccia reported that D-limonene given orally shortly before NNK administration inhibited lung neoplasias in female A/J mice [106]. In this study, the inhibition by D-limonene was attributed to its ability to modulate NNK activation. Ellagic acid and quercetin are common polyphenolic anti-oxidants in vegetables and fruits; both showed inhibitory activity against lung tumorigenesis in mice exposed to nitrosodiethylamine [107]. Although, data on the mechanisms were not available, the function of these compounds as anti-initiating agents was suggested. Ellagic acid was shown to induce G1 arrest and apoptosis in cancer cells [108]. It is plausible that these polyphenols, like tea catechins, are capable of inducing cell cycle arrest and apoptosis in mouse lung. Not all polyphenolic compounds are inhibitors of lung cancer. Resveratrol, a major active compound found in grape seeds, was shown to inhibit the activation of polyaromaric hydrocarbons by reducing expression of the CYP1A1 and 1B1 genes in human bronchial epithelial cells, having no effect lung tumorigenesis in A/J mice treated with BaP [109]. The lack of activity by resveratrol may be due to its limited tissue uptake in the target tissue, a common problem with most flavonoids.

Another flavonoid compound with potential in preventing lung cancer is deguelin, a rotenoid with activity to induce cell growth arrest and apoptosis in human bronchial epithelial cells, possibly by inhibiting PI3K/Akt signaling pathways [110, 111]. Its activity in cultured cells suggests that deguelin possesses anti-tumorigenic activity against lung cancer in animal models. Indeed, a recent study showed that deguelin is an effective inhibitor against tamoxifen and NNK-induced lung tumorigenesis in A/J mice [112]. Peperine, a compound found in pepper, was reported to reduce BaP induced lung tumorigenesis in mice by suppressing oxidative stress [113]. The inhibition of BaP-induced lung cancer in mice by diallyl sulfide, a garlic compound, appears to be associated with its inhibition of phase I enzyme for BaP activation and induction of glutathione transferase activity [114]. Diallyl sulfide also displays anti-proliferative activity against human lung cancer cells by inducing apoptosis [115, 116]. Genistein, a soy isoflavone, induced apoptosis in human cancer cells, including lung, possibly by inactivating nuclear factor kB [117]. However, it promoted lung carcinogenesis in a multi-organ carcinogenesis bioassay in rats [118]. These disparities possibly reflect its effective concentration in target tissue vs. cell culture. Indole-3-carbinol, occurring in cruciferous vegetables as glucobrassica, also showed chemopreventive activity for lung tumorigenesis in NNK-treated

A/J mice [119]. As a potent inducer of hepatic phase I enzymes, indole-3-carbinaol decreases the extrahepatic availability of NNK, thus negates its impact on lung. In a multi-organ carcinogenesis study, cacao liquor proanthocyanidin was found to reduce the incidence and multiplicity of lung carcinomas [120], although it is not known how this compound exerts its activity. A multi-organ tumor bioassay in mice showed that both incidence and multiplicity of lung adenomas and carcinomas were reduced significantly by lycopene, a tomato carotenoid [121]. Lycopene also protected against cigarette smoke-induced lung cancer in ferrets, possibly through induction of apoptosis via upregulated 1GFBP-3 and down-regulated phosphorylation of BAD [122]. Most of the phytoestrogens, including plant sterols, isoflavones and lignans, have not been studied in tumor bioassays in animals, but dietary intake of these compounds is reported to be linked to reduced risks of lung cancer in epidemiological studies [122, 123]. This family of compounds warrants more detailed evaluation in animal and cell culture studies.

While all of these compounds are studied in a less extensive manner as compared to cruciferous ITCs and tea polyphenols, the activities of these structurally diverse phytochemicals as inhibitors of lung carcinogenesis reinforce the notion that plants represent an extraordinarily rich source of potential chemopreventive agents. Many small molecules in plants other than the ones already being studied may be identified in the future for their chemopreventive activity against lung cancer.

References

1   Yang L, Parkin DM, Ferlay J, Li L, Chen Y: Estimates of cancer incidence in China for 2000 and projections for 2005. Cancer Epidemiol Biomarkers Prev 2005;14:243–250.
2   Parkin DM, Bray F, Ferlay J, Pisani P: Global Cancer Statistics, 2002, Pisani, CA. Cancer J Clin 2005;55:74–108.
3   Spira A, Ettiger D: Multidisciplinary management of lung cancer. N Eng J Med 2004;350: 379–392.
4   Mendez D, Warner KE: Adult cigarette smoking prevalence: declining as expected (not as desired). Am J Public Health 2004;94:251–252.
5   Institute of Medicine, Committee to Assess the Science Base for Tobacco Harm Reduction, and Board on Health Promotion and Disease Prevention: Clearing the Smoke: Assessing the Science Base for Tobacco Harm Reduction. National Academy Press, Washington, DC, 2001.
6   Greenwald P: Beta-carotene and lung cancer: a lesson for future chemoprevention investigations? J Natl Cancer Inst 2003;95:E1.
7   Hecht SS: Tobacco smoke carcinogens and lung cancer. J Natl Cancer Inst 1999;91:1194–1210.
8   Hecht SS: Cigarette smoking and lung cancer: chemical mechanisms and approaches to prevention. Lancet Oncol 2002;3:461–469.
9   Denissenko MF, Pao A, Tang M, Pfeifer GP: Preferential formation of benzo[a]pyrene adducts at lung cancer mutational hotspots in P53. Science 1996;274:430–432.
10 Hecht SS, Morse MA, Amin S, Stoner GD, Jordan KG, Choi C-I, Chung F-L: Rapid single dose model for lung tumor induction in A/J mice by 4-(methylnitrosamino)-1-(3-pyridyl)-1-butanone and the effect of diet. Carcinogenesis 1989;10:1901–1904.

11 Ross JA, Nelson GB, Wilson KH, Rabinowitz JR, Galati A, Stoner GD, Nesnow S, Mass MJ: Adenomas induced by polycyclic aromatic hydrocarbons in strain A/J mouse lung correlate with time-integrated DNA adduct levels. Cancer Res1995;55:1039–1044.

12 Hoffmann D, Rivenson A, Amin S, Hecht SS: Dose-response study of the carcinogenicity of tobacco-specific N-nitrosamines in F344 rats. J Cancer Res Clin Oncol 1984;108:81–86.

13 Shimkin MB, Stoner GD: Lung tumors in mice: application to carcinogenesis bioassay. Adv Cancer Res 1975;21:1–58.

14 Rehm S, Takahashi M, Ward JM, Singh G, Katyal SL, Henneman JR: Immunohisto-chemical demonstration of Clara cell antigen in lung tumors of bronchiolar origin induced by N-nitrosodiethylamine in Syrian golden hamsters. Am J Pathol 1989;134:79–87.

15 Schuller HM, Witschi HP, Nylen E, Joshi PA, Correa E, Becker KL: Pathobiology of lung tumors induced in hamsters by 4-(methylnitrosamino)-1-(3-pyridyl)-1-butanone and the modulating effect of hyperoxia. Cancer Res 1990;50:1960–1965.

16 Wang Y, Zhang Z, Kastens E, Lubet RA, You M: Mice with alterations in both p53 and Ink4a/Arf display a striking increase in lung tumor multiplicity and progression: differential chemopreventive effect of budesonide in wild-type and mutant A/J mice. Cancer Res 2003;63:4389–4395.

17 Witschi H: A/J mouse as a model for lung tumorigenesis caused by tobacco smoke: strengths and weaknesses. Exp Lung Res 2005;31:3–18.

18 Mauderly JL, Gigliotti AP, Barr EB, Bechtold WE, Belinsky SA, Hahn FF, Hobbs CA, March TH, Seilkop SK, Finch FL: Chronic inhalation exposure to mainstream cigarette smoke increases lung and nasal tumor incidence in rats. Toxicol Sci 2004;81:280–292.

19 Hutt JA, Vuillemenot BR, Barr EB, Grimes MJ, Hahn FF, Hobbs CH, March TH, Gigliotti AP, Seilkop SK, Finch GL, Mauderly JL, Belinsky SA: Life-span inhalation exposure to mainstream cigarette smoke induces lung cancer in B6C3F1 mice through genetic and epigenetic pathways. Carcinogenesis 2005;26:1999–2009.

20 Potter JD: Cancer prevention: epidemiology and experiment. Cancer Lett 1997;114:7–9.

21 Wattenberg LW: Inhibition of carcinogenic effects of polycyclic hydrocarbons by benzl isothio-cyanate and related compounds. J Natl Cancer Inst 1977;58:395–398.

22 Chung F-L, Juchatz A, Vitarius J, Hecht SS: Effects of dietary compounds on α-hydroxylation of N-nitrosopyrrolidine and N′-nitrosonornicotine in rat target tissues. Cancer Res 1984;44: 2924–2928.

23 Morse MA, Amin SG, Hecht SS, Chung F-L: Effects of aromatic isothiocyanates on tumorigenicity, O[6]-methylguanine formation, and metabolism of the tobacco-specific nitrosamine 4-(methylnitrosamino)-1-(3-pyridyl)-1-butanone in A/J mouse lung. Cancer Res 1989;49:2894–2897.

24 Morse MA, Wang C-X, Stoner GD, Mandal S, Conran PB, Amin SG, Hecht SS, Chung F-L: Inhibition of 4-(methylnitrosamino)-1-(3-pyridyl)-1-butanone-induced DNA adduct formation and tumorigenicity in the lung of F344 rats by dietary phenethyl isothiocyanate. Cancer Res 1989;49:549–553.

25 Jiao D, Eklind KI, Choi C-I, Desai DH, Amin SG, Chung F-L: Structure–activity relationships of isothiocyanates as mechanism-based inhibitors of 4-(methylnitrosamino)-1-(3-pyridyl)-1-butanone-induced lung tumorigenesis in A/J mice. Cancer Res 1994;54:4327–4333.

26 Zhang Y, Talalay P, Cho CG, Posner GH: A major inducer of anticarcinogenic protective enzymes from broccoli: isolation and elucidation of structure. Proc Natl Acad Sci 1992;89:2399–2403.

27 Zhang Y, Kensler TW, Cho CG, Ponser GH, Talalay P: Anticarcinogenic activities of sulforaphane and structurally related synthetic norbornyl isothiocyanates. Proc Natl Acad Sci 1994;91: 3147–3150.

28 Fahey JW, Haristoy X, Dolan PM, Kensler TW, Scholtus I, Stephenson KK, Talalay P, Lozniewski A: Sulforaphane inhibits extracellular, intracellular, and antibiotic-resistant strains of Helicobacter pylori and prevents benzo[a]pyrene-induced stomach tumors. Proc Natl Acad Sci 2002;99: 7610–7615.

29 Dinkova-Kostova AT, Jenkins SN, Fahey JW, Ye L, Wehage SL, Liby KT, Stephenson KK, Wade KL, Talalay P: Protection against UV-light-induced skin carcinogenesis in SKH-1 high-risk mice by sulforaphane-containing broccoli sprout extracts. Cancer Lett 2005; [epub ahead of print].

30 Yang Y-M, Conaway CC, Chiao JW, Wang C-X, Amin S, Whysner J, Dai W, Reinhardt J, Chung F-L: Inhibition of benzo(a)pyrene-induced lung tumorigenesis in A/J mice by dietary N-acetylcysteine

conjugates of benzyl and phenethyl isothiocyanates during the post-initiation phase is associated with activation of MAP kinases and p53 activity and induction of apoptosis. Cancer Res 2002; 62:2–7.

31 Conaway CC, Wang C-X, Pittman B, Yang Y-M, Schwartz JE, Tian D, McIntee EJ, Hecht SS, Chung F-L: Phenethyl isothiocyanate and sulforaphane and their N-acetylcysteine conjugates inhibit malignant progression of lung adenomas induced by obacco carcinogens in A/J mice. Cancer Res 2005;65:8548–8557.

32 Lin JM, Amin S, Trushin N, Hecht SS: Effects of isothiocyanates on tumorigenesis by benzo[a]pyrene in murine tumor models. Cancer Lett 1993;74:151–159.

33 Wattenberg LW: Inhibitory effects of benzyl isothiocyanate administered shortly before diethylnitrosamine or benzo[a]pyrene on pulmonary and forestomach neoplasia in A/J mice. Carcinogenesis 1987;8:1971–1973.

34 Chung FL, Kelloff G, Steele V, Pittman B, Zang E, Jiao D, Rigotty J, Choi C-I, Rivenson A: Chemopreventive efficacy of arylalkyl isothiocyanates and N-acetylcysteine for lung tumorigenesis in Fischer rats. Cancer Res 1996;56:772–778.

35 Morse MA, Eklind KI, Amin SG, Hecht SS, Chung F-L: Effects of alkyl chain length on the inhibition of NNK-induced lung neoplasia in A/J mice by arylalkyl isothiocyanates. Carcinogenesis 1989;10:1757–1759.

36 Morse MA, Eklind KI, Hecht SS, Jordan KG, Choi C-I, Desai DH, Amin SG, Chung F-L: Structure–activity relationships for inhibition of 4-(methylnitrosamino)-1-(3-pyridyl)-1-butanone lung tumorigenesis by arylalkyl isothiocyanates in A/J mice. Cancer Res 1991;51:1846–1850.

37 Matzinger SA, Crist KA, Stoner GD, Anderson MW, Pereira MA, Steele VE, Kelloff GJ, Lubet RA, You M: K-ras mutations in lung tumors from A/J and A/J × TSG-p53 F1 mice treated with 4-(methylnitrosamino)-1-(3-pyridyl)-1-butanone and phenethyl isothiocyanate. Carcinogenesis 1995;16:2487–2492.

38 El-Bayoumy K, Upadhyaya P, Desai DH, Amin S, Hoffmann D, Wynder EL: Effects of 1,4-phenylenebis(methylene)selenocyanate, phenethyl isothiocyanate, indole-3-carbinol, and d-limonene individually and in combination on the tumorigenicity of the tobacco-specific nitrosamine 4-(methylnitrosamino)-1-(3-pyridyl)-1-butanone in A/J mouse lung. Anticancer Res 1996;16:2709–2712.

39 Jiao D, Smith TJ, Yang CS, Pittman B, Desai D, Amin S, Chung F-L: Chemopreventive activity of thiol conjugates of isothiocyanates for lung tumorigenesis. Carcinogenesis 1997;18:2143–2147.

40 Witschi H, Espiritu I, Yu M, Willits NH: The effects of phenethyl isothiocyanate, N-acetylcysteine and green tea on tobacco smoke-induced lung tumors in strain A/J mice. Carcinogenesis 1998;19:1789–1794.

41 Nishikawa A, Furukawa F, Ikezaki S, Tanakamaru Z, Chung F-L, Takahashi M, Hayashi Y: Chemopreventive effects of 3-phenylpropyl isothiocyanate on hamster lung tumorigenesis initiated with N-nitrosobis(2-oxopropyl)amine. Jpn J Cancer Res 1996;87:122–126.

42 Milner JA, MCDonald SS, Anderson DE, Greenwald P: Molecular targets for nutrients involved with cancer prevention. Nutr Cancer 2001;41:1–16.

43 Ashendel CL: Diet signal transduction and carcinogenesis. J Nutr 1995;125:686S–691S.

44 Kong AN, Yu R, Hebbar V, Chen C, Owuor E, Hu R, Ee R, Mandlekar S: Signal transduction events elicited by cancer prevention compounds. Mutat Res 2001;480–481:231–241.

45 Keum YS, Jeong WS, Kong AN: Chemoprevention by isothiocyanates and their underlying molecular signaling mechanisms. Mutat Res 2004;555:191–202.

46 Dinkova-Kostova AT, Holtzdaw WD, Cole RN, Itoh K, Wakabayashi N, Katoh Y, Yamamoto M, Talalay P: Direct evidence that sulfhydryl groups of Keap1 are the sensors regulating induction of phase a2 enzymes that protect against carcinogens and oxidants: Proc Natl Acad Sci 2002;99:11908–11913.

47 Yu R, Chen C, Mo YY, Hebbar V, Owuor ED, Tan TH, Kong AN: Activation of mitogen-activated protein kinase pathways induces antioxidant response element-mediated gene expression via a Nrf2-dependent mechanism. J Biol Chem 2000;275:39907–39913.

48 Keum YS, Owuor ED, Kim BR, Hu R, Kong AN: Involvement of Nrf2 and JNK1 in the activation of antioxidant responsive element (ARE) by chemopreventive agent phenethyl isothiocyanate (PEITC). Pharm Res 2003;20:1351–1356.

49 Lee M, Hanson JM, Chu WA, Johnson JA: Phosphatidylinositol 3-kinase, not extracellular signal-regulated kinase, regulates activation of the antioxidant-responsive element in IMR-32 human neuroblastoma cells. J Biol Chem 2001;276:20011–20016.

50 Kang KW, Lee SJ, Park JW, Kim SG: Phosphatidylinositol 3-kinase regulates nuclear transloca-tion of NF-E2-related factor 2 through actin rearrangement in response to oxidative stress. Mol Pharmacol 2002;62:1001–1010.

51 Chen YR, Wang W, Kong AN, Tan TH: Molecular mechanisms of c-Jun N-terminal kinase-mediated apoptosis induced by anticarcinogenic isothiocyanates. J Biol Chem 1998;273:1769–1775.

52 Chen YR, Han J, Kori R, Kong AN, Tan TH: Phenylethyl isothiocyanate induces apoptotic signal-ing via suppressing phosphatase activity against c-Jun N-terminal kinase. J Biol Chem 2002;277:39334–39342.

53 Chaudhary PM, Eby MT, Jasmin A, Hood L: Activation of the c-Jun N-terminal kinase/stress-activated protein kinase pathway by overexpression of caspase-8 and its homologs. J Biol Chem 1999;274:19211–19219.

54 Yu R, Mandlekar S, Harvey KJ, Ucker DS, Kong AN: Chemopreventive isothiocyanates induce apoptosis and caspase-3-like protease activity. Cancer Res 1998;58:402–408.

55 Xu K, Thornalley PJ: Studies on the mechanism of the inhibition of human leukaemia cell growth by dietary isothiocyanates and their cysteine adducts in vitro. Biochem Pharmacol 2000;60:221–231.

56 Xiao D, Singh SV: Phenethyl isothiocyanate-induced apoptosis in p53-deficient PC-3 human prostate cancer cell line is mediated by extracellular signal-regulated kinases. Cancer Res 2002;62:3615–3619.

57 Huang C, Ma WY, Li J, Hecht SS, Dong Z: Essential role of p53 in phenethyl isothiocyanate-induced apoptosis. Cancer Res 1998;58:4102–4106.

58 Kuang YF, Chen YH: Induction of apoptosis in a non-small cell human lung cancer cell line by isothiocyanates is associated with P53 and P21. Food Chem Toxicol 2004;42:1711–1718.

59 Yang Y-M, Jhanwar-Uniyal M, Schwartz J, Conaway CC, Halicka HD, Traganos F, Chung F-L: N-acetylcysteine conjugate of phenethyl isothiocyanate enhances apoptosis in growth-stimulated human lung cells. Cancer Res 2005;65:8538–8547.

60 London SJ, Yuan J-M, Chung FL, Gao Y-T, Coetzee GA, Ross RK, Yu MC: Isothiocyanates, glutathione S-transferase M1 and T1 polymorphisms and lung cancer risk: a prospective study of men in Shanghai, China. Lancet 2000;356:724–729.

61 Spitz MR, Duphorne CM, Detry MA, Pillow PC, Amos CI, Lei L, de Andrade M, Gu X, Hong WK, Wu X: Dietary intake of ITC: evidence of a joint effect with GST polymorphisms in lung cancer. Cancer Epidemiol Biomarker Prev 2000;9:1017–1020.

62 Zhao B, Seow A, Lee EJ, Poh WT, Teh M, Eng P, Wang YT, Tan WC, Yu MC, Lee HP: Dietary ITC, GSTM1, t1, polymorphisms and lung cancer risk among Chinese women in Singapore. Cancer Epidemiol Biomarker Prev 2001;10:1063–1067.

63 Fowke JH, Chung FL, Jin F, Qi D, Cai Q, Conaway C, Cheng JR, Shu XO, Gao YT, Zheng W: Urinary ITC levels, Brassica, and human breast cancer. Cancer Res 2003;63:3980–3986.

64 Lin DX, Tang YM, Peng Q, Lu SX, Ambrosone CB, Kadlubar FF: GST null genotypes, broccoli and lower prevalence of colorectal adenomas. Cancer Epidemiol Biomarker Prev 1998;7:1013–1018.

65 Seow A, Yuan JM, Sun CL, Van Den Berg D, Lee HP, Yu MC: Dietary ITCs, GST polymorphisms and colorectal cancer risk in Singapore Chinese health study. Carcinogenesis 2002;23:2055–2061.

66 Shukla Y, Taneja P: Anticarcinogenic effect of black tea on pulmonary tumors in Swiss albino mice. Cancer Lett 2002;176:137–141.

67 Yang CS, Yang GY, Landau JM, Kim S, Liao J: Tea and tea polyphenols inhibit cell hyperprolifer-ation, lung tumorigenesis, and tumor progression. Exp Lung Res 1998;24:629–639.

68 Chung FL, Wang M, Rivenson A, Iatropoulos MJ, Reinhardt JC, Pittman B, Ho CT, Amin SG: Inhibition of lung carcinogenesis by black tea in Fischer rats treated with a tobacco-specific car-cinogen: caffeine as an important constituent. Cancer Res 1998;58:4096–4101.

69 Cao J, Xu Y, Chen J, Klaunig JE: Chemopreventive effects of green and black tea on pulmonary and hepatic carcinogenesis. Fundam Appl Toxicol 1996;29:244–250.

70 Katiyar SK, Agarwal R, Mukhtar H: Protective effects of green tea polyphenols administered by oral intubation against chemical carcinogen-induced forestomach and pulmonary neoplasia in A/J mice. Cancer Lett 1993;73:167–172.

71 Katiyar SK, Agarwal R, Zaim MT, Mukhtar H: Protection against N-nitrosodiethylamine and benzo[a]pyrene-induced forestomach and lung tumorigenesis in A/J mice by green tea. Carcinogenesis 1993;14:849–855.

72 Wang ZY, Agarwal R, Khan WA, Mukhtar H: Protection against benzo[a]pyrene- and N-nitrosodiethylamine-induced lung and forestomach tumorigenesis in A/J mice by water extracts of green tea and licorice. Carcinogenesis 1992;13:1491–1494.

73 Wang ZY, Hong JY, Huang MT, Reuhl KR, Conney AH, Yang CS: Inhibition of N-nitrosodiethylamine- and 4-(methylnitrosamino)-1-(3-pyridyl)-1-butanone-induced tumorigenesis in A/J mice by green tea and black tea. Cancer Res 1992;52:1943–1947.

74 Liao J, Yang GY, Park ES, Meng X, Sun Y, Jia D, Seril DN, Yang CS: Inhibition of lung carcinogenesis and effects on angiogenesis and apoptosis in A/J mice by oral administration of green tea. Nutr Cancer 2004;48:44–53.

75 Schuller HM, Porter B, Riechert A, Walker K, Schmoyer R: Neuroendocrine lung carcinogenesis in hamsters is inhibited by green tea or theophylline while the development of adenocarcinomas is promoted: implications for chemoprevention in smokers. Lung Cancer 2004;45:11–18.

76 Xu Y, Ho CT, Amin SG, Han C, Chung FL: Inhibition of tobacco-specific nitrosamine-induced lung tumorigenesis in A/J mice by green tea and its major polyphenol as antioxidants. Cancer Res 1992;52:3875–3879.

77 Shi S, Wang ZY, Smith TJ, Hong J-Y, Chen W-F, Ho C-T, Yang CS: Effects of green tea and black tea on 4-(methylnitrosamino)-1-(3-pyridyl)-1-butanone bioactivation, DNA methylation and lung tumorigenesis in A/J mice. Cancer Res 1994;54:4641–4647.

78 Yang G-Y, Wang ZY, Kim S, Liao J, Seril D, Chen X, Smith TJ, Yang CS: Characterization of early pulmonary hyperproliferation, tumor progression and their inhibition by black tea in a 4-(methylnitrosamino)-1-(3-pyridyl)-1-butanone-induced lung tumorigenesis model with A/J mice. Cancer Res 1997;57:1889–1894.

79 Luo D, Li Y: Preventive effect of green tea on MNNG-induced lung cancers and precancerous lesions in LACA mice. Hua Xi Yi Ke Da Xue Xue Bao 1992;23:433–437.

80 Luo SQ, Liu XZ, Wang CJ: Inhibitory effect of green tea extract on the carcinogenesis induced by asbestos plus benzo(a)pyrene in rat. Biomed Environ Sci 1995;8:54–58.

81 Landau J, Wang Z-Y, Yang G-Y, Ding W, Yang C-S: Inhibition of spontaneous formation of lung tumors and rhabdomyosarcomas in A/J mice by black and green tea. Carcinogenesis 1998;19:501–507.

82 Yang CS, Maliakal P, Meng X: Inhibition of carcinogenesis by tea. Annu Rev Pharmacol Toxicol 2002;42:25–54.

83 Paschka AG, Butler R, Young CY: Induction of apoptosis in prostate cancer cell lines by the green tea component, (–)-epigallocatechin-3-gallate. Cancer Lett 1998;130:1–7.

84 Ahn WS, Huh SW, Bae SM, Lee IP, Lee JM, Namkoong SE, Kim CK, Sin JI: A major constituent of green tea, EGCG, inhibits the growth of a human cervical cancer cell line, CaSki cells, through apoptosis, G(1) arrest, and regulation of gene expression DNA. Cell Biol 2003;22:217–224.

85 Isemura M, Saeki K, Kimura T, Hayakawa S, Minami T, Sazuka M: Tea catechins and related polyphenols as anti-cancer agents. Biofactors 2000;13:81–85.

86 Naasani I, Oh-Hashi F, Oh-Hara T, Fen WY, Johnston J, Chan K, Tsuruo T: Blocking telomerase by dietary polyphenols is a major mechanism for limiting the growth of human cancer cells in vitro and in vivo. Cancer Res 2003;63:824–830.

87 Chung JY, Huang C, Meng X, Dong Z, Yang CS: Inhibition of activator protein 1 activity and cell growth by purified green tea and black tea polyphenols in H-*ras*-transformed cells: structure-activity relationship and mechanisms involved. Cancer Res 1999;59:4610–4617.

88 Chung JY, Park JO, Phyu H, Dong Z, Yang CS: Mechanisms of inhibition of the Ras-MAP kinase signaling pathway in 30.7b Ras 12 cells by tea polyphenols (–)-epigallocatechin-3-gallate and theaflavin-3,3′-digallate. FASEB J 2001;15:2022–2024.

89 Lin JK, Liang YC, Lin-Shiau SY: Cancer chemoprevention by tea polyphenols through mitotic signal transduction blockade. Biochem Pharmacol 1999;58:911–915.

90 Liang YC, Chen YC, Lin YL, Lin-Shiau SY, Ho CT, Lin JK: Suppression of extracellular signals and cell proliferation by the black tea polyphenol, theaflavin-3,3′-digallate. Carcinogenesis 1999;20:733–736.

91 Cao Y, Cao R: Angiogenesis inhibited by drinking tea. Nature 1999;398:381.

92 Islam S, Islam N, Kermode T, Johnstone B, Mukhtar H, Moskowitz RW, Goldberg VM, Malemud CJ, Haqqi TM: Involvement of caspase-3 in epigallocatechin-3-gallate-mediated apoptosis of human chondrosarcoma cells. Biochem Biophys Res Commun 2000;270:793–797.

93 Rodriguez SK, Guo W, Liu L, Band MA, Paulson EK, Meydani M: Green tea catechin, epigallo-catechin-3-gallate, inhibits vascular endothelial growth factor angiogenic signaling by disrupting the formation of a receptor complex. Int J Cancer 2005; [epub ahead of print].

94 Lu H, Meng X, Yang CS: Enzymology of methylation of tea catechins and inhibition of catechol-O-methyltransferase by (−)-epigallocatechin gallate. Drug Metab Dispos 2003;31:572–579.

95 Lu H, Meng X, Li C, Sang S, Patten C, Sheng S, Hong J, Bai N, Winnik B, Ho CT, Yang CS: Glucuronides of tea catechins: enzymology of biosynthesis and biological activities. Drug Metab Dispos 2003;31:452–461.

96 Lu H: Mechanistic studies on the phase II metabolism and absorption of tea catechins. Toxicology 2002; Rutgers The State University of New Jersey, New Brunswick, 2002.

97 Meselhy MR, Nakamura N, Hattori M: Biotransformation of (-)-epicatechin 3-O-gallate by human intestinal bacteria. Chem Pharm Bull (Tokyo) 1997;45:888–893.

98 Hong J, Lu H, Meng X, Ryu JH, Hara Y, Yang CS: Stability, cellular uptake, biotransformation, and efflux of tea polyphenol (−)-epigallocatechin-3-gallate in HT-29 human colon adenocarcinoma cells. Cancer Res 2002;62:7241–7246.

99 IARC Working Group on the Evaluation of Carcinogenic Risks to Humans, Lyon: Coffee, tea, mate, methylxanthines, and methylglyoxal. IARC Monogr Eval Carcinog Risks Hum 1991;51: 1–530.

100 Yang CS, Wang ZY: Tea and cancer. J Natl Cancer Inst 1993;85:1038–1049.

101 Bonner MR, Rothman N, Mumford JL, He X, Shen M, Welch R, Yeager M, Chanock S, Caporaso N, Lan Q: Green tea consumption, genetic susceptibility, PAH-rich smoky coal, and the risk of lung cancer. Mutat Res 2005;582:53–60.

102 Nagano J, Kono S, Preston DL, Mabuchi K: A prospective study of green tea consumption and cancer incidence, Hiroshima and Nagasaki (Japan). Cancer Causes Control 2001;12:501–508.

103 Mendilaharsu M, De Stefani E, Deneo-Pellegrini H, Carzoglio JC, Ronco A: Consumption of tea and coffee and the risk of lung cancer in cigarette-smoking men: a case-control study in Uruguay. Lung Cancer 1998;19:101–107.

104 Hakim IA, Harris RB, Chow HH, Dean M, Brown S, Ali IU: Effect of a 4-month tea intervention on oxidative DNA damage among heavy smokers: role of glutathione S-transferase genotypes. Cancer Epidemiol Biomarkers Prev 2004;13:242–249.

105 Steele VE, Kelloff GJ: Development of cancer chemopreventive drugs based on mechanistic approaches. Mutat Res 2005;591:16–23.

106 Wattenberg LW, Coccia JB: Inhibition of 4-(methylnitrosamino)-1-(3-pyridyl)-1-butanone carcinogenesis in mice by D-limonene and citrus fruit oils. Carcinogenesis 1991;12:115–117.

107 Khanduja KL, Gandhi RK, Pathania V, Syal N: Prevention of N-nitrosodiethylamine-induced lung tumorigenesis by ellagic acid and quercetin in mice. Food Chem Toxicol 1999;37:313–318.

108 Narayanan BA, Geoffroy O, Willingham MC, Re GG, Nixon DW: P53/p21(WAF1/CIP1) expression and its possible role in G1 arrest and apoptosis in ellagic acid treated cancer cells. Cancer Lett 1999;136:215–221.

109 Berge G, Ovrebo S, Eilertsen E, Haugen A, Mollerup S: Analysis of resveratrol as a lung cancer chemopreventive agent in A/J mice exposed to benzo[a]pyrene. Br J Cancer 2004;91:1380–1383.

110 Lee HY: Molecular mechanisms of deguelin-induced apoptosis in transformed human bronchial epithelial cells. Biochem Parmacol 2004;68:1119–1124.

111 Lee HY, Oh SH, Woo JK, Kim WY, Van Pelt CS, Price RE, Cody D, Tran H, Pezzuto JM, Moriarty RM, Hong WK: Chemopreventive effects of deguelin, a novel Akt inhibitor, on tobacco-induced lung tumorigenesis. J Natl Cancer Inst 2005;97:1695–1699.

112 Chun K-H, Kosmeder JW 2nd, Sun S, Pezzuto JM, Lotan R, Hong WK, Lee HY: Effects of deguelin on the phosphatidylinositol 3-kinase/Akt pathway and apoptosis in premalignant human bronchial epithelial cells. J Natl Cancer Inst 2003;95:291–302.

113 Selvendiran K, Senthilnathan P, Magesh V, Sakthisekaran D: Modulatory effect of Piperine on mitochondrial antioxidant system in Benzo(a)pyrene-induced experimental lung carcinogenesis. Phytomedicine 2004;11:85–89.

114 Srivastava SK, Hu X, Xia H, Zaren HA, Chatterjee ML, Agarwal R, Singh SV: Mechanism of differential efficacy of garlic organosulfides in preventing benzo(a)pyrene-induced cancer in mice. Cancer Lett 1997;118:61–67.

115 Sakamoto K, Lawson LD, Milner JA: Allyl sulfides from garlic suppress the in vitro proliferation of human A549 lung tumor cells. Nutr Cancer 1997;29:152–156.

116 Wu XJ, Kassie F, Mersch-Sundermann V: The role of reactive oxygen species (ROS) production on diallyl disulfide (DADS) induced apoptosis and cell cycle arrest in human A549 lung carcinoma cells. Mutat Res 2005;579:115–124.

117 Li Y, Ahmed F, Ali S, Philip PA, Kucuk O, Sarkar FH: Inactivation of nuclear factor kappaB by soy isoflavone genistein contributes to increased apoptosis induced by chemotherapeutic agents in human cancer cells. Cancer Res 2005;65:6934–6942.

118 Seike N, Wanibuchi H, Morimura K, Wei M, Nishikawa T, Hirata K, Yoshikawa J, Fukushima S: Enhancement of lung carcinogenesis by nonylphenol and genistein in a F344 rat multiorgan carcinogenesis model. Cancer Lett 2003;192:25–36.

119 Morse MA, LaGreca SD, Amin SG, Chung F-L: Effects of indole-3-carbinol on lung tumorigenesis and DNA methylation induced by 4-(methylnitrosamino)-1-(3-pyridyl)-1-butanone (NNK), and on the metabolism and disposition of NNK in A/J mice. Cancer Res 1990;50:2613–2617.

120 Yamagishi M, Natsume M, Osakabe N, Okazaki K, Furukawa F, Imazawa T, Nishikawa A, Hirose M: Chemoprevention of lung carcinogenesis by cacao liquor proanthocyanidins in a male rat multi-organ carcinogenesis model. Cancer Lett 2003;191:49–57.

121 Kim DJ, Takasuka N, Nishino H, Tsuda H: Chemoprevention of lung cancer by lycopene. Biofactors 2000;13:95–102.

122 Liu C, Lian F, Smith DE, Russell RM, Wang XD: Lycopene supplementation inhibits lung squamous metaplasia and induces apoptosis via up-regulating insulin-like growth factor-binding protein 3 in cigarette smoke-exposed ferrets. Cancer Res 2003;63:3138–3144.

123 Schabath MB, Hernandez LM, Wu X, Pillow PC, Spitz MR: Dietary phytoestrogens and lung cancer risk. JAMA 2005;294:1550–1551.

Fungl-Lung Chung
Lombardi Comprehensive Cancer Center, Georgetown University Medical Center
3800 Reservoir Rd. LL level, Rm 128A, Box 571469
Washington, DC 20057 (USA)
Tel. +1 202 687 3021, Fax +1 202 687 1068
E-Mail flc6@georgetown.edu

Cho CH, Purohit V (eds): Alcohol, Tobacco and Cancer.
Basel, Karger, 2006, pp 288–304

....

Prevention of Tobacco Associated Morbidity and Mortality by a Vaccine against Nicotine

Erich H. Cerny[a], *Livia Franzini-Brunner*[b], *Thomas Cerny*[c]

[a]Chilka Limited, Lausanne, [b]Institut de Biochimie, Université de Lausanne, Lausanne, [c]Department of Oncology/Hematology, Kantonsspital St. Gallen, St. Gallen, Switzerland

Abstract

The finding that smoking increases significantly the risk for pulmonary and cardio-vascular diseases and subsequent death has been one of the most important findings of modern epidemiology. A large spectrum of behavioral and pharmacological treatment approaches have been developed over the last decades to help smokers quit their habit. The nicotine vaccine, which elicits antibodies against the nicotine molecule adds a new twist to this approach. The nicotine molecule itself is not recognized by the body as a foreign compound, because it is too small. The vaccine which is composed of a carrier molecule to which the nicotine molecule is chemically linked makes the nicotine molecule 'visible' to the immune system and elicits antibodies, which bind specifically to the nicotine molecules, once it has entered the body. The interaction of the nicotine molecule with its receptors is impended on the one side by steric hindrance through the huge immunoglobulin molecule binding to the nicotine and on the other side, by the fact, that antibodies cannot pass through the blood brain barrier due to their high molecular weight. Anti-nicotine vaccines suppress, therefore, the pharmacological effect of nicotine but not craving. The affordable expected cost of a vaccination combined with the long lasting effect of the vaccines and the absence of significant side effects in clinical trials published so far are important advantages of this approach. We review strategies used for the chemical synthesis of such vaccines as well as methods and results of animal experiments evaluating nicotine vaccines and report about promising results of the human clinical trials published so far.

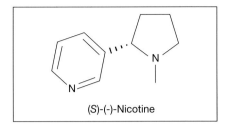

(S)-(-)-Nicotine

Fig. 1. Nicotine contains two rings with an asymmetric center, which occurs as the (*S*)-configuration in nature.

Nicotine Addiction and Classical Treatment Methods

Nicotine Addiction

The nicotine molecules have a close resemblance with the natural acetylcholine molecule (fig. 1) and interact with a broad range of cholinergic receptors in the body. The cholinergic receptors can be divided into two types, the muscarinic and the nicotinic receptors. The nicotinic receptors, which are particularly prevalent in the brain, the neuro-muscular junction and the autonomic nervous system, are composed of subtypes, which show for each tissue a particular composition. The long list of pharmacological effects due to nicotine includes in the central nervous system the releases of dopamine as well as inhibition of the enzymatic dopamine catabolism and in the periphery the discharge of epinephrine by the adrenal glands and a diminished output of insulin by the pancreas.

The success of the cigarette is based on the addictive properties of nicotine. Evolution has cleverly invented for the most elementary tasks of an individual, such as hunger, thirst or sex, the sensation of pleasure in order to assure that they are executed and created a structure in the brain where the information leading to the subjective sensation of pleasure travels. This is the weak spot in the design of the central nervous system, which drugs of abuse use to induce dependency. Mark Caron from Duke refers jokingly to evolution as the single villain in the development of addiction and describes addiction as a hijacking of a system, which evolution has developed for another purpose [1].

Today the following picture of the molecular basis of drug dependency emerges: In the fifties of last century B.F. Skinner and James Olds discovered a common neural pathway linking areas of the brain were the sensation of pleasure could be induced with the help of an experimental set up where they implanted electrodes into the brain of rats, which could administer their own electrical stimuli by pushing a lever if the electrode was placed at the right spot. The structure they identified and which is a highway of pleasure inducing neurons is called the medial forebrain bundle, which runs from the prefrontal cortex through the lateral hypothalamus to a small but important center known as the nucleus accumbens [2].

Work of Roy Wise in the sixties and seventies using selective blockers of the dopamine system was able to demonstrate that the application of such diverse euphoric stimuli as cocaine, heroin or sex let to secretion of dopamine in exactly the area of the nucleus accumbens.

The evaluation of dopamine transporter molecules by MG Caron and others in the eighties with the help of mice where genes were made non-functional (knock-out mice) showed that the dopamine receptors belong to the ubiquitous important class of G protein-coupled receptors. The cascade of events initiated by this class of receptors determines the specificity of a signal inside the cell. This molecular link ties the dopamine system to one of the most important molecular families mediating cellular communication and explains how the signal is transferred inside the cell.

Nicotine vaccines elicit antibodies against nicotine, which bind to its target once the nicotine molecule has entered the body, thereby abolishing any potential interaction of nicotine with its receptors. De Villiers et al. [3] demonstrated in 2002 as one would expect from theoretical grounds a significant reduction of dopamine levels in the nucleus accumbens after nicotine challenge in animals, which have been vaccinated as compared to control animals. This publication closes the circle by furnishing a proof for the efficacy of the nicotine vaccine up to the last element in the chain of events leading to dependence, namely dopamine secretion. Nicotine vaccines empty the traffic on the highway of pleasure because the antibodies binding to nicotine provide an efficient roadblock already outside the brain.

Classical Treatment Approaches for Smoking Cessation

The classical pharmacological treatment approach for drugs of abuse in general is based on the use of antagonists or agonists interacting with classes of a specific receptor of the drug or receptors for other signal transmitters involved in reinforcement loops connected with the drug. Antagonists (e.g. mecamylamine) for nicotine receptors interfere with the induction of the nicotine stimulation at the receptor, but induce at the same time withdrawal symptoms, which may lead to a reduction in compliance. Nicotine agonists such as nicotine substituting patches or smokeless tobacco on the other hand, replace the need for smoking by continuously delivering a nicotine-initiated stimulus to the brain. This treatment replaces one type of dependence by another. The true benefit of nicotine substitution therapies is the significant decrease in morbidity and mortality as compared to smoking. But this treatment form is less successful than one would hope for, because many smokers still prefer the cigarette [4]. The recent appearance of a partial nicotine agonist for smoking cessation treatment (Varenicline[TM] [5], Pfizer) is based on new research into the function of subtypes of receptor families and complements the existing treatments. The selective cannabinoid receptor 1 blocker Rimonabant[TM] is a promising drug in clinical evaluation, which has shown efficacy for smoking cessation in clinical trials [6]. The primary indication of this compound is for weight loss, which is of further benefit in smoking cessation treatments. The capacity of the brain to counter agonistic or antagonistic activity at a receptor family by down- or up-regulation of the number of receptors is a factor which influences the long-term therapeutic effect of many of those compounds. Smoking cessation leads, as one would expect from theoretical grounds to diminished dopamine levels in the nucleus accumbens of the smoker, which in turn may lead to depression. Anti-depressive agents such as bupropion treat depression due to diminished dopamine levels and have become a useful complement to smoking cessation treatments [7].

The literature shows that yearly 1–3% of smokers are able to quit by their own will, and professional counseling increases the 1 year smoking success rate to about 10%. But no smoking cessation method combining counseling with pharmacotherapy has consistently achieved a quit rate exceeding 25% after 6 month and the long-term success rates look even bleaker. There is therefore a need for more efficient forms of treatment.

The Historical Development of the Nicotine Vaccine

The development of immunodiagnostic tools such as RIA's and ELISA's in order to measure nicotine and cotinine in urine and blood let to the development of the first chemical method allowing the linkage of the nicotine molecule to a carrier protein for the production of antibodies (1973) [8].

Schmidt and Butler demonstrated in [9] the use of antibodies against digoxin in order to diminish the toxicity of digoxin in case of an overdose and found that antibodies can rip the drug of the receptor, if the drug is accessible to the specific antibodies. Working with opiates, Berkowitz and Spector [10] demonstrated the influence of anti-morphine antibodies on the analgesic effect of morphine in an animal model one year later.

Bonese et al. [11] studied heroin self-administration with the help of a rhesus monkey, which was immunized with a morphine-BSA conjugate. They found the animal at the beginning partially protected against a challenge with heroin (morphine antibodies cross react with heroin) but subsequent drug challenges overcame the protective effect and the authors concluded that this approach had no merit: 'This blockade been shown to be dose dependent and it can be overcome by high doses of the drug.' This publication by a prestigious group in the field has been the death knell for any further work for more than a decade. The same group of researchers has also evaluated a passive immunization model of opiate self-administration were they made a similar observation [12].

The first description of a vaccine against drugs of abuse including nicotine, which works by interrupting the vicious circle between drug consumption and drug stimulation is Swiss patent CH678394, 'Impfstoff und Immunserum gegen Drogen' [13] filed in 1990 by E.H. Cerny and which is assigned to Chilka Limited. The publication describes in addition to the antidrug vaccine the passive immunization in the case of a drug overdose, which is of no practical use in the case of nicotine dependence but useful for a drug such as cocaine. The anti-nicotine vaccines by Cytos AG, Xenova Group plc. as well as Nabi Inc. and of course Chilka Ltd. are based on the teachings, methods and principles of this patent.

Vaccines Against Nicotine, How They Work

Nicotine Vaccines Intercept Nicotine before it Reaches the Brain
The vaccine produces antibodies against the nicotine molecule, which in turn intercept it before it can reach the receptor. Antibodies are huge molecules equipped with binding sides having a form complementary to the epitope of the molecule to which they bind. The epitope of the target molecule fits into the binding groove of the antibody like a key into a lock and the non-covalent and therefore reversible binding between the complementary shapes of the electron clouds is based on hydrogen bridges, the van der Waals force, electrostatic force and hydrophobic interaction.

The rational behind the vaccine concept for the treatment of nicotine addiction is the interruption of the vicious circle between nicotine consumption and nicotine stimulation by binding of anti-nicotine antibodies to the nicotine molecule itself. Antibodies are too large a molecule to cross the blood brain barrier. As a consequence, the nicotine gets bound and concentrated outside the central nervous system. It is this different mechanism of action, which makes the nicotine vaccine an ideal complement to already existing forms of cessation therapy.

Why does the Nicotine Vaccine not Induce Serum Sickness?
A typical antigen such as a horse serum produces in the presence of antibodies against pathogenic immune complexes, which in turn produce diseases such as serum sickness and other immune complex initiated pathologies. Those large immune complexes are removed from circulation and deposited in various tissues [14, 15]. Very small molecules having typically only one epitope for simultaneous binding by an antibody on the other hand do not lead to the formation of immune complexes, which induce pathological processes in a mammal

because the formation of large immune complexes is based on the crosslinking of antibodies of antigenic molecules presenting more than one binding site for the antibodies. This fact allows the use of antibodies in the presence of the hapten for therapeutic purposes.

The Nicotine Vaccine Diminishes Immediate Influx of Nicotine in the Brain

Nicotine's journey from inhaling into the lungs to the central receptors in the brain takes about 10 s [16, 17]. It is this instant gratification which makes the molecule so addictive. Animals which have been vaccinated against nicotine bind the nicotine through the antibodies directed against nicotine, which circulate in blood and lymphatic fluid and are present in tissue. The nicotine is trapped by the antibodies in the peripheral blood and cannot reach the receptors in the central nervous system. Figure 4, taken from an earlier publication [18] by our group shows the distribution in the serum and the brain of a tritium labeled nicotine bolus 5 min after injection. The nicotine is injected into the tail vein and corresponds to the nicotine equivalent of 2 cigarettes in mice. As one would expect, a significant amount of the nicotine is bound in the serum of the vaccinated animals as compared to the non-vaccinated animal, but less than 10% of the dose can be found in the brain.

Nicotine Vaccine and Classical Vaccines: A Quantitative Comparison

A nicotine vaccine has to neutralize a hapten challenge in the order of 1 mg for every cigarette smoked, whereas the typical infectious disease challenge is limited in the case of a polio virus inocculum, e.g. to a quantity of a couple of hundred thousands or a million virus particles. We can perform a stoichiometric comparison of the two types of vaccines assuming for the sake of simplicity that one neutralizing antibody is needed for each polio virus, respectively, nicotine hapten. One molecular equivalent of nicotine weights 262 g and contains according to the Loschmidt number 6×10^{23} molecules. One milligram of nicotine has therefore about 2.3×10^{18} molecules which is about a thousand billion times more than the 10^6 polio virus of the infectious disease challenge. How does the immune system coop with such huge quantities of hapten molecules which need to by bound by antibodies?

Replacement of Specific Antibodies

The quantities of specific antibodies measured against nicotine are in the order of a few milligrams per liter in the case of successful immunization, which seems a huge quantity for the defense of an infectious invader but barely enough for anti-drug vaccines. Drug challenge studies with an accumulated drug dose above the stoichiometric binding capacity of the specific antibodies in the animals have shown a very good protective effect of the vaccine even in those cases. Such results have been reported in the case of significant cocaine challenge [19, 20] and with animals receiving high dose of continuous nicotine challenge with implantable nicotine pumps [18, 21].

The immune system has to replace antibodies lost by regular turnover, because the half live of immune globulins is only in the order of 20 days [22]. Clinical immunologists were the first to observe, that any depletion of specific antibodies leads to a very rapid replacement even under extreme conditions: Plasmapheresis is a procedure used for the detoxification of blood, which is based on separation of cells and plasma of blood in an extra corporal circuit. The plasma is replaced by a fluid substitute and the cells are again infused into the patient. Any specific antibodies lost during the procedure are rapidly replaced by the immune system to previous levels [23].

During the long apprenticeship of evolution the immune system has encountered infectious invaders, which have developed the capacity to produce exotoxins in quantities far in excess of their own mass, which the immune system had to neutralize. It seems therefore reasonable to speculate that other challenges encountered during evolution have prepared the immune system for the quantitative challenge necessary for a successful nicotine vaccine.

Antibodies as Jugglers of Nicotine

The interaction of the nicotine with the antibody is reversible and dependent mainly on the mean avidity of the polyclonal antibody population, antibody and nicotine concentration and distribution volume in the body as well as nicotine metabolism and excretion. One can imagine specific IgG antibodies as two pronged jugglers which are continuously binding and releasing the nicotine molecules, thereby prolonging the half live in vivo of the nicotine hapten by a factor of about 10. The interaction of the nicotine molecule with the specific antibody is only a fraction of the half live of the nicotine, which means that a given nicotine molecule can bind more than one time to a specific antibody before it is excreted or metabolized into a compound which the antibody does no more recognize. The retention factor due to those juggling antibodies can be very powerful. The conditions for long retention times of the hapten require the presence of a low hapten concentration, high antibody avidity and number as well as a long immunological stability of the hapten in the body (which means the epitope seen by the antibody is not metabolized): Schmidt et al. [24] have shown in a digoxin hapten model after 1 year in the serum of anti-digoxin vaccinated animals a concentration of the hapten, which is reached already after 12 h in non-vaccinated animals challenged with the same initial quantity. This corresponds to an increase in retention of the hapten by a factor of 700.

Nicotine Vaccines: Compliance, Side Effects and Price

Compliance: Nicotine antagonists as well as the nicotine vaccine induce withdrawal symptoms by suppression of the nicotine stimulation pushing the smoker to lighten up the next cigarette, a reflex which inspired Mark Twain to the remark: 'To cease smoking is the easiest thing I ever did. I ought to know, because I have done it a thousand time.' However, there is an important difference between antagonists and the nicotine vaccine: symptoms in the case of an antagonist will disappear if the medication is no more taken, the vaccine effect on the other hand is typically irreversible and of long duration (multiple month or years). Once a smoker has experienced that smoking does not lead to the usual stimulation and relieve of withdrawal symptoms, he will quit.

Side effects: An anti-nicotine vaccine could produce severe side effects if the antibodies against nicotine show immunological cross reactivity by binding to a molecule with partially similar structure and vital function such as, for e.g. acetylcholine. Competitive immunoassays performed by others and our group evaluating cross reactivity to the closest known biological molecules have shown no such activity exceeding background levels of the assay system used. Experimenting with different anti-nicotine vaccines in animal models for over a decade, and observing cohorts of mice in certain instances over a year, our group has never observed any type of vaccine related side effects except adjuvant induced symptoms at the site of injection. Similarly, side effects in clinical trials reported below are almost exclusively limited to the site of injection (swelling, pain, itching and altered skin coloration) and known to be related to the use of adjuvant substances such as alum. No incidence of severe side effects has been reported so far in any of the clinical trials.

Price: Until now there are no vaccines against nicotine or any other drug of abuse on the market and there exists therefore no reference price. The technical complexity of producing the nicotine vaccine is limited and represents only a small percentage in the calculation of the price to the potential end user. Any smoking cessation treatment will not only involve the cost for the vaccine. The treatment needs a close follow up and counselling by qualified health care personal, which will further add to the final bill of an anti-vaccine treatment. This part of the treatment is labor intensive, requires qualified personal and will be costlier than the price of the vaccine itself. Smoking substitution medications in developed countries exceed monthly costs of 100 US dollars.

Smoking itself is not exactly cheap either: a smoker consuming 2 packages a day at a price of more than 3 US dollars a package spends over 2,000 US dollars a year. In view of the many unknowns, it seems too early to speculate about pricing of anti-nicotine vaccines, but the vaccine should be a bargain if compared to the annual cost of cigarettes to an average smoker.

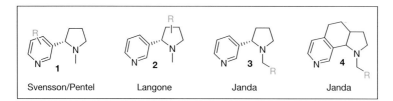

Fig. 2. The nicotine haptens employed so far by several groups can be classified into 3 types, depending on the site of attachment of the linker on the nicotine molecule. Svensson's and Pentel's group introduce the linker on the pyridine ring. Langone et al. attach the linker to the pyrrolidine ring of 3'-hydroxymethylnicotine. Janda et al. attach alkyl linkers to the pyrrolidyl nitrogen.

Efficiency: Experimentation in animal models show that the different nicotine vaccines used to diminish efficiently the initial in stream of nicotine into the brain, which is of crucial importance for the dependence inducing potential of nicotine. The same data show, furthermore, a protective effect even after multiple nicotine challenges and over long periods of time, data which are corroborating by the stream of results coming from the ongoing clinical trials. But nothing replaces the careful collection of clinical data over significant periods of time and it is too early to judge the therapeutic efficiency of those vaccines and their final value as smoking cessation treatments. Studies with very long observation times are particularly important in the smoking cessation field in view of the fact, that a smoker may even after a decade of abstinence restart with his smoking habit.

The Chemical Synthesis of Nicotine Vaccines:

Nicotine contains two rings with an asymmetric center, which occurs as the (*S*)-configuration in nature (fig. 1).

The haptens employed so far by several groups can be classified into 3 types, depending on the site of attachment of the linker on the nicotine molecule (fig. 2).

Svensson's [3, 25], Hieda's [26] and Pentel's group [27] used 2-aminonicotine and 6-aminonicotine (**1**, R = NH$_2$) to introduce the linker on the pyridine ring. Langone et al. attached the linker to the pyrrolidine ring of 3'-hydroxymethylnicotine (**2**, R = CH$_2$OH) [28, 29]. This type of hapten is perhaps the most widely recognized, studied and accepted for anti-nicotine antibodies production. Janda and coworkers [30, 31] prepared nicotine and cotinine based haptens of type 3, with alkyl linkers attached to the pyrrolidyl nitrogen in order to present correct mimics of the naturally occurring molecules to the immune system, in terms of stereochemistry and unaltered ring systems. They studied also conformationally constrained nicotine haptens like 4 [32].

Table 1. Compiles information concerning published coupling chemistries as well as carrier proteins used in the conjugate for nicotine vaccines. K_d = nicotine-antibody dissociation constant; K = average antibody affinity constant, calculated according to R. Müller's method [41], NR: No reference data for affinity or dissociation constant.

Structure of hapten + carrier protein	Affinity of raised antibodies against nicotine	Reference
	NR	[33, 34, 40]
	NR	[33, 34]
	NR	[33, 34]
	NR	[33, 34]
	$K = 0.5–1.6 \times 10^8 \, M^{-1}$	[29]
	$K_d = 37 \pm 23 \times 10^{-9} \, M$	[36]

Castro et al. [33, 34] compared haptens of structure **2** bearing two different types of linker, a rigid and short one (*para*-aminobenzamide-nicotine) and a longer and flexible one (ε-aminocapramide-nicotine). He observed no significant differences.

Table 1 summarizes the results obtained so far with different haptens and carrier proteins. The conjugate in row 6 of table 1 has been used by Cytos to

Table 1. (continued)

Structure of hapten + carrier protein	Affinity of raised antibodies against nicotine	Reference
	$K = 2.2 \pm 2.7 \times 10^8 \, M^{-1}$ $K_d = 4.5 \times 10^{-9} \, M$	[27]
	NR	[38]
	$K = 2.4 \pm 1.6 \times 10^7 \, M^{-1}$	[26]
	$K_d = 2–3 \times 10^{-7} \, M$	[30]
	$K_d = 1.0 \pm 0.10 \times 10^{-6} \, M$	[32]
	$K_d = 0.60 \pm 0.10 \times 10^6 \, M$	[32]
	$K_d = 0.2–50 \, mM$	[40]

rEPA = recombinant exoprotein from *pseudomonas aeruginosa*; AAS = aminoacids; BSA = bovine serum albumin; KLH = keyhole limpet hemocyanin; VLP = virus-like particle.

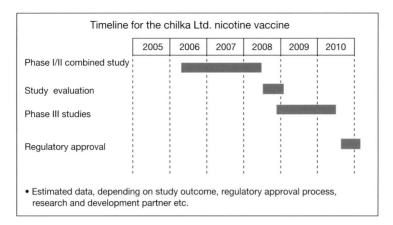

Fig. 3. The estimated time table until market launch of the Chilka Ltd. nicotine vaccine. (Researchers familiar with clinical studies are aware that many events may interfere with established planning.)

prepare the vaccine CYT002NicQb, which just completed phase II clinical trials [35, 36]. The conjugate in the row below, NicVax, is undergoing phase II clinical trials for Nabi Biopharmaceuticals.

Table 1 compiles information concerning conjugates synthesized for nicotine vaccines (coupling chemistries, carrier proteins, affinity constants [37]).

Vaccines in Preclinical or Clinical Evaluation

Chilka Ltd.
Chilka Limited, British Virgin Islands, is preparing a combined phase I/II clinical trial which is designed as a placebo controlled efficacy study using detection of nicotine metabolites in serum and urine and CO in exhaled air as objective evaluation criteria of treatment success. A prospective time line for the development of the Chilka nicotine vaccine is shown in figure 3. Chilka has shown in a mouse model that intra-nasal immunizations can elicit significant *IgA* levels of nicotine antibodies as measured in serum and saliva. This is a useful observation because nicotine enters the body through the lungs and the superficial lining of the lungs contains significant amounts of IgA antibodies. Figure 4 comparing intra-nasal (group IM 1) and subcutaneous vaccination (IM 2) demonstrates a comparable protective effect as measured by the reduction in radio labeled nicotine in the brain. Based on long-term data in the mouse model it seems reasonable to assume that the interval between the initial vaccine treatment a latter booster shots will be at least a year and probably even longer.

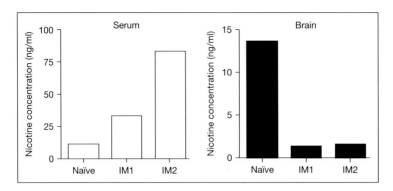

Fig. 4. Distribution in the serum and the brain of tritium labeled nicotine bolus: the figure shows the fate of a nicotine bolus injected into the tail vein corresponding to the equivalent of 2 cigarettes (600 ng in a mice of 20 g) in mice sacrificed 5 min after injection. As one would expect, a significant amount of the nicotine is bound in the serum of the vaccinated animals as compared to the naïve animal, but less than 10% of the dose can be found in the brain. (IM1: 3 × i.n., IM2 3 × s.c., serum of 5 animals is pooled for each data point.) Originally published in [18].

Cytos AG

Cytos AG, Schlieren, Switzerland, presented data of the company's phases I and II clinical trial in May 2005 at the annual meeting of the American Society of Clinical Oncology (abstract 1008) [36].

The vaccine is a virus like particle based conjugate on which the nicotine derivative is linked covalently, Alum is used as an adjuvant and the vaccine is applied intra muscular injection (i.m.). The phase I study performed in England was designed as a single-center parallel group dose comparison study with 40 healthy non-smokers. The phase II study performed in Switzerland evaluated over 300 smokers in a multi-center placebo controlled efficacy study using continuous abstinence from smoking in months 4–6 after first vaccination as primary end-point. The vaccine showed an excellent safety profile, fair immunogenicity and most importantly an excellent correlation between antibody levels and smoking cessation. Cytos projects conservatively that phase III/1 and III/2 studies will take from 2007 to 2009 and that regulatory approval by the Health Authorities will put the launch of the product into the year 2010.

Nabi Biopharmaceuticals Inc.

Nabi Biopharmaceuticals initiated in 2002, a placebo-controlled, double-blind phase I clinical trial of a single dose of NicVAX™ in healthy, nonsmoker volunteers showing that a single dose of vaccine resulted in a rapid immune response and generated nicotine-specific antibodies. The vaccination was well tolerated not requiring therapeutic intervention in any patient.

In January 2003, Nabi Biopharmaceuticals initiated a phase I/II placebo-controlled, double-blind clinical trial of NicVAX™ in smokers, ex-smokers and non-smokers in the Netherlands. Evaluating the development of specific nicotine antibody levels and safety of the vaccine in study participants. This trial demonstrated that multiple injections of NicVAX were safe, well-tolerated and generated substantial amounts of nicotine specific antibodies. In August 2003, Nabi initiated with NicVAX™ in the US a randomized, placebo-controlled double-blind phase II clinical trial with 63 smokers who have expressed a desire to quit smoking. Nabi Inc., Boca Raton, USA, has reported clinical phase II data in September 2004 on its NicVax anti-nicotine vaccine showing promising results.

Xenova Group plc.

Xenova Group plc., Great Britain, uses a vaccine based on nicotine linked to a recombinant Cholera B toxin subunit carrier. Aluminium hydroxide gel is used as adjuvant and the vaccine is applied i.m. Xenova has shown in 2001–2002 during phase I study, the immunogenicity and inoccuity of their vaccine. A second double-blind, randomized, placebo-controlled trial starting 2003 and involving 60 smokers evaluated doses of 50, 250 and 1000 μg of the vaccine. Results of this second trial announced during July 2004 reported a good safety profile for the two lower doses. As far as efficacy is concerned the company reports: 'Although not designed to test the effect of TA-NIC™ on smoker's quit rates, there was a clear reduction across all groups receiving the vaccine in terms of those smokers who voluntarily quit during the 12-week period or self-reported a reduction in smoking pleasure compared to those receiving the placebo'.

Xenova has started in June 2004 two dose escalation phase II studies of an anti-cocaine vaccine based on the same cholera toxin B subunit carrier as the above anti-nicotine vaccine (presented at the College on Problems of Drug Dependence 66th Annual Scientific Meeting in Puerto Rico, June 12–17) and obtained promising results in those studies. A cocaine vaccine is particularly challenging if one considers that the average acute dose of cocaine consumption is in the order of 60–250 mg [37] as compared to 1 or 2 mg of nicotine inhaled from a cigarette.

In view of the large market for efficient smoking cessation treatments it is no surprise that besides the above mentioned companies a number of other research groups have published results on preclinical evaluation of nicotine vaccines [25, 31, 38, 39, 40].

Nicotine Vaccines: Five Year from Now

As stated above Chilka Ltd. sees a launch of their own anti-nicotine vaccine not before 2010.

Cytos AG which prepares now phase III clinical trials has communicated an estimated launch date of their nicotine vaccine not before 2010. Nabi Inc. has communicated that additional data of the phase II clinical trial of the nicotine vaccine will be published in 2005 but the company has not yet articulated any estimated date of product launch for the nicotine vaccine. Xenova Group plc. either has not yet communicated a date for their nicotine vaccine product launch. The recent friendly take over of the company by a well-financed group may add additional funds to the promising nicotine vaccine project and speed up development.

The time lines until product launch by the above companies may seem rather conservative in view of continuing the shortening of product development and approval times. In view of the competitive situation one may hope that the reality will for once surpass the promises.

Some Thoughts About the Future of the Nicotine Vaccines

The familiar health warnings on cigarette packages appeared in the USA for the first time in 1965 and the number of smokers in this country has been cut in half since then. The measure has been seconded efficiently by banning of radio and televisions advertisement since 1971, smoking prohibition in public transportation since 1990 and a legal challenge to the tobacco industry as a whole by state and federal governments since 1994. From today's perspective the main motivation for the average person to quit smoking in developed countries is based in addition to the fear of declining health due to smoking and the knowledge of the addictive properties of nicotine by the increasing restrictions on places were smoking is still tolerated as well as ever increasing prices of tobacco products. Are those measures so successful because they were based first and mainly on information and only auxiliary on constraint?

The appearance of the nicotine vaccine as a new treatment option to the already impressive arsenal of medications for smoking cessation adds some new elements besides the low price of a treatment and the long duration of its effect. The vaccine may also shift some of the responsibility for ruining a smokers health from the cigarette producer to the smoker. In the presence of a vaccine, which efficiently blocks the pharmacological effect of nicotine, a smoker not only decides if and when he wants to start smoking despite all the risks implied, but he has in the future additionally the possibility to start blocking the pharmacological stimulation of nicotine whenever he likes.

The fear of getting hooked to cigarette smoking and to suffer from a smoking related damage to health is a powerful criterion related to the decision to quit or not to start. The appearance of a vaccine, which may limit the addictive potential of smoking could lower the perception of smoking related risk to health and incite people who would otherwise refrain from smoking to start smoking. Could this mean that the nicotine vaccine will favor a trend where frustrated smokers quit and young smokers replace them because they loose their fear of getting irreversibly hooked? It will be a task of the medical community to continue to inform the young objectively about the risks implied in such a new treatment option.

References

1   Taubes G: in HHMI Bulletin 2001;26–29.
2   Wise RA: Neurobiology of addiction. Curr Opin Neurobiol 1996;6:243–251.
3   de Villiers SH, et al: Active immunization against nicotine suppresses nicotine-induced dopamine release in the rat nucleus accumbens shell. Respiration 2002;69:247–253.
4   Peters MJ, Morgan LC: The pharmacotherapy of smoking cessation. Med J Aust 2002;176: 486–490.
5   Coe JW, et al: Varenicline: an alpha4beta2 nicotinic receptor partial agonist for smoking cessation. J Med Chem 2005;48:3474–3477.
6   Cohen C, Kodas E, Griebel G: CB1 receptor antagonists for the treatment of nicotine addiction. Pharmacol Biochem Behav 2005;81:387–395.
7   Hall SM, et al: Cost-effectiveness of bupropion, nortriptyline, and psychological intervention in smoking cessation. J Behav Health Serv Res 2005;32:381–392.
8   Ausubel FM: Short Protocols in Molecular Biology: A Compendium of Methods from Current Protocols in Molecular Biology Hoboken, N.J., Wiley, 2002.
9   Schmidt DH, Butler VP Jr: Immunological protecion against digoxin toxicity. J Clin Invest 1971;50:866–871.
10 Berkowitz B, Spector S: Evidence for active immunity to morphine in mice. Science 1972;178: 1290–1292.
11 Bonese KF, Wainer BH, Fitch FW, Rothberg RM, Schuster CR: Changes in heroin self-administration by a rhesus monkey after morphine immunisation. Nature 1974;252:708–710.
12 Killian A, Bonese K, Rothberg RM, Wainer BH Schuster CR: Effects of passive immunization against morphine on heroin self-administration. Pharmacol Biochem Behav 1978;9:347–352.
13 Cerny EH: Impfstoff und Immunserum gegen Drogen. Swiss patent CH678394 1990.
14 Cochrane CG: The role of immune complexes and complement in tissue injury. J Allergy 1968;42: 113–129.
15 Dixon FJ, Cochrane CG: The pathogenicity of antigen-antibody complexes. Pathol Annu 1970;5: 355–379.
16 Benowitz NL: Pharmacology of nicotine: addiction and therapeutics. Ann Rev Pharmacol Toxicol 1996;36:597–613.
17 Zevin S, Gourlay SG, Benowitz NL: Clinical pharmacology of nicotine. Clin Dermatol 1998;16:557–564.
18 Cerny EH, et al: Preclinical development of 'a vaccine against smoking'. Onkologie 2002;25: 406–411.
19 Carrera MR, et al: Suppression of psychoactive effects of cocaine by active immunization. Nature 1995;378:727–730.
20 Fox BS, et al: Efficacy of a therapeutic cocaine vaccine in rodent models. Nat Med 1996;2: 1129–1132.

21 Hieda Y, Keyler DE, Ennifar S, Fattom A, Pentel PR: Vaccination against nicotine during continued nicotine administration in rats: immunogenicity of the vaccine and effects on nicotine distribution to brain. Int J Immunopharmacol 2000;22:809–819.

22 Hassig A: Intravenous immunoglobulins: pharmacological aspects and therapeutic use. Vox Sang 1986;51:10–17.

23 Unanue ER, Benacerraf B: Textbook of Immunology, Williams & Wilkins, 1984, p 188.

24 Schmidt DH, Kaufman BM, Butler VP Jr: Persistence of hapten-antibody complexes in the circulation of immunized animals after a single intravenous injection of hapten. J Exp Med 1974;139: 278–294.

25 Lindblom N, et al: Active immunization against nicotine prevents reinstatement of nicotine-seeking behavior in rats. Respiration 2002;69:254–260.

26 Hieda Y, et al: Active immunization alters the plasma nicotine concentration in rats. J Pharmacol Exp Ther 1997;283:1076–1081.

27 Satoskar SD, et al: Tissue-dependent effects of immunization with a nicotine conjugate vaccine on the distribution of nicotine in rats. Int Immunopharmacol 2003;3:957–970.

28 Langone JJ, Gjika HB, Van Vunakis H: Nicotine and its metabolites. Radioimmunoassays for nicotine and cotinine. Biochemistry 1973;12:5025–5030.

29 Bjercke RJ, et al: Stereospecific monoclonal antibodies to nicotine and cotinine and their use in enzyme-linked immunosorbent assays. J Immunol Methods 1986;90:203–213.

30 Isomura S, Wirsching P, Janda KD: An immunotherapeutic program for the treatment of nicotine addiction: hapten design and synthesis. J Org Chem 2001;66:4115–4121.

31 Carrera MR, et al: Investigations using immunization to attenuate the psychoactive effects of nicotine. Bioorg Med Chem 2004;12:563–570.

32 Meijler MM, Matsushita M, Altobell LJ 3rd, Wirsching P, Janda KDA: New strategy for improved nicotine vaccines using conformationally constrained haptens. J Am Chem Soc 2003;125: 7164–7165.

33 Castro A, McKennis H, Monji N, Bowman ER: Semi-rigid and flexible linkages in antibody production for determination of nicotine. Biochem Arch 1985;1:205–214.

34 Castro A, Monji N, Ali H, Bowman ER, McKennis H: Characterization of antibodies to nicotine. Biochem Arch 1985;1:173–183.

35 Cerny T: Anti-nicotine vaccination: where are we? Recent Results Cancer Res 2005;166:167–175.

36 Cornuz J, Jungi F, Cerny T, Klingler K, Mueller P: In American Society of Clinical Oncology, ASCO 2005 Annual Meeting, Orlando, Florida, 2005.

37 Llosa T: The standard low dose of oral cocaine: used for treatment of cocaine dependence. Subst Abus 1994;15:215–220.

38 Sanderson SD, et al: Immunization to nicotine with a peptide-based vaccine composed of a conformationally biased agonist of C5a as a molecular adjuvant. Int Immunopharmacol 2003;3: 137–146.

39 Matsushita H, Noguchi M, Tamaki E: Conjugate of bovine serum albumin with nicotine. Biochem Biophys Res Commun 1974;57:1006–1010.

40 Dickerson TJ, Yamamoto N, Janda KD: Antibody-catalyzed oxidative degradation of nicotine using riboflavin. Bioorg Med Chem 2004;12:4981–4987.

41 Muller R: Determination of affinity and specificity of anti-hapten antibodies by competitive radioimmunoassay. Methods Enzymol 1983;92:589–601.

Prof. Thomas Cerny, MD, Professor, Head
Department of Oncology/Hematology
Kantonsspital St. Gallen
CH–9000 St. Gallen (Switzerland)
Tel. +41 71 494 1061, E-Mail thomas.cerny@kssg.ch

Author Index

Subject Index

iron combined pro-oxidant actions,
see also Iron metabolism
acetaldehyde production, *see*
Acetaldehyde
alcohol dehydrogenase 31
breast cance risk 121–124
CYP1A2 31
CYP2E1 31, 32
CYP3A4 31
reactive oxygen species production,
see Reactive oxygen species
trends of use in United States 19, 20
Alcohol dehydrogenase (ADH)
alcohol metabolism 31
gene polymorphisms 33, 51, 52, 122,
123, 146
high-activity Caucasians 56, 57
microbial enzyme in mouth 53, 54
pancreatic stellate cell expression 113
retinoid substrates and alcohol
competition 142, 145, 146
Aldehyde dehydrogenase (ALDH)
deficiency in Asians 56, 57
gene polymorphisms 34, 51, 79,
123, 146
retinoid substrates and alcohol
competition 142, 145, 146
Aldehyde oxidase, acetaldehyde
metabolism 122
Angiogenesis, nicotine induction 261, 262
AP-1, smoking and alcohol effects 150,
151
Apoptosis
S-adenosylmethionine regulation in
hepatocytes 169
nicotine inhibition 260, 261

B cell, nicotine immunosuppression 263
Bladder cancer, smoking risks 199
Breast cancer
alcohol association case-control and
cohort studies 25, 120
breast density and alcohol consumption
128
DNA methylation and one-carbon
metabolism 128–130
epidemiology 18, 119, 120

estrogen metabolism and alcohol effects
126, 127
ethanol-induced oxidative stress 125,
126
ethanol metabolism in breast 121–124
p53 mutational spectrum 130, 131
Buccal pouch model, experimental
modification by cigarette smoke and
constituents in hamster model 245

Caffeine, lung cancer chemoprevention
220
Cancer, *see also* specific cancers
carcinogenesis stages 141
DNA methylation patterns
CpG island de novo methylation 7
gene mutation role 8, 9
hypomethylation 7, 8
overview 6, 7, 40, 41
environmental causes 2–4
gene alterations 5, 6
genetic susceptibility 4, 5
history of theories 1, 2
β-Carotene, *see* Retinoids
Cellular retinol-binding protein (CRBP),
alcohol effects 147, 148
Cervical cancer, smoking risks 200
Colorectal cancer (CRC)
acetaldehyde role
bacterial metabolism 70, 71
metabolism in colorectal mucosa
68–70
toxicity 67, 68
alcohol association
animal studies 65, 66
case-control and cohort studies 24,
25, 64, 65
CYP2E1 induction 71, 72
dietary factors 3, 72, 73
epidemiology 17, 18
fiber in prevention 3, 4
folate in prevention 4
smoking
association studies 198, 199, 232
experimental modification by cigarette
smoke and constituents in rodent
models 243, 244

Genistein, lung cancer chemoprevention 280

Glucocorticoids, lung cancer chemoprevention 219

Glutathione, ethanol depletion 98

Helicobacter pylori, see Gastric cancer

Hepatocellular carcinoma, *see* Liver cancer

Hepatocyte growth factor (HGF), methionine adenosyltransferase induction 168

Hepcidin, iron metabolism 180

HFE, liver cancer mutations 183, 184

Hydroxyl ethyl radicals (HER), DNA adducts 37, 38

Indole-3-carbinol, lung cancer chemoprevention 280, 281

Inflammatory bowel disease, smoking effects 232

Iron
  abundance in tissues 178
  carcinogenicity 181
  ethanol combined pro-oxidant actions
    liver carcinogenesis 182–184
    prospects for study 185, 186
  Fenton reactions 176, 177
  free radical generation 176–178
  lipid peroxidation 177, 178
  metabolism 178, 180

Jun N-terminal kinase (JNK)
  cruciferous vegetable isothiocyanate effects 275
  smoking and alcohol effects 150, 151

Larynx cancer
  alcohol association case-control and cohort studies 23, 24, 49
  epidemiology 16
  smoking association studies 50, 196, 198
  synergistic effects of alcohol and smoking 50

Leukemia, smoking risks 199, 200

D-Limonene, lung cancer chemoprevention 280

Lipid peroxidation
  DNA-peroxide adducts 38, 39
  iron role 178

Liver cancer
  S-adenosylmethionine management 169, 170
  alcohol
    association case-control and cohort studies 24, 96, 97
    cell signaling pathway effects 99–101
    hepatocellular carcinoma progression effects 100–102
    liver damage mechanisms 97–99
  antioxidants in prevention 102, 103
  epidemiology 16, 17, 95, 96
  HFE mutations 183
  iron and ethanol pro-oxidant actions in carcinogenesis 182–184
  methionine adenosyltransferase role 164
  nuclear factor-κB activation 184, 185
  p53 mutations 181, 182
  risk factors 96
  smoking
    experimental modification by cigarette smoke and constituents in rodent models 240
    risks 233

Lung cancer
  β-adrenergic receptor signaling 213–217
  carcinogens 206, 207
  chemoprevention
    animal models 271, 272
    CARET trial 217–219
    cruciferous vegetable isothiocyanates
      animal studies 273, 274
      cell studies 273, 275, 276
      human studies 276
    deguelin 280
    epidemiology studies of phytochemicals 272, 273
    garlic 280
    genistein 280
    indole-3-carbinol 280, 281
    D-limonene 280
    tea studies
      animal studies 276, 277
      cell studies 278